Classical Mechanics and Quantum Mechanics
An Historic-Axiomatic Approach

Authored by

Peter Enders

Taraz State Pedagogical University, Kazakhstan

Classical Mechanics and Quantum Mechanics: An Historic-Axiomatic Approach

Author: Peter Enders

eISBN (Online): 978-1-68108-449-7

ISBN (Print): 978-1-68108-450-3

need for a court order if at any point you breach any terms of this License Agreement. In no event will any delay or failure by Bentham Science Publishers in enforcing your compliance with this License Agreement constitute a waiver of any of its rights.

3. You acknowledge that you have read this License Agreement, and agree to be bound by its terms and conditions. To the extent that any other terms and conditions presented on any website of Bentham Science Publishers conflict with, or are inconsistent with, the terms and conditions set out in this License Agreement, you acknowledge that the terms and conditions set out in this License Agreement shall prevail.

Bentham Science Publishers Ltd.
Executive Suite Y - 2
PO Box 7917, Saif Zone
Sharjah, U.A.E.
Email: subscriptions@benthamscience.net

BENTHAM SCIENCE

Dedication

Dedicated to my parents, Lieselott and Gerhart Enders.

CONTENTS

LIST OF FIGURES

LIST OF TABLES

FOREWORD

The title 'Classical Mechanics and Quantum Mechanics. An historic-axiomatic approach' promises to merge two topics, which are usually separately treated and presented. The one – history – systematises former sciences, while the other one – axiomatic – systematises contemporary sciences. Axiomatization, however, is simultaneously – implicitly or explicitly – also an *interpretation* of former science. One intention of the book is to uncover the implicit interpretations and make them available for a reinterpretation of former and current theories. It provides an axiomatic of Schrödinger's wave mechanics on the basis of Euler's axiomatic of classical mechanics and its generalisation by Planck and Einstein.

Usually, the results of the drastic paradigmatic turn introduced by Heisenberg and Schrödinger are superimposed upon the prehistory of quantum mechanics. This overshadows the common features of Euler's axiomatic and quantum mechanics.

First of all, there is no violation of Newton's 1st axiom, if "stationary state of straight uniform motion" and "external force" are replaced with 'stationary state of constant total energy' ('energetic state') and 'external cause', respectively.

LAW I (modified). Every body perseveres in its energetic state, unless it is compelled to change that state by an external cause.

This generalisation is an important part of Planck's and Einstein's approach to quantum theory.

Thus, this book is not just another textbook about quantum mechanics as it presents quite a novel, axiomatic path from classical to quantum physics and also a novel approach to the history of physics, which is rethought.

Generally speaking, there are two sorts of rethinking, first, rethinking or reinterpretation of theories and, second, rethinking of the path towards a theory. Famous examples of the former sort are the creations of the calculus by Newton and Leibniz and of quantum theory by Heisenberg and Schrödinger. This book rethinks the path to Schrödinger's wave mechanics and the interpretation of it, notably, the discretisation of energy instead of frequency (the latter one being the classical case) and the use of the classical expressions for the potential and kinetic energies.

There is a nice classical analogue to quantum jumps, which plagued Planck and Schrödinger quite a lot[1], but was, on principle, settled already in Schrödinger's 'Third Communication' 1926.

Schrödinger's back-transition from wave to classical mechanics is reinterpreted and complemented.

The approach presented refers to the common basis formed by physical conservation laws and mathematical representation of physical relations between the *energy and the states of a system* by differential equations, whose solutions are continuous space functions (Laguerre functions for the hydrogenium atom in Schrödinger's 'First Communication', Hermitean functions for the harmonic oscillator in his 'Second Communication').

In contrast to Planck himself, Einstein most consequently applied Planck's quantum hypothesis. Classical theory is completely replaced with quantum theory, where the fundamental conservation laws, notably, the energy law, are retained in a new representation. In his 1907 pioneering investigations of the specific heat of solids, he concluded, "that the number of stationary states a microscopic system can assume is smaller than that of the bodies of our everyday experience." This observation defines quantization to be a *selection problem* rather than an "eigenvalue problem" (Schrödinger), the latter denoting the mathematics of *classical* standing waves. Peter Enders does not only claim, but carefully analyses and explains that paradigm. He demonstrates, how the problems are rooted in history and shows to the reader that they can adequately be represented and even convincingly be solved as he supplies the reader with numerous quotations and hints to relevant sources, respectively.

Thus, this book presents quite a novel, axiomatic path from classical to quantum physics. The novelty begins with an original sketch of classical mechanics, which rests on Euler's and Helmholtz's rather than Newton's or Hamilton's axiomatics. Special attention is paid to the commons rather than to the differences between classical and quantum mechanics. Schrödinger's 1926 forgotten demands on quantization are taken seriously and are finally fulfilled. The Schrödinger equation is derived without any special assumptions about the nature of quantum systems, such as interference and superposition, or the existence of a quantum of action, h.

---

[1] "...an act of despair", Planck wrote to Wood. "If we are going to stick to this damned quantum-jumping, then I regret that I ever had anything to do with quantum theory." (after Schrödinger, 'Are there quantum jumps?')

The use of the classical expressions for the potential and kinetic energies within quantum physics is justified.

Another doubtless benefit of this textbook is its extensive reference to original texts. This includes many details not entering contemporary representations of classical mechanics, although being essential for understanding quantum physics. Another benefit is that it addresses not only students and scientists, but also teachers and historians; it sheds new light on the history of ideas and notions. The level of mathematics is seldom higher than that of the common (Riemannian) integral. Basic notions and quantities are carefully introduced.

The author has preserved and even improved all advantages of his preceding book on the foundations of quantum mechanics. They have been highlighted by readers as follows.

"From Newton to Planck and even more. The mix of exacting physics, biography and history exhibits its quite own charm and quickly fascinates the reader, in particular, when nowadays physical equations and the historical literature are so masterly interwoven as the paramount historical figures with their epochal works." (Carsten Hansen, review on buchkatalog.de)

"I would like to express my thanks to the author for this book. It contains an impressive and comprehensible derivation and representation of quantum physics. Due to his approach, I have eventually found an approach that is acceptable for me and not purely formal. The way from classical to quantum physics is impressively understood. Using a profound knowledge of the historical literature, including the original texts, the Schrödinger equation is derived through an extension of Euler's and Helmholtz's representations of classical mechanics to non-classical systems. It is the concentration on the commons rather than on the differences between classical and quantum systems that makes this approach reasonable. Moreover, this makes the interpretation and meaningfulness of the quantum-mechanical models and concepts clearly visible."[2]

Dr. Dr. Dieter Suisky
Humboldt University of Berlin
Germany

---

[2] https://www.amazon.de/klassischen-Physik-Quantenphysik-historisch-kritische%20Anwendungsbeispielen/dp/3540250425/ref=sr_1_3?s=books_&ie=UTF8&qid=1440617257&sr=1-3&keywords=enders+physik

PREFACE

The Preface provides the reader with a very general overview over the goal and the approach of this book, including the new features against previous publications of its author. It emphasizes the need for studying the original texts and stresses, in particular, Schrödinger's discarded requirements to any path from classical to quantum mechanics. Further keys are, (i), Newton's forgotten notion of state, (ii), Euler's unknown axiomatic of classical mechanics, (iii), Helmholtz's concept of configurations, (iv), Helmholtz's less known foundations of energy conservation, (v), Einstein's overlooked idea of quantization as selecting the quantum states of a system out of its classical states (discrete *versus* continuous energy spectrum). Inspired by Gödel's first incompleteness theorem, the path from classical to quantum mechanics is sought for and found through a question, which can be posed, but *not* be answered within classical mechanics. The answer to this question provides a *novel* definition of quantum mechanics. Quantum mechanics is not intuitive, but classical mechanics is not either.

"... so oft ich mich mit den grundlegenden Arbeiten unserer großen Meister unmittelbar vertraut machte, hatte ich einen Gewinn an Einsicht und Verständnis zu verzeichnen, der weit über das hinaus ging, was aus den sekundären Quellen, den Lehrbüchern und dergleichen zu entnehmen war." [1]

"Es ist dies [der Mangel der sonst bereits erfolgreichen Vermittlung wissenschaftlicher Kenntnisse] das Fehlen des historischen Sinnes und der Mangel an Kenntniss jener großen Arbeiten, auf welchen das Gebäude der Wissenschaft ruht." [2]

---

[1] Wilhelm Ostwald (1853 - 1932), *Johann Wilhelm Ritter*, 1894; quoted in W. Hollmann, *Die Zeitschriften der exakten Naturwissenschaften in Deutschland*, 1937, p. 8 (after Th. Hapke, *100 Jahre Ostwald's Klassiker der exakten Wissenschaften 1889 - 1989*, 2003) – Engl.: every time when I have acquainted myself directly with the fundamental work of our great masters, I have registered a gain of insight and understanding, which exceeded by far that, what could be extracted from the secondary sources, the text books and the like.

[2] Ostwald, 1899, after R. Zott, *Über Wilhelm Ostwalds wissenschaftshistorische Beiträge zum Problem des wissenschaftlichen Schöpfertums*, 1999, p. 16, fn. 17. See also Wilhelm Engelmann (1808 - 1878), *Ankündigung [Announcement]*, in: Ostwalds Klassiker der exakten Wissenschaften, No. 121, 1911. – Engl.: It is this [the deficit of the otherwise already successful imparting of scientific knowledge] the missing of the historic sense and the deficit of knowledge of those great works, on which the building of science rests.

"Wer über die handwerkliche Handhabung von Physik hinausgehen will, muss sich zwingend mit der Frage befassen, was physikalisches Denken ausmacht und wie es entstanden ist."[3]

Why another book about quantum mechanics, and why a physical rather than a philosophical treatise about the relationship between classical (CM[4]) and quantum mechanics (QM[5])? Because it treats both CM and QM such, that their commons are stressed, while their differences are developed in a natural and smooth manner. This serves not only the unity of physics, but also the understanding of both.

The basic features of CM needed here are exposed along Newton's[6] and Helmholtz's[7] *original* texts and complemented by Euler's forgotten axiomatic of CM[8]. The following transition to QM is guided by,

- Einstein's 1907 idea of quantization as selection problem,

- Schrödinger's 1933 thoughts about the relationship between CM and QM,

- Schrödinger's 1926 requirements to any transition from CM to QM.

For the sake of the unity of physics, Hertz had required to represent CM such, that all other branches of physics can be derived from it ('Hertz's program'[9]). The feasibility of Hertz's vision will be shown.

Special attention is paid to carefully treating the basic notions. For the notions are the tools of thinking, so that the accuracy of science is by no means better than the accuracy of the notions used. One of the best-known historical examples is the confusion of energy and force.

---

[3] Wilfried Kuhn (1923 - 2009), *Ideengeschichte der Physik*, 2016, cover – Engl.: In order to exceed workmanship of physics, one must concern himself with the question, what constitutes physical thinking and how it is developed.

[4] CM means CM of discrete mass points.

[5] QM refers to matrix mechanics, wave mechanics and path-integral approach, respectively.

[6] Sir Isaac Newton (1643 - 1727), *Philosophiae Naturalis Principia Mathematica*, 1687…1726

[7] Hermann Ludwig Ferdinand von Helmholtz (1821 - 1894), *Über die Erhaltung der Kraft*, 1847; *Vorlesungen über die Dynamik discreter Massenpunkte*, 1911

[8] Leonhard Euler (1707 - 1783), *Anleitung zur Naturlehre*, ca. 1750, publ. only in 1862

[9] Heinrich Rudolf Hertz (1857 - 1894), *Die Prinzipien der Mechanik in neuem Zusammenhange dargestellt*, 1910, *Preface*

Although dealing with basic topics of CM and QM and being self-containing, this book does not represent complete introductions into CM respectively QM as there is few account for their experimental bases.

Despite of presenting a non-conventional approach to QM, it presents new results. Among others, for a free particle, a *square-integrable* wave function is *derived*, and the methodologically interesting similarities between its paradigm 'quantization as selection problem' and Mittelstaedt's reconstruction of QM[10] are considered.

One of the most striking – and often considered to be "the *only* mysterious"[11] – feature of quantum physics is the 'interference of particles'. Its similarities to the interference of classical waves led to the term 'wave-particle duality'. This duality is often taken as starting point for introducing quantum physics. On the other hand, the notions 'wave' and 'particle' are classical ones, so that the apparent duality arises from the assignment of classical properties to non-classical objects.

Thus, this book aims at an *axiomatic* foundation of non-relativistic QM. For this, it reconsiders the historical development, where it concentrates on aspects, which have been discarded, although being crucial for reaching that goal.[12] It presents an *axiomatic* derivation of the stationary and time-dependent Schrödinger equations. "Axiomatic" means, that features like quantum of action, wave-particle dualism, indistinguishability and uncertainty are *deduced* rather than postulated.

The possibility of that is usually disbelieved. Admittedly, it became possible only through exploiting largely discarded results, notably, Newton's notion of stationary state as well as Helmholtz's approach to the energy conservation law and concept of configurations. Even more important is Euler's *forgotten* axiomatic of CM. In contrast to Newton's axioms, it is *not* tied to the motion along trajectories. This makes it possible to abandon the trajectories *without* touching the axiomatic.

Usually, quantum physics is perceived as being counter-intuitive. However, one can hardly expect QM to be intuitive, when even common CM is not. For instance,

---

[10] Peter Mittelstaedt (1929 – 2014), *Rational Reconstructions of Modern Physics*, 2011; see also *The Problem of Interpretation of Modern Physics*, 2011

[11] http://www.feynmanlectures.caltech.edu/III_01.html#Ch1-S7

[12] Due to the axiomatization, vague notions lose their mystic character, *cf* David Hilbert (1862 - 1943), Lothar Wolfgang Nordheim (1899 - 1985) & John von Neumann (1903 - 1957), *Über die Grundlagen der Quantenmechanik*, 1927; see also James Clerk Maxwell (1831 - 1879), *Matter and Motion*, 1991, p. xi, and Stephen G. Brush (*1935), *Kinetische Theorie. Vol. I*, 1970, Introduction.

- if Newton's 1st axiom (Galileo's law of relativity[13]) were intuitive, already Aristotle had stated it[14];

- if Newton's 2nd axiom (the change of the momentum vector is proportional to the vector of the external force applied) were intuitive, it would be taught in school (instead of the over-simplification 'force = mass × acceleration');

- if Newton's description of the force of gravity in the *Definitions* of the *Principia* were intuitive, the notion of field would have been ascribed to him rather than to Faraday[15];

- if Newton's representation of CM were intuitive, it had not taken 100 years to accept it, and the confusion between 'force' and 'energy' had been settled much earlier, *etc*.

Moreover, as observed by the pioneers of QM (notably, Bohr 1913, Heisenberg 1925, Schrödinger 1926), there is *no smooth* way from Newtonian or canonical CM to QM. On the contrary, there *is* a *smooth* way from Eulerian CM to QM, and this way will be described here.

But what is the motivation for changing from CM to QM? Historically, or empirically, it was the inability of CM to explain certain experimental results. Axiomatically, it is an intrinsic limitation of CM. Such a limitation is suggested by Gödel's incompleteness theorem[16], *viz*, to pose questions, which can be stated, but not be answered within CM.

To be specific, let me recall the following. The set of configurations a classical-mechanical conservative system is able to assume is limited by the energy law. For instance, it makes the motion of an oscillator like a pendulum or spring to be bounded by the two turning points. Therefore, the question reads,

How the mechanics of oscillators *without* turning points looks like?

This question can be posed, but *not* be answered within CM. For answering it, Helmholtz's analysis of the relationships between (momentum) configurations and

---

[13] Galileo Galilei (1564 - 1642), *Dialogo sopra i due massimi sistemi del Mondo, Tolemaico, e Copernicano*, 1632
[14] Aristotle (384 - 322 BCE), *Physics* – for a fairly account of sources and comments, see Wikipedia.
[15] See Enders, *Precursors of force fields in Newton's 'Principia'*, 2010.
[16] Kurt Friedrich Gödel (1906 - 1978), *Über formal unentscheidbare Sätze der Principia Mathematica und verwandter Systeme, I.*, 1931, Theorem VI – I am *not* claiming that Gödel's theorem applies to CM.

(kinetic) potential energy is extended to the classically *forbidden* sets of (momentum) configurations. For an oscillator, these are the sets beyond the turning points. This will lead us to the stationary, time-independent Schrödinger equation. The non-stationary, time-dependent Schrödinger equation will be obtained through generalizing Euler's principles of stationary-state change for classical bodies, first, to classical conservative systems and, then, to quantum-mechanical systems.

When treating the symmetry of quantum systems, the approach of this book tells immediately, that stationary quantum state quantities assume the symmetries of the corresponding classical quantities. Combining that with a generalization of Helmholtz's explorations about the relationships between forces and energies leads to a *non*-classical, *quantum* class of interactions, *viz*, to the gauge invariance of the Schrödinger equation and to the (Ehrenberg-Siday-)Aharonov-Bohm effect. In turn, this suggests an approach to the unification of classical mechanics and electromagnetism.[17]

Thus, this book is about the *axiomatic* derivation of the Schrödinger equation and its *non*-classical solution rather than about its interpretation.[18] That derivation corroborates Schrödinger's view of $|\psi(x)|^2$ to represent a weight [density] function (1926, *4th Commun.*, § 7) and reveals $x_{ch}|\psi(x,t)|^2V(x,t)$ and $p_{ch}|\phi(p,t)|^2T(p,t)$ to be the *effective* potential and kinetic energies, respectively, 'seen' by a quantum particle (x_{ch} and p_{ch} being characteristic extensions). This view is compatible with the quantum-logical interpretation, that QM is just the description of the outcome of measurements of quantum objects by means of quantum methods.

The mathematics used is not more complicated than the Fourier transform[19]. The physical content is accessible even for high-school pupils. For one goal of this book is to free both teachers and pupils from the fear against quantum physics. It should be used as a companion to lectures that provide experimental results having forced the physicists a century ago to search for a novel description of atoms. These experiments provided the facts the novel, *non*-classical mechanics had to describe – on the other hand, they gave virtually *no* hints about the way, how to go

---

[17] Enders, *Towards the Unity of Classical Physics*, 2009

[18] For a short (and necessarily incomplete) survey, see Daniel F. Styer (*1955), Miranda S. Balkin, Kathryn M. Becker, Matthew R. Burns, Christopher E. Dudley, Scott T. Forth, Jeremy S. Gaumer, Mark A. Kramer, David C. Oertel, Leonard H. Park, Marie T. Rinkoski, Clait T. Smith & Timothy D. Wotherspoon, *Nine formulations of quantum mechanics*, 2002.

[19] Jean-Baptiste Joseph Baron de Fourier (1768 - 1830), *Mémoire sur la propagation de la chaleur dans les corps solides*, 1807; *La Théorie Analytique de la Chaleur*, 1822, Ch. IX

axiomatically from CM to QM. An axiomatic way is sought – last but not least – for the sake of the unity of physics.

This book represents a complete revision of my 2006 book *From Classical Physics to Quantum Physics*. Selection and arrangement of the material as well as many reasonings have been largely changed. The exposition of CM concentrates on the needs for axiomatically deriving QM. The chapters on solid-state theory (except Bloch's theorem and phonons) are omitted as not being related to the axiomatic issues.[20] Accordingly, the citations and quotations have been revisited and updated. The problems have been adapted and extended; an asterisk '*' marks suggestions for own research (and I would be happy to participate in it!). Moreover, symmetry is dealt with in an own part, in order to highlight the benefits of this approach and to scope with new results. The ideas on field quantization in § 9.2.5 of the 2006 book are revised and exposed in much more detail in a new part, too. The index has been extended accordingly.

Major changes concern also the account for Mittelstaedt's recent book *Rational Reconstructions of Modern Physics* and the representation of Newton's axioms, of the notion of state, of the selection problems, of the representation of the energy of a QM system, and of the recursion relations. The Fourier transform between the wave functions in position and momentum representations is now established by means of *a single* (new) argument. For *free* spinless particles in *whole* space, $\mathbb{R}^3$, *square-integrable* wave functions are *derived*.[21] Chapter 1 about the conservation and changes of stationary quantum states has been completely rewritten. Chapter 5 'Aequat causa effectum' presents not previously published results about a common feature of the symmetries of classical and quantum systems. Section 4.2 with the completion of Schrödingers transition from QM to CM by means of coherent states is new, too. There is a new Chapter 2 about the energy law as foundation of CM. It discusses Planck's 1909 treatment in view of Euler's forgotten results in the *Anleitung* and Carlsons 2016 approach. Subsection 2.3.2 about energy and extension in momentum space is improved by means of the hodograph. The treatment of tunneling has been extended to Bohmian formulae. In connection with that, there is a new Section 3 about the benefits of our approach for treating conservations of weight (probability) and energy.

---

[20] In the meanwhile, these issues have been exposed in a comparable manner in Tom Lancaster (*1979) & Stephen J. Blundell (*1969), *Quantum Field Theory for the Gifted Amateur*, 2014.

[21] Few of that developments have been sketched in my short review *Quantization as Selection rather than Eigenvalue Problem*, 2013.

Sections not being necessary for understanding the essentials of this book are marked with an asterisk (*).

First of all, and again, I feel most indebted to my friend and colleague Dr. Dieter Suisky. He has discovered the power of Euler's axiomatic of CM for tackling QM and special relativity, and he has extracted from Einstein's 1907 pioneering paper on the specific heat of solids[22] the paradigm 'quantization as selection problem'[23]. I am also still indebted to all the colleagues mentioned in my foregoing book. The discussions with Prof. Sigurd Schrader and his co-workers at the Technical University of Applied Sciences Wildau as well as with my friend and former colleague Dr. Michael Erdmann have encouraged me to change substantially the order as well as the manner of exposition of the material to the present form.

Moreover, I thank Dr. Christian Baumgarten for useful hints, Prof. Shawn Carlson for the discussions about his "Principle of Dynamic State Equilibrium" prior publication[24], Prof. Maurice A. de Gosson for elucidating various aspects of quantization, Dr. Pekko Kuopanportti for hints on the energy density, Prof. Peter Mittelstaedt for explanations to his book *Rational Reconstructions of Modern Physics* and Prof. Michael Müller-Preußker[†] for his hints about gauge theory. The discussions with Prof. Günther Nimtz have very much clarified my picture of the time-dependent 'tunnel effect'. I feel indebted to Prof. Harry Paul for earlier discussions and for sending me his contribution about the black-body radiation, Prof. Lev Petrovich Pitaevskii for insightful remarks on the energy density, Prof. Joseph Rosen for his comments on the relationship between symmetry, cause and action, and Prof. Robert H. Swendsen for discussions on (in)distinguishability and for providing texts prior publication. Dr. Eugene V. Stefanovich's work[25] has

---

[22] Einstein, *Die Plancksche Theorie der Strahlung und die Theorie der spezifischen Wärme*, 1907

[23] Dieter Suisky (*1944), *Über eine Differenz in der Begründung des Wirkungsprinzips bei Maupertuis und Euler*, 1999; Suisky & Enders, *Leibniz's foundation of mechanics and the development of 18th century mechanics initiated by Euler*, 2001; Enders & Suisky, *Über das Auswahlproblem in der klassischen Mechanik und in der Quantenmechanik*, 2004; *Quantization as selection problem*, 2005; Suisky & Enders, *Dynamische Begründung der Lorentz-Transformation*, 2005; Suisky, *The Newton - Leibniz controversy on space and time and the development of mechanics by Euler and Einstein*, 2006; *Zur methodologischen Bedeutung von Euler's Begründung der Mechanik*, 2006; *Euler's early relativistic theory*, 2007, *On the post-Newtonian period in the development of mechanics*, 2007; Heinz Lübbig (*1942) & Suisky, *From the origin of forces to the origin of mass. Euler's algorithm for the definition of inert mass reconsidered*, 2007; Suisky, *Euler's mechanics as a unified theory of matter and motion*, 2007; *Euler as Physicist*, 2009; *Are there elements of Leibniz's theory in Newton? On the different shapes of Newton's 2nd Law*, 2012

[24] Shawn Carlson, *A novel formalism of classical mechanics*, to be publ.; *Why Not Energy Conservation?*, 2016

[25] Stefanovich (*1965), *Relativistic Quantum Dynamics. A Non-Traditional Perspective on Space, Time, Particles, Fields, and Action-at-a-Distance*, 2004/2014; *Relativistic Quantum Theory of Particles. I: Quantum Electrodynamics, II. A Non-Traditional Perspective on Space, Time, Particles, Fields, and Action-at-a-Distance*, 2015

influenced this book much more than being visible, and thus I feel highly indebted to him for his numerous additional explanations. Prof. Norbert Straumann has pointed to the Hilbert space formalism as a reason for the fact that Schrödinger's own criticism of his derivation of wave mechanics has eventually been discarded. Prof. Johannes Richter has provided valuable explanations about localization in ideal crystals (flat-band systems).

My friend and schoolmate Dr. Wolfgang Peters posed many stimulating philosophically driven questions and gifted me Torretti's book[26]. My friends and former colleagues Dr. Andreas Förster and Andreas Rothenberg have made several valuable proposals for the text.

Moreover, numerous suggestions and hints to references I have obtained from the newsgroup sci.physics.foundations. The teaching at the Institute of Physics, Mathematics and Informatics at the Abai University Almaty, Kazakhstan, was a source of rethinking several issues.

I am also indebted to anonymous referees for pointing to various weaknesses in the draft. I hope to have eliminated (most of) the mixing of introducing and advanced paragraphs.

Fig. 1. Science should make fun (source: ericweisstein.com).

Despite of the varying quality of its articles, Wikipedia has become a most valuable reference to biographical and bibliographical data. This makes every euro I have

---

[26] Robert Torretti, *The Philosophy of Physics*, 1999

donated to it to be worth it.[27] Thus, I feel highly indebted to all the enthusiasts and projects making the wisdom of the past and present available in the web.[28]

This book has been typeset using TeXstudio and pdfLaTeX. I thank the German TeX user group Dante e.V. (http://www.dante.de) as well as the many enthusiasts on the www TeX help pages for their support.

Peter Enders

CONFLICT OF INTEREST

The author declares no conflict of interest, financial or otherwise.

ACKNOWLEDGEMENTS

My beloved parents, Dr. Lieselott Enders[29] and Dr. Gerhart Enders[30], gifted me a universal humanistic education and *the freedom of independent thinking*. My faithful wife, Galina Nurtasinowa, has shared the up and downs of writing, again.

---

[27] Basically, I second (Paul-)Michel Foucault's (1926 - 1984) criticism of scientific discourse in *Les mots et les choses. Une archéologie des sciences humaines*, 1966. The number of hits of a web site is by no means a measure of its scientific quality. And I try to underestimate neither the affect of a real book upon its readers, nor the value of a searchable document. – The addresses for free download provided in this book are, of course, far from being complete. Commercial sources are discarded, because 30 $ or € for a single paper is considered not to be a fair price.

[28] Special thank is due to Eric Wolfgang Weisstein (*1969), who has created and maintains mathworld.wolfram.com/ as well as scienceworld.wolfram.com/.

[29] Friedrich Beck (*1927) & Klaus Neitmann (*1954) (Eds.), *Brandenburgische Landesgeschichte und Archivwissenschaft. Festschrift für Lieselott Enders zum 70. Geburtstag*, 1997; Neitmann, Beck, Heinrich Kaak (*1950), Frank Göse (*1957), Jan Peters (1932 - 2011) & Wolfgang Neugebauer (*1953), *Lieselott Enders in memoriam. Das archiv- und geschichtswissenschaftliche Werk im Rückblick und im Ausblick*, 2011; P. Enders, *Veni – Vidi – Cassavi. Methodologische Gespräche zwischen Historikerin und Physiker*, 2011

[30] Hans Wolfgang Gerhart Enders (1924 - 1972), *Zur Kassation von Akten statistischer Dienststellen*, 1954; *Archivverwaltungslehre*, 1963…2004; P. Enders, *Gerhart Enders als Wissenschaftler*, 2015

CHAPTER 1

Introduction

Abstract: The Introduction provides the reader with a more detailed overview over the goal and the approach of this book, including the new features against previous publications. Special attention is paid to the importance of notions and to the relationship between classical and quantum mechanics. Another key topic is Schrödinger's forgotten 1926 requirements to any path from classical to quantum mechanics. The approach of this book fulfills all four of them, and this will help to overcome several difficulties of interpretation. Mittelstaedt's "reconstruction" of quantum mechanics is discussed in some detail, because there are several similarities to that approach. The Introduction concludes with the key lines along which this book is organized.

Keywords: Axiomatic, Classical mechanics, Mittelstaedt, Quantization, Quantum mechanics, Reconstruction, Relationship between classical mechanics and quantum mechanics, Schrödinger, State, Stationary state.

1.1. FEW GENERAL NOTIONS

"On y verra de ces fortes de demonstrations, qui ne produisent pas une certitude aussi grande que celles de Geometrie, & qui mesme en different beaucoup, puisque au lieu que les Geometres prouvent leurs Propositions par of Principes certains & incontestables, icy les Principes se verifient par les conclusions qu'on en tire; la nature de ces choses ne souffrant pas que cela se fasse autrement."[1]

"Durch die Untersuchungen über die Grundlagen der Geometrie wird uns die Aufgabe nahegelegt, nach diesem Vorbilde diejenigen physikalischen Disciplinen axiomatisch zu behandeln, in denen schon heute die Mathematik eine hervorragende Rolle spielt; dies sind in erster Linie die Wahrscheinlichkeitsrechnung und die Mechanik."[2]

---

[1] Christiaan Huygens (1629 - 1695), *Traité de la Lumière*, 1678/1690, Preface. – Shortly, in mechanics, the demonstrations are not of that great certitude as those of geometry. Geometers prove their propositions by fixed and incontestable principles. The principles of mechanics are verified by the conclusions drawn from them. The nature of these things does not allow for doing it otherwise. – See also Fokko Jan Dijksterhuis (*1965), *Lenses and Waves. Christiaan Huygens and the Mathematical Science of Optics in the Seventeenth Century*, 2005, in particular, *Ch. 5. 1677–1679 — Waves of Light. The road to the wave theory and the transformation of geometrical optics*, pp. 159-211.

[2] Hilbert, *Mathematische Probleme*, 1900, Sect. 6 – Engl.: "The investigations on the foundations of geometry suggest the problem: To treat in the same manner, by means of axioms, those physical sciences in which

The notions are the tools of thinking. Hence, the precision of thinking is by no means better than the precision of the notions it uses. A well-known example is the intermingling of the notions 'force' and 'energy'. For this, special attention will be paid to the 'labor of notion'.

1.1.1. Physics

> "Mathematics may be compared to a mill of exquisite workmanship, which grinds you stuff of any degree of fineness; but, nevertheless, what you get out depends upon what you put in; and as the grandest mill in the world will not extract wheat-flour from peascod, so pages of formulae will not get a definite result out of loose data."[3]

> "Physical Science is that department of knowledge which relates to the order of nature, or, in other words, to the regular succession of events.

> The name of physical science, however, is often applied in a more or less restricted manner to those branches of science in which the phenomena considered are of the simplest and most abstract kind, excluding the consideration of the more complex phenomena, such as those observed in living beings." (Maxwell, Matter and Motion, 1877, § 1)

Actually, the subject of this book is still narrower, *viz*, the classical and quantum mechanics of material systems of point-like bodies and particles, respectively.

1.1.2. Space Time Matter

Body: "A body is what which fills a place."[4]

Particle: "A body so small that, *for the purpose of our investigation*, the distances between its different parts may be neglected, is called a material particle." (Maxwell, *loc. cit.* § 6)

---

mathematics plays an important part; in the first rank are the theory of probabilities and mechanics." (Hilbert, *Mathematical Problems*, revis. transl. by David E. Joyce)

[3] Thomas Henry Huxley (1825 - 1895), *Geological Reform*, 1869 (quoted after Am. J. Phys. 58 (1990) 12, p. 1172)

[4] Newton, *De Gravitatione*, Def. 2 – This unfinished manuscript represents a milestone on Newton's way from Descartes to his *Principia*. It deals also with fundamental aspects of mechanics that are not accounted for in the *Principia*, because they play no role within *celestial* mechanics, for which the *Principia* were originally designed.

For the sake of clarity, I will use the word 'particle' for quantum particles and retain the word 'body' for classical particles.

As this book deals with the non-relativistic theory, Newton's concepts of relative (!) space and time (*Principia*, Definitions, Scholium) are adopted.

Further basic notions are the following ones.

Place: "A place is a part of space that a thing fills adequately." (Newton, *De Gravitatione*, Def. 1)

Rest: "Rest is remaining in the same place." (*Ibid.*, Def. 3)

Motion: "Motion is change of place." (*Ibid.*, Def. 4)

Internal and external relations or actions: "All relations between one part of this system and another are called Internal relations or actions. Those between the whole or any part of the system and bodies not included in the system are called External relations or actions." (Maxwell, *loc. cit.*, § 3)

Configuration: "[The] assemblage of relative positions is called the *Configuration* of the system." (*Ibid.*, § 4)

1.1.3. Extensive and Intensive Variables

"If you wish to converse with me," said Voltaire, "define your terms." How many a debate would have been deflated into a paragraph if the disputants had dared to define their terms! This is the alpha and omega of logic, the heart and soul of it, that every important term in serious discourse shall be subjected to the strictest scrutiny and definition. It is difficult, and ruthlessly tests the mind; but once done it is half of any task.[5]

A variable whose values depend on the quantity of substance under study is called *extensive*. Examples are volume, energy, mass, charge.[6]

---

[5] William (Will) James Durant (1885 - 1981), *The Story of Philosophy*, 2nd ed., Ch. 2, Pt. 3, p. 68
[6] *Cf* http://scienceworld.wolfram.com/physics/ExtensiveVariable.html.

A variable which is independent of the quantity of material present is called *intensive*. Examples are mass and charge densities, pressure.[7]

There are ordinary and characteristic intensive variables.[8] *Ordinary* intensive variables are the ratio of extensive variables. All densities belong to this type. *Characteristic* intensive variables characterize the direction of interaction processes and the corresponding equilibria when they cease. Examples are negative pressure, voltage, chemical potential.

Rules for intensive and extensive variables[9]:

Rule 1: The product of an intensive and an extensive variables is an extensive variable (example: momentum, see Problem 1.2)

Rule 2: All conserved quantities are extensive variables (example: energy). The reverse is not true (example: entropy).

Rule 3: Intensive variables may add when weighted appropriately (example: mass-weighted velocity).

Rule 4: The change of a conserved quantities equals the product of an characteristic intensive variable and of the change of an extensive variable (example: change of mass-weighted velocity).

Problem 1.1. *Show, that the velocity is an intensive variable!*

Problem 1.2. *Show, that the momentum is an extensive variable!*

1.2. WHAT IS QUANTUM MECHANICS?

"The calculation of eigenvalues is a problem of great practical and theoretical importance. Here are two very different types of application: in the dynamics of structures it is essential to know the resonance frequency of the structure… Another class of fundamental applications

---

[7] *Cf* http://scienceworld.wolfram.com/physics/IntensiveVariable.html.

[8] *Cf* Szücs Ervin (1930 - 2008), *Dialógusok a müszaki tudományokról* [Dialogues on Technical Processes], 1971, First and Second Talks.

[9] *Cf* Szücs, *loc. cit.*, Talks 1, 2, 4.

is related to the determination of the critical value of a parameter for the stability of a dynamical system…"[10]

What exactly is 'quantum mechanics' (QM)?

Literally, the mechanics of quanta.

Thus, what are quanta?

The first quanta in the sense of nowadays quantum physics are Planck's "energy elements"[11]. In order to provide his guessed radiation formula with a physical foundation, Planck postulated, that the energy of standing electromagnetic waves in resonators assumes discrete rather than continuous values, as in classical resonators, such as strings, pipes and cavities.

Many definitions refer to 'tiny': tiny bodies/particles, systems or actions. However, the action of a body at rest is zero, and quantum effects also occur on a macroscopic scale, *e.g*, in lasers and in superconductors.

For this, the interference phenomena of particles are much more striking. It is impossible to explain them in any classical way and has been considered to contain "the only mystery" of QM[12]. In the Ehrenberg-Siday-Aharonov-Bohm experiment[13], additionally, the interference pattern depends on the *static longitudinal* component of the vector potential, which does *not* act upon classical charges.

In bypassing I mention the observation, that oil drops on the surface of a liquid can behave in a quantum-like manner.[14] This, however, does not disprove the impossibility of classical explanation. It is rather an analogy, where physically different phenomena can be dealt with by means of the same mathematical formalism. Such analogies are seducing, but misleading. Speeding ahead, I notice

---

[10] Françoise Chatelin, *Eigenvalues of Matrices*, 1995, *Preface*, p. xi

[11] Max Planck (1858 - 1946, Nobel Award 1918), *Zur Theorie des Gesetzes der Energieverteilung im Normalspektrum*, 1900

[12] Richard Phillips Feynman (1918 - 1988), Robert Benjamin Leighton (1919 - 1997) & Matthew Linzee Sands (1919 - 2014), *The Feynman Lectures on Physics*, 2006, Vol. 3, Sect. 1. It seems, that entanglement is also a "mystery", see Section 4.5.

[13] Werner Ehrenberg (1901 - 1975) & Raymond Eldred Siday (1912 - 1956), *The Refractive Index in Electron Optics and the Principles of Dynamics*, 1949; Yakir Aharonov (*1932) & David Josef Bohm (1917 - 1992), *Significance of Electromagnetic Potentials in the Quantum Theory*, 1959; *Further considerations of electromagnetic potential in the quantum theory*, 1961 – here, Aharonov & Bohm point to the earlier prediction in the Ehrenberg & Siday paper.

[14] Pavel S. Kamenov & I. D. Christoskov, *Interference of classical particles on the surface of a liquid*, 1989

that, analogously, the discreteness of the energy spectrum of quantum systems should *not* be described by means of the same mathematics that leads to the discreteness of the wavelength of standing waves in resonators.

The interference of waves is described through the addition of amplitudes, say, $u(\vec{r}, t)$, rather than the addition of intensities, $|u(\vec{r}, t)|^2$. For this, interference phenomena of particles have been interpreted in assigning some amplitudes to them.

These amplitudes are "associated with an entire motion of a particle as a function of time, rather than simply with a position of the particle at a particular time", as in (Wentzel[15]-Dirac[16]-)Feynman's path integral method[17]. Here, the 'amplitude' is the wave function in the form $\exp(iS/\hbar)$, S being the classical action. Aside from the fact, that Schrödinger has *expressis verbis* refused that connection[18], the use of the *classical* expressions for the kinetic and potential energies for the description of *non*-classical motion lacks justification (*cf* Schrödinger's 4th reqirement quoted in the over-next subsection).

Another basic possibility to cope with the absence of classical locations, at which a particle EITHER *is*, OR *is not*, consists in the introduction of propositions about physical observables, where a particle *may* be at a given position – quantum logic[19]. This leads to a description of states by rays in Hilbert spaces[20]. However, the use of the *classical* expressions for the operators needs justification, again.

[15] Gregor Wentzel (1898 - 1978), *Zur Quantenoptik*, 1924

[16] Paul Adrien Maurice Dirac (1902 - 1984, Nobel Award 1933), *The Lagrangian in Quantum Mechanics*, 1933

[17] Feynman, *Space-Time Approach to Non-Relativistic Quantum Mechanics*, 1948; Feynman & Albert Roach Hibbs (1924 - 2003), *Quantum Mechanics and Path Integrals* (emended by D. F. Styer), 2005

[18] Erwin Rudolf Josef Alexander Schrödinger (1887 - 1961), *Quantisierung als Eigenwertproblem. Zweite Mitteilung*, 1926, p. 489, fn. 1

[19] The originating paper is Garrett Birkhoff (1911 - 1996) & John von Neumann (János Neumann Margittai / Neumann János Lajos, 1903 - 1957), *The logic of quantum mechanics*, 1936. Charles Francis, *Quantum Logic*, 2010, provides a representation that is tailored to the needs of physicists, who wish quickly catch upon.

[20] Hilbert, Nordheim & v. Neumann, *Über die Grundlagen der Quantenmechanik*, 1927; see also Hermann Klaus Hugo Weyl (1885 - 1955), *Gruppentheorie und Quantenmechanik*, 1928; v. Neumann, *Mathematische Grundlagen der Quantenmechanik*, 1932

Further phenomena contradicting classical physics concern the stability of the atoms and the discreteness of the atomic spectral lines; few more examples will be presented in Chapter 5 below.

1.3. CLASSICAL *VERSUS* QUANTUM MECHANICS

> CM = "yes, we can" – QM = "no, we can't"
> CM = Beethoven symphony, QM = Webern symphony[21]

> "Instead of viewing classical mechanics as a theory that had been replaced, he [Dirac] saw it as a theory that should continue to be developed, modified, and extended. He took no part of physics to be a permanent achievement, correct for all time."[22]

From the very beginning (Planck, 1900), the development of quantum physics is indivisibly connected with the discussion of its relationship to classical physics. And even after the overwhelming success and general acceptance of quantum theory, it had to be stated, that

> "It is in principle impossible, however, to formulate the basic concepts of quantum mechanics without using classical mechanics."[23]

Indeed, all formulations of QM known to me exploit the classical expressions for position, momentum, kinetic and potential energies, and so on with the only justification that 'it works'.

Thus, there are various ways to lay down the foundations of QM. They differ, first, in their view on the *relationship between QM and CM*.

Some approaches stress the differences between CM and QM, while other ones – like this book – seek for their commons.

On discussing the notion of path for classical and for quantum particles, Schrödinger concluded,

---

[21] Ignacio de Gispert Pastor, *What is the basic difference between classical mechanics and quantum mechanics?*, 2012

[22] Alisa Bokulich, *Open or Closed? Dirac, Heisenberg, and the Relation between Classical and Quantum Mechanics*, 2004

[23] Lev Davidovich Landau (1908 - 1968, Nobel Award 1962) & Evgeny Mikhailovich Lifshitz (1915 - 1985), *Course of Theoretical Physics. 3. Quantum Mechanics*, 1959

"We are faced here with the full force of the logical opposition between an

EITHER – OR (point mechanics)

and a

BOTH – AND (wave mechanics)

This would not matter much, if the old system were to be dropped entirely and to be replaced by the new one.

Unfortunately, this is not the case."[24]

Schrödinger's contemplation means, in particular, that solely a deeper analysis of the roots and foundations of CM provides the keys for understanding QM.

Second, the various ways to lay down the foundations of QM differ in *how* to use CM.

The pioneers of quantum theory understood CM as being necessary, but not sufficient. For instance, Bohr assumed,

"(1) That the dynamical equilibrium of the [atomic] systems in the stationary states can be discussed by help of the ordinary mechanics, while the passing of the systems between different stationary states cannot be treated on that basis.[25]

(2) That the latter is followed by the emission of a homogeneous radiation, for which the relation between the frequency and the amount of energy emitted is the one given by Planck's radiation theory."[26]

Consequently, *additional* assumptions have been made[27], *e.g,*

- to restrict the energy spectrum to *discrete* values (Bohr, 1918), such as *nhv*

---

[24] Schrödinger, *The fundamental idea of wave mechanics,* Nobel Lecture 1933. – This both – and occurs also in the path integral representation of QM (Feynman, *Space-Time Approach to Non-Relativistic Quantum Mechanics,* 1948), *cf* Heinz Lübbig, *Das Wirkungsprinzip von Maupertuis und Feynmans Wegintegral der Quantenphase,* 1999.

[25] We will see, that this statement holds true for Newton's, but *not* for Euler's representation of CM.

[26] Niels Hendrik David Bohr (1885 - 1962, Nobel Award 1922), *On the Constitution of Atoms and Molecules,* 1913, p. 7; see also *The quantum theory of line spectra,* 1918, § 1.

[27] For a critical review of modern textbook treatments, see John S. Briggs (*1942) & Jan-Michael Rost, *On the derivation of the time-dependent equation of Schrödinger,* 2001, pp. 698f.

(Planck, 1900, Einstein, 1907) or $\frac{n}{2} h\nu$ (Bohr, 1913);

- to quantize the phase space into cells of *finite* size: $\iint dqdp = h$ (Planck, 1913, eq. (210));

- to "*distinguish*" (Heisenberg, 1977) or to "*select*" (Pauli, 1926[28]) the values $n\hbar$ of the action integral, $\oint pdq$;

- to suppose the existence of h and the absence of classical paths (Heisenberg, 1925)[29];

- to replace the trajectories, $x(t)$, with transition amplitudes, x_{mn} (*ibid.*)[30];

- to suppose the existence of h and of a wave function being the solution of an eigenvalue problem (Schrödinger, 1926);

- to assume, that the classical expressions can be used for quantum equations or in the Hilbert space that emerges from quantum logic;

- to suppose the existence of transition probabilities[31], which obey the Chapman-Kolmogorov equation (see Section 12.2).

However, *none* of those approaches satisfies all four of Schrödinger's requirements quoted below. In particular, the change of the meaning of kinetic and potential energies is not accounted for. This contributes to the difficulties of interpreting QM and leads to misunderstandings like the 'tunnel effect', as we will see in Section 10.1.[32]

---

[28] See also Albert Messiah (1921 - 2013), *Quantum Mechanics*, 1999, Ch. I, § 15.

[29] Einstein concluded the necessity to abandon the classical equations of motion from the invalidity of the equipartition theorem within quantum statistic, see *Bietet die Feldtheorie Möglichkeiten für die Lösung des Quantenproblems?*, 1923, p. 360, fn. 1.

[30] Bartel Leendert van der Waerden (1903 - 1996) rightly mentions the *contradiction* between, (i), Heisenberg's statement, that the trajectory, $\vec{r}(t)$, of a quantum particle is not observable [actually, within QM, it even does not exist], and, (ii), the appearance of the position variable, $\vec{r}$, as 'observable' (*Sources of Quantum Mechanics*, 1967).

[31] Feynman, *Space-Time Approach to Non-Relativistic Quantum Mechanics*, 1948; nowadays known as path integral method, see, *e.g*, Feynman & Hibbs, *Quantum Mechanics and Path Integrals*, 2005; Hagen Kleinert (*1941), *Path Integrals in Quantum Mechanics, Statistics, Polymer Physics, and Financial Markets*, 2009

[32] In contrast to classical bodies moving along trajectories, a quantum particle with energy E can 'move' through areas with potential energy $V(x) > E$. I will explain this behavior through the fact, that the physically

Furthermore, there are two opposite claims about the role CM is playing for QM.

1. CM is the limit case of QM, which thus does not need to refer to CM;

2. QM needs CM for its foundation.

This book supports the second claim. It hence does not follow the point of view, that CM cannot be fully understood without using QM[33].

Moreover, there are at least six manners of going from CM to QM.[34]

1. Pose additional assumptions upon the motion of bodies/particles, in order to obtain a discrete set of stationary states; best known are the quantum conditions used before 1925[35];

2. Assume a wave equation for modeling the wave-like behavior of small particles;

3. Introduce additional mechanisms; for instance, let vacuum fluctuations smear out the classical trajectories[36];

4. Reinterpret classical formulae[37];

5. Find principles, which cannot or not completely be implemented within CM[38];

6. Represent CM such, that other branches of physics can be derived from it: Hertz's program[39].

---

relevant potential energy is not the classical potential energy, $V(x)$, but the non-classical expression $x_E|\psi_E(x)|^2V(x)$, x_E being a characteristic extension and $\psi_E(x)$ the time-independent wave function.

[33] See, *e.g*, Fritz (Friedrich Arnold) Bopp (1909 - 1987), *Grundlagen der klassischen Physik in gegenwärtiger Sicht*, 1971, p. 98.

[34] See also Styer, Balkin, Becker, Burns, Dudley, Forth, Gaumer, Kramer, Oertel, Park, Rinkoski, Smith & Wotherspoon, *Nine formulations of quantum mechanics*, 2002.

[35] For a comprehensive review of the historical development, see Jagdish Mehra (1931 - 2008) & Helmut Rechenberg (*1937-2016), *The Historical Development of Quantum Theory*, 1999ff.

[36] Edward Nelson (*1932), *Feynman Integrals and the Schrödinger Equation*, 1964; *Derivation of the Schrödinger Equation from Newtonian Mechanics*, 1966; L. Fritsche & M. Haugk, *A new look at the derivation of the Schrödinger equation from Newtonian mechanics*, 2003.

[37] See, for instance, Maurice A. de Gosson (*1948) & Basil Hiley (*1935), *Imprints of the Quantum World in Classical Mechanics*, 2010.

[38] Alon E. Faraggi (*1961) & Marco Matone, *The Equivalence Postulate of Quantum Mechanics*, 1999; their path from CM to QM appears to fail for the state at rest as that state is absent within QM.

[39] Hertz, *Die Prinzipien der Mechanik in neuem Zusammenhange dargestellt*, 1910, *Preface*

This book follows the sixth way: Hertz's program. It concentrates on the commons rather than on the differences between CM and QM. Basing on Euler's axiomatic of CM, the axiomatic is kept at a minimum. In contrast to Newton's one, Euler's axiomatic allows for *deductive* generalizations of CM. Examples are special-relativistic mechanics[40] and the (gravito-)electromagnetic equations (see Subsection 17.2.5) as well as their linear generalizations[41] like that by Proca[42] and Yukawa[43]. This is related to the methodological point of view, that any (branch of) science is determined by both its subject and its methods of investigation.

As a matter of fact, a recent observation demonstrates an even closer relationship between CM and QM than known before: The Schrödinger equation for a non-relativistic spin-less particle is in a certain sense equivalent to Hamilton's equations of motion.

> "There is a one-to-one and onto correspondence between Hamiltonian flows generated by a [classical] Hamiltonian H and strongly continuous unitary one-parameter groups satisfying Schrödinger's equation with Hamiltonian operator $\widehat{H} = H(x, -i\hbar\nabla, t)$ obtained from H by Weyl quantization."[44]

At once, de Gosson & Hiley meritoriously stress, that this represents a derivation of the time-dependent Schrödinger equation (p. 7), but *not* a derivation of QM from CM (p. 4). QM is not just an algorithm for certain formal operations on classical equations.[45] That correspondence describes another imprint left by QM in CM. – For me, it describes another possibility to go from CM to QM *without any* additional postulates. However, as it is a purely mathematical way,

- the physical meaning of the wave function remains undetermined;

- the change of the meaning of the potential and kinetic energies is obscured.

---

[40] Suisky & Enders, *Dynamische Begründung der Lorentz-Transformation*, 2005; Suisky, *Euler's mechanics as a unified theory of matter and motion*, 2007

[41] Enders, *Towards the Unity of Classical Physics*, 2009; *Physical, metaphysical and logical thoughts about the wave equation and the symmetry of space-time*, 2011. The existence of electrical charges is not obtained this way, of course.

[42] Alexandru Proca (1897 - 1955), *Sur la théorie ondulatoire des électrons positifs et négatifs*, 1936. – In Michel Gondran, *The Proca equations derived from first principles*, 2009, "first principles" means the correspondence principle as to replace E with $i\hbar\,\partial/\partial t$ and $\vec{p}$ with $-i\hbar\nabla$. This presupposes QM, in contrast to Hertz's program.

[43] Hideki Yukawa (1907 - 1981, Nobel Award 1949), *On the interaction of elementary particles I*, 1935

[44] de Gosson & Hiley, *Imprints of the Quantum World in Classical Mechanics*, 2010

[45] See also George Whitelaw Mackey (1916 - 2006), *The Relationship Between Classical and Quantum Mechanics*, 1998, p. 106 (cited after de Gosson & Hiley, *loc. cit.*, p. 4).

Of course, Newton, Euler and Helmholtz did not have any imagination of quantum physics. However, their representations of CM are much more general than usually taught as they contain many elements that are applicable beyond the realm of CM. As we will see lateron, Newton's and Euler's notions of [stationary] state is much closer to QM as well as to classical and quantum statistical mechanics than the nowadays notion of [*non*-stationary] state. Euler's principles of change of stationary states hold still true in QM. Helmholtz's approach to the mechanical energy conservation law using Leibniz's theorem on the conservation of living force facilitates the way to QM as it can largely be formulated without trajectories.

The quantization of kinematic relations (Heisenberg, 1925) or of classical equations of motion (Born, Heisenberg & Jordan, 1925f.) do *not* provide the stationary-state quantities[46]. Schrödinger's recurrence to Hamilton's analogies between CM and optics[47] led to the difficulties quoted above. The common treatments of Hamiltonian mechanics concentrate on his principles of least action[48] and on his equations of motion. On the contrary, this book stresses the role of the Hamiltonian as a generalization of Newton's *stationary*-state function (the momentum vector of a free body) to the situation of a body subject to an external force field. As a matter of fact, this represents one part of the bridge to atomic mechanics searched for by Schrödinger.

1.4. SCHRÖDINGER'S REQUIREMENTS TO ANY QUANTIZATION

"I've found that while many people are seeking ever more complex solutions to ever more abstract issues, a lot of the basics are ignored."[49]

Schrödinger begins the first of his four pioneering communications *Quantization as Eigenvalue Problem*[50] with the enthusiastic claim,

"§ 1. In dieser Mitteilung möchte ich zunächst an dem einfachsten Fall des (nichtrelativistischen und ungestörten) Wasserstoffatoms zeigen, dass die übliche Quantisierungsvorschrift sich durch eine andere

---

[46] *Cf* van der Waerden, *Sources of Quantum Mechanics*, 1967, *Introduction*

[47] Sir William Rowan Hamilton (1805 - 1865), *On a General Method in Dynamics*, 1834; *Second Essay on a General Method in Dynamics*, 1835

[48] See, *e.g*, the classic Cornelius Lanczos (Kornél Löwy, Kornél Lánczos; 1893 - 1974), *The Variational Principles of Mechanics*, 41970.

[49] John B. Merryman, quoted after Le Tat Dieu, einsteinerrs.com/newtons-laws.html

[50] Schrödinger, *Quantisierung als Eigenwertproblem*, 1926 – For a reprint of Schrödinger's letters about his derivation of wave mechanics, see Karl von Meyenn (*1937), *Eine Entdeckung von ganz außerordentlicher Tragweite: Schrödingers Briefwechsel zur Wellenmechanik und zum Katzenparadoxon*, 2011.

Forderung ersetzen lässt, in der kein Wort von "ganzen Zahlen" mehr vorkommt. Vielmehr ergibt sich die Ganzzahligkeit auf dieselbe natürliche Art, wie etwa die Ganzzahligkeit der Knotenzahl einer schwingenden Saite. Die neue Auffassung ist verallgemeinerungsfähig und rührt, wie ich glaube, sehr tief an das wahre Wesen der Quantenvorschriften."[51]

Since then, this enthusiasm dominates the perception of wave mechanics, *although* Schrödinger himself posed quite *contrary* requirements as given below.

Planck[52] welcomed the "natural", or 'intuitive' aspect, *i.e*, the (erroneous) impression, that Schrödinger's wave mechanics is much closer to CM than the matrix mechanics that had been invented by Heisenberg[53] one year before and quickly developed by him, Born and Jordan[54]. The still highly influential Lorentz[55] immediately realized the significance of Schrödinger's approach and, at once, the difficulties of its interpretation. Einstein, the work of which on the wave-particle duality of light radiation[56] had laid the foundation for de Broglie's 'matter waves'[57],

---

[51] Schrödinger, *1st Commun.*, 1926, p. 361 – Engl.: § 1. In this article I would like to show – using the simplest case of the (non-relativistic and unperturbed) hydrogen atom – that the usual rule for quantization can be replaced by another requirement in which there is no longer any mention of 'integers'. In fact, the integral property follows in the same natural way that, say, the number of nodes of a vibrating string must be an integer. The new interpretation can be generalized and, I believe, strikes very deeply into the true nature of the quantization rules.

[52] After Martin Jesse Klein (1924 - 2009), *Introduction*, in: Albert Einstein (1879 - 1955, Nobel Award 1921), *Letters on Wave Mechanics*, 2011

[53] Werner Karl Heisenberg (1901 - 1976, Nobel Award 1932), *Über quantentheoretische Umdeutung kinematischer und mechanischer Beziehungen*, 1925

[54] Max Born (1882 - 1970, Nobel Award 1954) & Pascual Jordan (1902 - 1980), *Zur Quantenmechanik*, 1925; Born, *Zur Quantenmechanik der Stoßvorgänge*, 1926; *Zur Wellenmechanik der Stoßvorgänge*, 1926; Born, Heisenberg & Jordan, *Über Quantenmechanik II*, 1926 (the 'Three men paper')

[55] Hendrik Antoon Lorentz (1853 - 1928, Nobel Award 1902), *Letter to Schrödinger*, in: Einstein, *Letters on Wave Mechanics*, 2011, letter 19

[56] Einstein, *Über einen die Erzeugung und Verwandlung des Lichtes betreffenden heuristischen Gesichtspunkt*, 1905; *Zur Theorie der Lichterzeugung und Lichtabsorption*, 1906; *Zum gegenwärtigen Strand des Strahlungsproblems*, 1909; *Über die Entwicklung unserer Anschauungen über das Wesen und die Konstitution der Strahlung*, 1909; *Strahlungs-Emission und -Absorption nach der Quantentheorie*, 1916; *Zur Quantentheorie der Strahlung*, 1916; *Bietet die Feldtheorie Möglichkeiten für die Lösung des Quantenproblems?*, 1923; *Anmerkung zu S. N. Boses Abhandlung Plancks Gesetz und Lichtquantenhypothese*, 1924

[57] Louis-Victor-Pierre-Raymond, 7th duc de Broglie (1892 - 1987, Nobel Award 1929), *Recherches sur la théorie des quanta*, 1924

misread Schrödinger's equation and proposed to Schrödinger a correction that was actually the original equation.[58]

Pauli wrote to Jordan,

> "Ich glaube, dass diese Arbeit mit zu dem Bedeutendsten zählt, was in letzer Zeit geschrieben wurde. Lesen Sie sie sorgfältig und mit Andacht."[59]

In contrast, Heisenberg complained, that

> "Schrödinger really simply throws overboard everything in quantum theory; namely, the photoelectric effect[60], the Franck[-Hertz] collisions[61], the Stern-Gerlach effect[62], *etc*. Then it isn't hard to make a theory."[63]

Feynman, finally, is told to have stated, that the Schrödinger equation cannot be derived, as it emerged out of Schrödinger's head.[64]

Unfortunately, the success of Schrödinger's eigenvalue mathematics as well as the discussions about the meaning of the wave function, ψ, and of its modulus-squared, $|\psi|^2$, made Schrödinger's own criticism of his eigenvalue formalism to be discarded.[65]

---

[58] Both Einstein and Schrödinger never agreed with the 'Copenhagen interpretation' of quantum mechanics, see Klein, *Introduction, loc. cit.*; Norbert Straumann (*1936), *Schrödingers Entdeckung der Wellenmechanik*, 2001, p. 23.

[59] Wolfgang Ernst Pauli (1900 - 1958, Nobel Award 1945), *Letter to P. Jordan*, 12. April 1926; in: Arnold Johannes Wilhelm Sommerfeld (1868 - 1951), *Wissenschaftlicher Briefwechsel*, 1979, Vol. I, Doc. [131], quoted after Straumann, *Schrödingers Entdeckung der Wellenmechanik*, 2001. – Engl.: I believe, that this paper belongs to the most important one that has been written recently. Read it carefully and with devotion.

[60] Einstein (1905) had explained it by applying Planck's quantum hypothesis for light in resonators to light rays in free space.

[61] James Franck (1882 - 1964, Nobel Award 1925) & Gustav Hertz (1887 - 1975, Nobel Award 1925), *Über Zusammenstöße zwischen Elektronen und Molekülen des Quecksilberdampfes und die Ionisierungsspannung desselben*, 1914

[62] Walter Gerlach (1889 - 1979) & Otto Stern (1888 - 1969, Nobel Award 1949), *Der experimentelle Nachweis der Richtungsquantelung im Magnetfeld*, 1922; *Das magnetische Moment des Silberatoms*, 1922

[63] Quoted after Klein, *Introduction, loc. cit.*

[64] Richard Feynman (1918 - 1988, Nobel Award 1965), after M. Komma, *Mathematische Behandlung der Schrödingergleichung*, 2003. Komma rightly comments, that *all* theories emerge in heads.

[65] Remarkably enough, Schrödinger has formulated these requirements *during* his development of wave mechanics, and this the more as they criticize *his own* work.

1. The "quantum equation" should "carry the quantum conditions in itself" (*Second Commun.*, pp. 511f.)[66];

2. There should be a special mathematical method for solving the stationary Schrödinger equation, which accounts for the *non*-classical character of the quantization problem, *i.e*, which is *different* from the classical methods for calculating the eigenmodes of strings, resonators and so on[67];

3. The derivation should uniquely decide, that the energy rather than the frequency values are discretized, since frequency discretization is a *classical* phenomenon (*ibid.*, pp. 511, 519)[68];

4. The use of the *classical* expressions for the potential and kinetic energies should be justified (*Fourth Commun.*, p. 113).

The approach presented here is the only one I am aware of that observes these requirements.

Thus, quantization is not really an eigenvalue problem. Indeed, quantization is *much more* than the classical 'discretization as eigenvalue problem' found, *e.g*, in strings and resonators.

Historically, the embedding of QM into the theory of Hilbert spaces[69] has sanctified the boundary conditions.

---

[66] As a matter of fact, the boundary conditions belong *not* to the differential equation, but to the differential *operator*. Hence, the quantum conditions should be independent of boundary conditions. Like de Broglie's condition, which interprets Bohr's 'quantum orbits' as standing waves (*Thèses*, 1924), they are *classical* conditions (*cf* also Heisenberg, *Die Geschichte der Quantentheorie*, 1977).

[67] Schrödinger (*ibid.*, p. 512) remarks, that the usual solution methods for eigenvalue problems are indirect ones in that first the general solution is constructed and that only then, its parameters are adjusted such, that the boundary conditions are fulfilled. I will show that *the quantization is independent of the boundary conditions*.

[68] In his calculations, Schrödinger has replaced the frequency, ν, with E/h. Otherwise, he had obtained $H\psi = h\nu\psi$, whereby ν would be discretized. Wavelength and frequency are *not* dynamical variables like position and momentum, and *not* stationary-state variables like total energy and angular momentum.

[69] Hilbert, Nordheim & von Neumann, *Über die Grundlagen der Quantenmechanik*, 1927; see also Weyl, *Gruppentheorie und Quantenmechanik*, 1928; von Neumann, *Mathematische Grundlagen der Quantenmechanik*, 1932

1.5. MITTELSTAEDT'S "RECONSTRUCTION" OF QUANTUM MECHANICS *

Dedicated to Peter Mittelstaedt (1932 - 2014)

Our approach is methodologically related to Mittelstaedt's "reconstruction" of quantum mechanics[70]. Both ones draw relevant conclusions from the history of physics and consider CM *not* to be just a limit case of QM.

Mittelstaedt "reconstructs" QM through "abandoning or relaxing the various not justified metaphysical ontological hypotheses contained in" Newtonian mechanics (p. x). These hypotheses are (p. 7),

Conjecture O(C)[1] There exists an absolute time. It establishes an universal temporal order of two or more events, it provides a universal measure of time and it explains the concept of simultaneity of two spatially separated events.[71]

Conjecture O(C)[2] There exists an absolute space. It explains the concepts of absolute motion and absolute rest. Euclidean Geometry applies to this absolute space.[72]

Conjecture O(C)[3] There are individual and distinguishable objects.[73] These objects cannot only be named and identified at a certain instant of time, but also re-identified at any later time. Generally they possess also a temporal identity.

Conjecture O(C)[4] These objects possess elementary properties in the following sense. An elementary property, P_λ, refers to an object system such, that either P_λ or the counter property, $\bar{P}_\lambda$, pertains to the system. Furthermore, objects are subject to the law of thoroughgoing determination according to which "if all predicates are taken together with their contradictory opposites then one of each pair of contradictory opposites must belong to it" (Immanuel Kant).

---

[70] Mittelstaedt, *Rational Reconstructions of Modern Physics*, 2011; see also *The Problem of Interpretation of Modern Physics*, 2011

[71] Notice, that Newton has stressed the fact, that physical measurements are done within *relative* time (*Principia*, Definitions, Scholium).

[72] Notice, that Newton has stressed the fact, that physical measurements are done within *relative* space (*ibid.*).

[73] This formulation suggests, that (in)distinguishability is a property of bodies/particles. It will be shown below, that – on the contrary – (in)distinguishability is a property of *states*.

Conjecture O(C)[5] For objects of the external objective reality, the causality law holds without any restriction. There is an unbroken causality.[74]

Conjecture O(C)[6] For objects of the external objective reality, the law of conservation of substance holds without any restriction.

Mittelstaedt eliminates or relaxes,

• conjecture **O(C)[1]** for obtaining Minkowski[75] rather than Galileo space-time[76];

• conjecture **O(C)[1,2]** for obtaining the pseudo-Riemannian geometry of the general theory of relativity;

• conjectures **O(C)[3...6]** for reconstructing non-relativistic QM.

He admits,

"The coupling between matter, the source of the gravitational field, and the geometry of space-time, cannot be obtained merely by eliminating or relaxing convenient conjectures $O(C)^k$ of classical mechanics or classical ontology, respectively..." (pp. 8f.)

This favors Euler's approach to derive the dynamics from the properties of the objects under consideration.

Mittelstaedt's reconstruction of QM concentrates on the discussion of certain variants of quantum logic. Quantum logic leads to a Hilbert space, but not to the concrete form of the operators to be used therein. Hence, it does not fulfill Schrödinger's 4th requirement quoted in the foregoing section.

---

[74] This, perhaps, refers to causality in the sense of Laplace's demon (Pierre Simon Marquis de Laplace (1749 - 1827), *Essai Philosophique sur la Probabilité*, 1814, *Introduction*). As a matter of fact, the time-dependent Schrödinger equation represents a *deterministic* evolution equation for the wave function. The latter is related to the probability of the outcome of experiments. The fact, that the quantum-mechanical predictions of those outcomes is formulated in terms of probabilities, has seduced many researchers to abandon causality. Strictly speaking, however, 'broken causality' implies 'broken laws', and 'broken laws' implies 'broken science'. In other words, science is impossible without causality, *cf* Enders, *Physical, metaphysical and logical thoughts about the wave equation and the symmetry of space-time*, 2011.

[75] Hermann Minkowski (1864 - 1909), *Raum und Zeit*, 1908

[76] Strictly speaking, there is *no* 'Galilean space-time', because the Galileo transform cannot be represented by means of a symmetric metric tensor, see Enders, *loc. cit.*

1.6. KEY LINES OF THIS BOOK

Thus, this approach benefits from Mittelstaedt's one and at once differs from it in that,

  i. it uses not only Newton's, but also Euler's axiomatic of CM, because, here, generalizations can be made *without* touching the axiomatic;

 ii. it does not exploit metaphysical/ontological arguments, but poses a physical question, which is allowed, but cannot be answered within CM, and which thus transcends CM.

Accordingly, this book is organized according to the following Newton-Eulerian key steps:

Conservation of stationary states – change of stationary states – motion.

Moreover,

• Helmholtz's treatment of energy conservation induces a *selection problem* between possible and impossible configurations of classical systems. This selection is related to the following point.

• Planck (1900) and Einstein (1907) discriminate between classical systems and atomic aggregates through the fact, that the latter ones exhibit a *less* amount of stationary states than those.[77] This idea is exploited for solving the stationary Schrödinger equation as a *selection* rather than as an eigenvalue problem. This way, Schrödinger's 2nd requirement quoted in Section 1.4 above will be fulfilled.

• The energy law itself does not imply the existence of turning points for an oscillator. This allows for the question, which is the mechanics of oscillators *not* exhibiting classical turning points and thus assuming BOTH classically allowed AND classically forbidden configurations?

---

[77] Planck resorts to Ludwig Eduard Boltzmann (1844 - 1906), *Über die Beziehung zwischen dem zweiten Hauptsatze der mechanischen Wärmetheorie und der Wahrscheinlichkeitsrechnung respektive den Sätzen über das Wärmegleichgewicht*, 1877. This seminal paper – featuring the entropy of a thermodynamic state to be proportional to the logarithm of the probability of that state – is now available in English: *On the Relationship between the Second Fundamental Theorem of the Mechanical Theory of Heat and Probability Calculations Regarding the Conditions for Thermal Equilibrium*, 2015.

The answer to this questions results in a non-classical representation of the total energy, the stationary Schrödinger equation being a consequence of it.

- Schrödinger's imagination of the meaning of the modulus-squared wave function, $|\psi|^2$, as "a kind of weight function" (*4th Commun.*, § 7) is recovered and clarified. Moreover, *square-integrable* wave functions for (interaction-) *free* particles emerge.

- The Eulerian principles of stationary-state change for classical bodies can be reformulated in terms of quantum-mechanical quantities, *without* additional requirements. Due to that, the time-dependent Schrödinger equation can be *axiomatically* derived, too. A nice byproduct is the Hellmann-Feynman theorem[78] in Section 11.5.

- Using Newton's and Euler's rather than Lagrange's and Laplace's notions of state, the symmetry of classical stationary-state quantities (total energy, total momentum, total angular momentum) is exploited to found the symmetry of quantum quantities, notably that of many-body wave functions.

- The symmetry of classical systems as the basis of the symmetry of quantum systems is extended to the old rule 'aequat causa effectum'.

The dualism between bodies (particles) and fields (waves) in classical physics is deliberately *not* simply transferred to quantum physics. For both aspects needs to be *synthesized*[79]. The interference experiments with mesoscopic particles like fullerenes[80] and even larger molecules[81] show a rather continuous transition between quantum and classical mechanics.[82]

---

[78] Hans Gustav Adolf Hellmann (1903 - 1938), *Einführung in die Quantenchemie*, 1937; Feynman, *Forces in molecules*, 1939; see also David Carfi, *The Pointwise Hellmann-FeynmancTheorem*, 2010.

[79] See, *e.g*, Heisenberg, *Die physikalischen Prinzipien der Quantentheorie*, 1929; de Broglie, *Certitude et incertitude de la science*, 1966.

[80] Markus Arndt (*1965), Olaf Nairz, Claudia Keller, Gerbrand van der Zouw & Anton Zeilinger (*1945), *Wave–particle duality of C_{60} molecules*, 1999.

[81] Stefan Gerlich, Sandra Eibenberger, Mathias Tomandl (*1984), Stefan Nimmrichter, Klaus Hornberger, Paul J. Fagan, Jens Tüxen, Marcel Mayor (*1965) & Arndt, *Quantum interference of large organic molecules*, 2011

[82] The recently evoked influence of gravitation on this transition (Igor Pikovski, Magdalena Zych, Fabio Costa, Časlav Brukner (*1967), *Universal decoherence due to gravitational time dilation*, 2015) sounds seducing, but is far beyond the scope of this book.

Part I

Classical-Mechanical Systems

"Die Aufgabe … ist diese, … eine vollkommen bestimmte Zusammen-stellung der Gesetze der Mechanik anzugeben, welche mit dem Stande unserer heutigen Kenntnis verträglich ist, welche nämlich in Beziehung auf den Umfang dieser Kenntnis weder zu eng ist, noch zu weit."[83]

In a non-mechanistic sense, Hertz aims to represent CM such, that other branches of physics can be derived from it through sensible generalizations. I will call this task *Hertz's program*. It is a program towards the *practical* unity of physics. This book is an example for its feasibility. Another example is the derivation of special-relativistic mechanics through lifting an implicit assumption in Euler's derivation of Newton's equation of motion.[84]. Helmholtz's explorations of the relationship between forces and energies can be generalized toward classical electromagnetism and related equations, see Subsection 17.2.5.

This book presents CM such, that its generalization to QM can be done in a smooth manner, *i.e.*, without any assumptions like waves or Planck's constant. The most important corner stones are, (i), the notion of state and, (ii), Euler's axiomatic of CM.[85]

---

[83] Heinrich Rudolf Hertz (1857 - 1894), *Die Prinzipien der Mechanik*, 1894, *Vorwort des Verfassers*, [2]1910, pp. XIX f. – Engl.: "The problem which I have endeavoured…is the following:– To fill up the existing gaps and to give a complete and definite presentation of the laws of mechanics which shall be consistent with the state of our present knowledge, being neither too restricted nor too extensive in relation to the scope of this knowledge. The presentation must not be too restricted: there must be no natural motion which it does not embrace. On the other hand it must not be too extensive: it must admit of no motion whose occurrence in nature is excluded by the state of our knowledges." (English ed., 1899, *Author's Preface*, pp. xxi f.)

[84] Suisky & Enders, *Dynamische Begründung der Lorentz-Transformation*, 2005

[85] For the historic references I have used – among others – Károly Simonyi (1916 - 2001), *Kulturgeschichte der Physik*, 1990; István Szabó (1906 - 1980), *Geschichte der mechanischen Prinzipien und ihrer wichtigsten Anwendungen*, 1996; René François Dugas (1897 - 1957), *History of Mechanics*, 1955; Ernst Waldfried Josef Wenzel Mach (1838 - 1916), *Die Mechanik in ihrer Entwicklung*, 1988; historically conscious representations like Sommerfeld's lectures; dictionaries of scientists like David, Ian, John & Margaret Millar, *The Cambridge Dictionary of Scientists*, 1996; general dictionaries like Josef Naas (1906 - 1993) & Hermann Ludwig Schmid (1908 - 1956), *Handbuch der Mathematik*, 1974; last but not least, the Internet.

CHAPTER 2

Newton-Eulerian Axiomatic: The Notion of State

Abstract: This chapter deals with the all-important notion of state as it is central to *all* parts of physics. For this book, most important are, on the one hand, the notions of state by Newton and Euler – on the other hand, the notion of state by Lagrange and Laplace used nowadays. Surprisingly enough, Newton and Euler's notions of state are closer to quantum mechanics than the nowadays' one. Newton's axiomatic contains solely the conservation of states (1st and 3rd axioms) and the change of states (2nd axiom). In contrast, the equation of motion is *not* part of the axiomatic. Euler's axiomatic contains solely the conservation of states. The change of state (and subsequently the equation of motion) is to be examined according to the problem under consideration. In contrast to Newton's axiomatic and correcting Bohr's corresponding claim, Euler's principles of state change for classical bodies are formulated such, that they can be translated even to quantum-mechanical motion. Their power is also demonstrated through an alternative derivation of Hamilton's equation of motion.

Keywords: d'Alembert's principle, Axiomatic, Dynamics, Euler's principles of stationary-state change, Hamiltonian, Inertia, Newton's axioms, State, State function, Statics, Stationary state.

The notion of state is central to *all* parts of physics. Its concrete content changes from branch to branch and even during the history of physics, so that it is impossible to assign a general definition to it. For this, I will confine myself on the differences between Newton's and Euler's notions of state on the one hand, and Lagrange's and Laplace's notion of state used nowadays on the other hand. These differences are of paramount importance not only for the relations between CM and QM, but also for statistical mechanics.[1] Accordingly, Newton's and Euler's forgotten notion of state will be sketched in somewhat more detail.

In contrast to the Lagrangian, the Hamiltonian represents a generalized Newtonian (stationary-)state function. This makes a novel, more direct way from Newton's to Hamilton's mechanics possible.

Referring to Newton's and Euler's line of reasoning, this chapter is structured along the line.

---

[1] Enders, *Equality and Identity and (In)distinguishability in Classical and Quantum Mechanics from the Point of View of Newton's Notion of State*, 2008; *Gibbs' Paradox in the Light of Newton's Notion of State*, 2009

Conservation of stationary state – change of of stationary state – motion.

2.1. STATIONARY STATES

According to the line of reasoning just written down at the end of the foregoing paragraph, I begin with the notion 'stationary state' and its fundamental role for this book.

2.1.1. On the Fundamental Role of Stationary States

The first three tasks of atom mechanics were,

1. the explanation of the stability of the atoms,

2. the explanation of the discreteness of the atomic spectral lines,

3. the calculation of the intensities of the atomic spectral lines.

The first two tasks refer to the *stationary* states of the atoms. This favors Euler's axiomatic of CM, again, see Table **2.1**.

Table 2.1. Euler's axiomatic of CM is closer to QM than Newton's axiomatic of CM ('states' mean stationary states).

Point	Newtonian CM *vs* QM	Eulerian CM *vs* QM
Conservation of states	common	common
Change of states	*different*	*common*
Motion	different	different

The set of possible stationary states of a system determines the set of possible changes of stationary states, therefore, its possible motions.

Problem 2.1. *Prove the last statement!*

2.1.2. Descartes', Newton's and Euler's Notions of (Stationary) State

Descartes, Newton and Euler[2] refer the notion of state to the interaction*less* (mechanical) body. Any change of state needs an *external cause*. Without such a cause, the state of a body is that of being at rest (law of sufficient reason, see Subsection 2.2.1), or of being in straight uniform motion (Galileo's law of inertia).

This (stationary!) state is quantified by the 'quantity of motion'. The latter one equals

- the *modulus* of the momentum vector (Descartes),

- the momentum vector itself (Newton),

- the 'living force' (Leibniz),

- the velocity vector (Euler),

respectively. The 'quantity of motion' is at once the stationary-state variable.

In contrast to the other ones, Newton's stationary-state variable, the momentum vector, $\vec{p} = m\vec{v}$ (*Principia*, Definition 2), admits the following immediate physical interpretation. The total mass, m, of a body, is the result of the masses of its parts. This mass distribution be described through the mass density, $\rho_m(\vec{r}, t)$.

- If the body consists of N atoms (Newton) of masses m_j ($j = 1 \ldots N$) at (changing) positions $\vec{r}_j(t)$, we have $\rho_m(\vec{r}, t) = \sum m_j \delta(\vec{r} - \vec{r}_j(t))$.

- If the body is infinitely divisible (Euler), $\rho_m(\vec{r}, t)$ is a smooth scalar field inside the body and zero outside.

In both cases, $\vec{j}_m = \rho_m \vec{v}$ is the mass current density, *i.e*, the *actual* flow of matter. This hydrodynamic point of view is corroborated by the fact, that the motion of bodies considered here is convective.

It is crucial, that the position vector does *not* enter the state descriptions above. For during straight uniform motion, the position changes *without* external action.[3]

---

[2] René Descartes (1596 - 1650), *Principia philosophiae*, 1644; *Le Monde*, 1664; Newton, *Principia*, 1687; Euler, *Anleitung zur Naturlehre*, 1750
[3] *Cf* Carl Friedrich Freiherr von Weizsäcker (1912 - 2007), *Aufbau der Physik*, 2002, p. 235.

In modern terms, this state is a stationary state, but the set of variables describing it is different from the nowadays notion of stationary state, see Table **2.2** below.

That arguing has been abandoned during the 2nd half of the 18th century. The fundamental equations are no longer the equations of conservation and of change of stationary states, but the equations of motion.[4] As a consequence, the position of a body is part even of the *stationary*-state description. Unfortunately, the name 'state' has not been changed.

This *hidden* paradigm change has *in*creased the distance between CM and QM. This surprising observation is elucidated in Table **2.2**.

Table 2.2. Variables describing the stationary states of a free (interaction-less) classical point-like body resp. free, spinless quantum particle.

Free body	Newton	Euler	Lagrange, Laplace	Hamilton	QM
Variables	$\vec{p}(t) = \vec{p}(0)$	$\vec{v}(t) = \vec{v}(0)$	$\vec{r}(t)$, $\vec{v}(t)$	$\vec{r}(t)$, $\vec{p}(t)$	$\vec{k} = \frac{1}{\hbar}\vec{p}(0)$
No. of variables	3	3	6	6	3
$\vec{r}(t)$	plays *no* role		*does* play a role		absent

The number of stationary-state variables is *the same* (3) within QM and Newton's and Euler's CM, but *different* (6) within nowadays CM (Lagrange/Laplace, Hamilton). Similarly, Newton's and Euler's descriptions of the stationary states of a Coulomb system is much closer to QM than Lagrange's and Hamilton's descriptions. This concerns not only the number of stationary-state variables (3 and 6, respectively, again), but also the role of the initial conditions and the spatial domain involved, see Table **2.3**.

---

[4] Joseph Louis, comte de Lagrange (Giuseppe Ludovico De la Grange Tournier, 1736 - 1813), *Mécanique Analytique*, 1788; Laplace, *Essai Philosophique sur la Probabilité*, 1814; Hamilton, *On a General Method in Dynamics*, 1834; *Second Essay on a General Method in Dynamics*, 1835

Table 2.3. Newton's and Euler's descriptions of the stationary states of a Coulomb system is closer to Bohr's atom model than Lagrange's and Hamilton's descriptions.

Coulomb System	**Newton, Euler**	**Lagrange, Hamilton**	**Bohr 1913**
Variables	E, L, L_z	$\vec{r}(t), \vec{v}(t)$ resp. $\vec{p}(t)$	$n \leftrightarrow E, l \leftrightarrow L, m \leftrightarrow L_z$
No. of variables	3	6	3 (quantum numbers)
Initial conditions	all initial conditions yielding the same orbit are equivalent	determine the state at all subsequent times	no trajectories $\rightarrow$ no $\vec{r}(0), \vec{p}(0)$
Description of	whole orbit, time-*in*dependent	path in phase space, time-dependent	whole orbital, time-*in*dependent

For this, I will build QM on classical stationary states.

Problem 2.2 (Momentum *vs* velocity). *Discuss the advantages of the momentum as (stationary-)state variable, when compared with the velocity! Hint: See also Subsection 1.1.3!*

Problem 2.3 (SI unit for charge*). *Comment on the fact, that the SI unit for mass is a basis unit, while that of the electrical charge is not!*

Problem 2.4 (Pressure*). *Describe pressure! Hint: See also Subsection 1.1.3!*

2.2. CONSERVATION OF STATIONARY STATE

The best known statements about the conservation of stationary states are Newton's 1st axiom and Bohr's 1913 postulate, that interaction-less atoms assume stable states. For this, it is reasonable to start with fundamental statements about the conservation of stationary states.

2.2.1. The Law of Sufficient Reason. Galilean Relativity

"A perfectly round ball never ceases to move on its own."[5]

Von Kues claimed that the Lord is everywhere and every-time in the same manner.[6] In agreement with that view, Newton presupposes, that,

1. Space and time are homogeneous and isotropic (*Principia*, Definitions, Scholium).

2. A body does not change its motion on its own, without external reason (*De Gravitatione*, p. 32, Def. 5; *Principia*, Definition 3).

Then, by virtue of the law of sufficient reason, a free body at rest never ever starts to move, because, (i), there is no 'moving force', and, (ii), no place or direction is preferred over the other ones.

For constant mass, 'constant momentum' implies constant velocity. Axiom *N1* (see Subsection 2.2.3) makes no difference between rest and motion at constant velocity. This is Galilean relativity (Galileo, *Dialogo*): The physical processes in two equal systems are the same, if the two systems move with constant velocity each relatively to another.[7]

Problem 2.5 ('Cusanian' space-time*). *Von Kues' claim referred to at the beginning of this subsection, that the Lord is everywhere and every-time in the same manner, suggests the geometry of space-time to be Euclidean*[8]. *This means, that the differential line element,* ds, *is given through the generalized Pythagorean theorem*[9]

$$ds^2 = dt^2 + dx^2 + dy^2 + dz^2 \ (c = 1) \tag{2.1}$$

---

[5] Nikolaus Kardinal von Kues (Nicolaus Cusanus / de Cusa, 1400 - 1464), *De ludo globi* – see also Robert Wilhelm Hermann Rompe (1905 - 1993) & Hans-Jürgen Treder (1928 - 2006), *Nikolaus von Kues als Naturforscher*, 1979.

[6] After Regine Kather (*1955), *Gott ist der Kreis, dessen Mittelpunkt überall ist ... Von der Dezentrierung der Erde und der Unendlichkeit des Universums bei Nikolaus von Kues und Giordano Bruno*, 2004.

[7] In bypassing, let me mention that – in contrast to the space of positions – the space of velocities is not a vector space, but an affine space, or – if including infinitively large velocities – a projective space, *cf* Rock Brentwood, *Baez's Week 250, The Meaning of Relativity and Affine Spaces*, 2007.

[8] Euclid of Alexandria (ca. 300 BC), *Elements*; *cf* also Enders, *Physical, metaphysical and logical thoughts about the wave equation and the symmetry of space-time*, 2011.

[9] In bypassing I note, that there is a proof by Einstein, which shows much of his thinking, see Steven Strogatz, *Einstein's First Proof*, 2015.

Which motion is described through a straight line in this 'Cusanian' space-time?

Problem 2.6 (Minkowski and 'Van Leunen' space-times*). *According to special relativity, the geometry of space-time is not Euclidean, but Minkowskian (*Raum und Zeit, *1908, I).*

$$ds^2 = dt^2 - dx^2 - dy^2 - dz^2 \qquad (2.2)$$

Van Leunen considers the infinitesimal space-time step, ds, *to be an infinitesimal proper time step,* $d\tau$. *Then, the infinitesimal coordinate time step,* dt, *displays "a Pythagoras format"*[10].

$$dt^2 = d\tau^2 + dx^2 + dy^2 + dz^2 \qquad (2.3)$$

Discuss straight uniform motion within Minkowski space time (2.2) and within 'van Leunen' space time (2.3)!

2.2.2. Descartes' "Main Rules" of Motion*

We owe Descartes analytical geometry with the system of straight perpendicular coordinates that bears his name. Its use has become so self-evident, that it is rather strange that it has not been invented earlier.

Less generally known is Descartes' representation of mechanics being embedded in his natural philosophy. Despite of its shortcomings, it has posed a standard for the subsequent generations from Huygens over Leibniz and Newton till Euler.

Descartes' "main rules" for the motion of bodies read (*Le Monde*, 1664, Ch. VII),

Rule D1 Each individual part of matter always continues to remain in the same state unless collision with others constrains it to change that state.

Rule D2 When one of these bodies pushes another, it cannot give the other any motion except by losing as much of its own at the same time; nor can it take away from the other body's motion unless its own is increased by as much.

Rule D3 When a body is moving, even if its motion most often takes place along a curved line and…can never take place along any line that is not in some

---

[10] Hans van Leunen (*1941), *Quaternionic versus Maxwell based differential calculus*, 2016

way circular, nevertheless each of its individual parts tends always to continue its motion along a straight line.

The emphasis on the stationary states and their conservation is obvious. As a matter of fact, conservation is more easily accessible to philosophical speculation than change (Descartes did not perform experiments).

Rule *D1* expresses Galilean inertia. It should be noted, that it is by no means trivial. Real ballistic curves through air are surprisingly close to the prediction of the ancient impetus theory[11].

In rule *D2*, "[quantity of] motion" means the product of mass, m, and *modulus*, v, of the velocity vector, $\vec{v}$, therefore, the *modulus*, $p = mv$, of the momentum vector, $\vec{p} = m\vec{v}$ (*Principia Philosophiae*, 1644, II, No. 36). As a consequence, not all of Descartes' seven rules for the elastic impact (*ibid.*, pp. 46ff.) are correct. Huygens (*De motu corporum ex percussione*, 1669) took into account the direction of the motion and hence obtained a complete description of the elastic impact. The momentum *vector*, $\vec{p} = m\vec{v}$, became Newton's 'quantity of motion' (*Principia*, Definition 2) and stationary-state variable.

In contrast, Leibniz and – *expressis verbis* – du Châtelet[12] considered the 'living force', mv^2, to measure the quantity of motion. (The factor $1/2$ in the kinetic energy, $mv^2/2$, becomes relevant only, if one balances it with other energies.) Huygens knew, that *both* total momentum and total living force are conserved during an elastic impact, though he did not express his results in this form[13].

Rule *D3* is illustrated in Fig. **2.1**[14].

Problem 2.7. *Check the conservation of, (i), $\vec{p}_{tot} = m_1\vec{v}_1 + m_2\vec{v}_2$ and, (ii), $(m_1v_1 + m_2v_2)$ for all characteristic cases of the central elastic impact of two unequal bodies $(m_1 \neq m_2)$! (To find the characteristic cases is part of the problem.)*

[11] See Michael McCloskey, *Intuitive Physics*, 1983; Thomas Sonar (*1958), *Die Entwicklung der Ballistik von Aristoteles bis Euler,* 2008; both cited after Raúl Rojas (*1955), *Impetustheorie des Fuß balls*, 2014.
[12] Gabrielle Émilie Le Tonnelier de Breteuil, Marquise du Châtelet-Laumont (1706 - 1749), *Institutions de Physique*, 1740
[13] As a matter of fact, the discussion went on for long, see, notably, Immanuel Kant (1724 - 1804), *Gedanken von der wahren Schätzung der lebendigen Kräfte ...*, 1746.
[14] https://newsletter.westfalia.de/go/5/29J0J5D7-292RG4K6-29DEX2XB-I5J18M1.php.

Fig. 2.1. The sparks of a disc grinder move according to Descartes' rule *D3*.

2.2.3. Newton's 1st, the Galilean Axiom

> "The *Principia* is a scientific work and not a bible. One should study it and ponder, admire – yes! – but not swear on it. One finds in it novelties and repetitions, an elegant perfection, but also mistakes, enlightening conciseness and superfluous detours, extraordinary requirements on rigor, but also incompleteness of the logic, the clearance of earlier posed hypotheses and the introduction of unexplained new assumptions."[15]

Newton's main work, *Philosophiae Naturalis Principia Mathematica*, has laid down major parts of the foundations of modern physics. The history of its reception[16] is at once a history of novel developments, which reaches far beyond Newton and lasts till today[17].

---

[15] Clifford Ambrose Truesdell (1919 - 2000), *Essays in the History of Mechanics*, 1968, p. 88 (quoted after Simonyi, *Kulturgeschichte der Physik*, 1990, p. 296).

[16] See, in particular, François Marie Arouet de Voltaire (1694 - 1778), *Élements de philosophie de Newton*, 1738; *Sammlung verschiedener Briefe des Herrn von Voltaire, die Engelländer und andere Sachen betreffend*, 1747; after Horst-Heino von Borzeszkowski (*1940) & Renate Wahsner (*1938), *Newton und Voltaire*, 1980.

[17] *Cf* Subrahmanyan Chandrasekhar (1910 - 1995), *Newton's Principia for the Common Reader*, 1995; Newton, *The Principia*, 1999, is a new translation with a 370(!) pages *Introduction* by Isaac Bernard Cohen

The "axioms or laws of motion" form a crucial part of Newton's representation of CM. Nowadays, they are taught even in high-schools, although *at variance to* their original formulation. More important, however, is the fact, that they prevent any step-by-step generalization of CM. In order to understand, *why*, I will consider them in somewhat more detail.

In doing so, I follow Liebscher[18] in sequence and naming of the axioms. This exhibits the following advantages:

- It reflects the decreasing abstractness of the axioms;

- it groups the statements about conserved quantities together;

- the two qualitative statements occur before the quantitative one;

- the 1st axiom does not need a coordinate system; this may indicate a superiority of the special-relativistic geometry of space-time[19] against the different treatment of space and time by the Galileo transform[20];

- the expression "weighting of velocity" allows for introducing the inert rest mass as well as relativistic generalizations.

On the other hand, Liebscher's wording – being guided by the *kinematic* aspects of special relativity – does *not* contain the notion of state. For this, I combine his and Newton's original formulations accordingly.

Furthermore, because Newton's 2nd and 3rd axioms deal with the momentum, I follow Sommerfeld (*Mechanik*, §1) to write the 1st axiom in terms of the momentum, too. Notice, that this makes the inertia to completely disappear in the axioms. For an alternative, see Euler's approach outlined in the over-next subsection.

Axiom N1 (Galilean axiom) "Every body perseveres in its stationary state at rest or of constant momentum vector, except insofar as it is compelled to change its state by external forces."

(1914 - 2003) – if not otherwise indicated, I quote this translation. New translations into German appeared in 1988 by Ed Dellian (*1939) and in 1999 by Volker Schüller.

[18] Dierck-Ekkehard Liebscher (*1940), aip.de/ lie/; see also *Relativitätstheorie mit Zirkel und Lineal*, 1977, Ch. 3.

[19] Minkowski, *Raum und Zeit*, 1908

[20] See also Enders, *Physical, metaphysical and logical thoughts about the wave equation and the symmetry of space-time*, 2011.

Thus, Newton begins his axiomatic with the *force-free* body. Moreover, he states (although only indirectly), that the stay of a body in a stationary state is a basic property of the bodies. The force of inertia (*Principia*, Definition 3) is *not* mentioned[21].

Axiom *N1* defines the stationary state for a single body and expresses the principle of causality. If the stationary-state quantities are properly chosen (for instance, the total energy), it applies to an extremely wide class of systems.

The position, $\vec{r}$, of a body is *not* a stationary-state quantity, because it is *not* constant in the stationary state of straight uniform motion.

2.2.4. Newton's 3rd, the Newtonian Axiom

Axiom N3 (Newtonian axiom) To any action there is always an action of equal
　　quantity, but opposite direction such, that the total momentum vector
　　remains the same.

This slight reformulation of the original text is corroborated by Newton's

Corollary 3 "The quantity of motion, which is determined by adding the motions
　　made in one direction and subtracting the motions made in the opposite
　　direction, is not changed by the action of bodies on one another".

Consider two bodies interacting solely among themselves. Newton's presuppositions of the foregoing subsection generalize straightforwardly to them as they build a closed system. For this *system* as a whole, space and time are homogeneous and isotropic, again; and it cannot change its motion as a whole on its own.

More generally speaking, internal interactions do not affect the stationary-state quantities, such as total momentum, energy and angular momentum. The latter ones are natural generalizations of the individual momentum, energy and angular momentum of a single body.

Like axiom *N1*, axiom *N3* is not necessarily bound to the motion along trajectories. This makes its generalization possible, where the concrete representation of the momentum will change, of course.

---

[21] This is by no means self-evident and not yet contained in Newton's earlier work, *cf* Richard Samuel "Sam" Westfall (1924 - 1996), *Isaac Newton*, 1996, pp. 214f.; Gernot Böhme (*1937), comment 13 to Newton, *De Gravitatione et aequipondio fluidorumde gravitatione…*, 1988.

2.2.5. Euler's Axioms

> "[Euler] ist auch durch die Universalität and die spekulative Tiefe seines Denkens ausgezeichnet, welches unablässig darauf ausging, die forschende and konstruierende Wissenschaft nach ihrem philosophischen Gehalt sich zum Bewusstsein zu bringen. …Seine physikalischen and mathematischen Arbeiten …sind durchweg von dem Bestreben getragen, die Grundbegriffe zur höchsten Klarheit zu bringen."[22]

As we will see in the next section, the axiomatic status of Newton's 2nd axiom makes it virtually impossible to discard it – and also Newton's equation of motion – when seeking generalizations of CM. For this, the historical developments of relativistic mechanics as well as quantum mechanics were not smooth generalizations of Newtonian mechanics.

In contrast, Euler's axiomatic of CM[23] is *not* limited by those constraints. Solely the 'internal principles of motion': the conservation of stationary-state quantities (like Newton's axioms *N1* and *N3*) have got an axiomatic status. The 'external principles of motion': the change of stationary-state quantities under external forces (like axiom *N2*), represent merely problems to be solved accordingly to the settings under consideration.

Remarkable enough, Hertz holds quite a similar view.

> "**308** Wir betrachten es als die Aufgabe der Mechanik, aus den von der Zeit unabhängigen Eigenschaften materieller Systeme die in der Zeit verlaufenden Erscheinungen derselben und ihre von der Zeit abhängigen Eigenschaften abzuleiten."[24]

---

[22] Friedrich Ueberweg (1826 - 1871), *Grundriß der Geschichte der Philosophie*, Third Part: *Die Philosophie der Neuzeit bis zum Ende des XVIII. Jahrhunders* (completely revised ed. by Max Frischeisen-Köhler (1878 - 1923) & Willy Moog (1888 - 1935)), 1924, p. 462 – Engl.: [Euler] is also distinguished through the universality and the speculative depth of his thinking, which diligently strove for bringing to awareness the philosophical content of the researching and constructing science. …His physical and mathematical work…is throughout carried by the aspiration to bring the basic concepts to highest clearness.

[23] Euler, *Gedancken von den Elementen der Cörper*, 1746; *Réflexions sur l'espace et le temps*, 1748; *Recherches sur l'origine des forces*, 1750; *Harmonie entre les principes généraux de repos et de mouvement de M. de Maupertuis,* 1751; *Découverte d'un nouveaux principe de mécanique*, 1752; *Lettres à une princesse d'Allemagne*, 1768; *Anleitung zur Naturlehre*, written c. 1750, publ., however, only in 1862.

[24] Hertz, *Die Prinzipien der Mechanik in neuem Zusammenhange dargestellt*, 1910, p. 162 – Engl.: "308. We consider the problem of mechanics to be to deduce from the properties of a material system which are

> **"309 Grundgesetz** Jedes freie System beharrt in seinem Zustande der Ruhe oder der gleichförmigen Bewegung in einer geradesten Bahn.
>
> Systema omne liberum perseverare in statu suo quiescendi vel movendi uniformiter in directissimam."[25]

Hertz (*loc. cit.* Art. 323, p. 167) stresses, that his formulation deliberately closely follows Newton's one.

> "Corpus omne perseverare in statu suo quiescendi vel movendi uniformiter in directum, nisi quatenus a viribus impressis cogitur statum illum mutare." (Newton, *Principia*, 1686, Lex. I, p. 12)

The reader is encouraged to explore Hertz's ideas in more detail, in particular, in view of its general approach to mechanical motion and of the absence of forces, as within general relativity and in QM (see eq. (5.3) below).

2.2.6. The General Properties of the Bodies

In Euler's representation of CM, the linear motion of bodies is founded on,

- the general properties of the bodies;

- the conservation and changes of the stationary states.

(*cf* Euler, *Anleitung*). The general properties of the bodies are,

1. The *extension*; its measure being the volume of a body (*ibid.*, Ch. 2);

2. The *movability*; it has no measure (*ibid.*, Ch. 3);

3. The *steadfastness*; its measure being the (inert) mass of a body (*ibid.* § § 31, 33f.)[26];

---

independent of the time those phenomena which take place in time and the properties which depend on time." (*The Principles of Mechanics Presented in a New Form*, 1899, p. 144)

[25] Hertz, *ibid.* – Engl.: "309. Fundamental Law. Every free system persists in its state of rest or of uniform motion in a straightest path." (*ibid.*)

[26] Euler avoids the notion *inertia*, because the force of inertia is an "inherent property of matter" (Newton, *Principia*, Definition 3). In order to expel ghosts from science, Euler denies bodies to have got inherent properties.

4. The *impenetrability*; it has no measure (*ibid.*, Ch. 5).[27]

2.2.7. Euler's Axioms

The existence of stationary states is postulated in the following axioms.

Axiom E0 Every body is EITHER resting OR moving.[28]

Axiom E1 A body preserves its state at rest, unless an external cause sets it in motion (Euler, *Anleitung*, § 26).[29]

Axiom E2 A body preserves its state of straight uniform motion, unless an external cause changes that state (*ibid.*, § 28).[30]

Axiom *E0* means, that the axioms *E1* and *E2* are not independent each of another; they exclude each another and, at once, they are *in harmony* each with another[31]. This represents a dialectical harmony in the sense, that the one is meaningless without the other one.[32] The *coincidentia oppositorum* is the kernel of the scientific method of Cusanus.[33]

There are two obvious differences to Newton's axioms *N1* and *N3*.

1. It is the external cause (in particular, the external force), which changes the stationary state of a body, not the body itself.

2. The cases of rest and of straight uniform motion are addressed separately.

For the latter one, I just wish to mention the following observations ("rest" being referred to a given reference system, of course).

---

[27] Newton considers "extension, hardness, impenetrability, movability and force of inertia" to be the general properties of bodies (*Principia*, Book III, 3rd Rule).

[28] *Tertium non datur*, *cf* Nicolas (Nicole de) Malebranche (1638 - 1715), *Entretiens sur la métaphysique*, 1688, Dialogue 7 (8), p. 71 (2007 transl.)

[29] Strictly speaking, *E1* is not an axiom, but a consequence of the theorem of sufficient reason, see Subsection 2.2.1.

[30] *Cf* Galileo's law of inertia in Subsection 2.2.1 above.

[31] *Cf* Euler, *Harmonie entre les principes généraux de repos et de mouvement de M. de Maupertuis*, 1751.

[32] For instance, there is no need to mention 'finite', if there is no 'infinite', *etc.*, see Georg Wilhelm Friedrich Hegel (1770 - 1831), *Wissenschaft der Logik*, 1812-1816.

[33] von Kues (Cusanus), *De docta ignorantia*, 1488, Book I – During a ship journey, he observed, that sky and earth come together at the horizon. The same was felt and painted by Henri Émile Benot Matisse (1869 - 1954).

- According to Malebranche (*De la recherche de la vérité*, 1712), 'nothing' (no motion, no speed) is not the limit case of 'something existing' (motion, finite amount of speed).

- Dirac (*The Principles of Quantum Mechanics*, 1930) has argued for the existence of an absolute motion.

- The distinction between rest and motion plays an important role in the transform properties of the Hamilton-Jacobi equation.[34]

- Is statics merely a special case of dynamics? (In Subsection 2.5.2 I will consider dynamics as extension of statics; see also Problem 2.9.)

Problem 2.8 (Impenetrability - "the essence of the bodies"). *Show, that the impenetrability represents "the essence of the bodies" (Euler,* Anleitung, *Ch. 5) as it includes the other three general properties: extension, movability and steadfastness (inertia)!*

Problem 2.9 (Electrostatics*). *Electrostatics is often represented as limit case of electrodynamics. Does this exhaust it completely?*

Problem 2.10 (A little bit topology*). *Assume space and the loci of the bodies to be topological point sets. Show, that impenetrable bodies are to be represented by* open *point sets! Which consequences does this have for the touching of bodies?*

2.2.8. Derivation of Newton's 3rd Axiom

In order to illustrate the power of Euler's axiomatic, let me sketch his arguing for Newton's 3rd axiom.

All forces are supposed to originate from the impenetrability of the bodies.[35] The impenetrability has no measure. Hence, the forces that two bodies exert one upon another are opposite in direction, but equal in magnitude (see also Problem 2.12).

$$\vec{F}_{21} = \frac{m_2 \vec{v}_2}{t} = -\vec{F}_{12} = -\frac{m_1 \vec{v}_1}{t} \qquad (2.4)$$

This is equivalent to the conservation of total momentum as postulated in axiom *N3*.

---

[34] Faraggi & Matone, *The Equivalence Postulate of Quantum Mechanics*, 1999
[35] Newton ascribes only the occurrence of pressure to the impenetrability, see *De Gravitatione*, Def. 9.

$$m_1 \vec{v}_1 + m_2 \vec{v}_2 = \vec{0} \tag{2.5}$$

Euler's derivation of Newton's 2nd axiom will be given within the next section.

Problem 2.11. *Show, that the Eulerian arguing above complies with the axiom*

"Bodies that touch each another exert the same pressure upon another."[36]

Problem 2.12 (Minimum action). *Show that the magnitude of a force is just as large as being necessary and sufficient to prevent the mutual penetration of two bodies and hence its action being minimum!* Hint: *Discuss the meaning of the notion 'action', consulting Maupertuis' 'principle of least action' and related work*[37]*!*

2.3. CHANGE OF STATIONARY STATE

2.3.1. Newton's 2nd, the Huygensian Axiom

"Das erste Gesetz beschreibt, was los ist, wenn nichts los ist, das zweite, wenn es eine Einwirkung von außen gibt, und das dritte, wie auf diese Einwirkung reagiert wird."[38]

Axiom N2 The change of the momentum vector is proportional to the external force vector, $\vec{F}_{ext}$.

In Definition 8 of the *Principia*, the moving force is defined as momentum change per time unit. Therefore, this axiom means

$$d\vec{p} = n\vec{F}_{ext}dt \tag{2.6}$$

---

[36] Newton, *De Gravitatione*, p. 36, Axiom 2
[37] Pierre Louis Moreau de Maupertuis (1698 - 1759), *Accord de différentes Loix de la Nature qui avoient jusqu'ici paru incompatibles*, 1744; Euler, *Harmonie entre les principes généraux de repos et de mouvement de M. de Maupertuis*, 1751; *Découverte d'un nouveaux principe de mécanique*, 1752; *Dissertation sur le principe de la moindre action*, 1753; de Broglie, *Recherches sur la théorie des quanta*, 1924; Lanczos, *The Variational Principles of Mechanics*, 1970; Szabó, *Geschichte der mechanischen Prinzipien und ihrer wichtigsten Anwendungen*, 1987; Suisky, *Über eine Differenz in der Begründung des Wirkungsprinzips bei Maupertuis und Euler*, 1999
[38] Tobias Henz & Gerald Langhake, *Pfade durch die Theoretische Mechanik 1*, 2016, p. 43 – Engl.: The first law describes, what happens when nothing happens, the second, if there is an external action, and the third, what is the reaction on that action.

where n adjusts different systems of units of measurement. In cgs and SI units, $n = 1$[39].

The *ex*ternal (w.r.t. the body) force, $\vec{F}_{ext}$, represents the reason for the change of the stationary state and needs not to be presupposed. This circumstance becomes fully clear within Euler's treatment, see next subsection.

Axiom *N2* is bound to motion along trajectories. This makes it *im*possible to generalize Newton's axiomatic to cases, where there are no trajectories, notably, to QM.

Problem 2.13 ($\vec{F}=m\vec{a}$). *Discuss the differences between Newton's original formulation of axiom N2 and the widely formulation 'force equals mass time acceleration'!*

Problem 2.14 (Axiom *N3**). *Discuss the fact, that axiom N3 requires a counteraction of the body upon $\vec{F}_{ext}$!*

Problem 2.15 (Parallelogram of forces). *Justify the parallelogram of forces!* Hint: *Consider first two forces of equal magnitude along the same line and assume, that they, (i), add when being parallel, and, (ii), cancel each another when being directed each against another! (Why these assumptions are admissible?) Then, decompose two unequal forces, the directions of which are perpendicular each to another, into forces of equal amount!*[40]

Problem 2.16 (Action and reaction*). *The sentence by Henz & Langhake quoted at the beginning of this subsection suggests a temporal sequence of action and reaction. Explore this idea for explaining the absence of immediate action-at-distance!*

2.3.2. Euler's Principles of Stationary-State Change for a Single Body

> "The very existence of the general principles of mechanics is their justification."[41]

---

[39] In most cases in this book, 'change' is understood as change in time, where 'time' is the scale (intensity) of change as given by physical processes ('clocks'), such as day, month, year.

[40] *Cf* Siméon Denis Poisson (1781 - 1840), *Traité de Mécanique*, 1811, Vol. 1, pp. 11-19; Engl.: *A Treatise of Mechanics*, 1842, Vol. 1, pp. 36-42; after Marc Lange (*1963), *Why do forces add vectorially? A forgotten controversy in the foundation of classical mechanics*, 2011; see also Lange, *A Tale of Two Vectors*, 2009.

[41] Lanczos, *The Variational Principles of Mechanics*, 1986, p. vii

As mentioned above, Newton's axiomatic of CM bounds the manner of stationary-state change to the motion along trajectories. In contrast, Euler describes the stationary-state change (and the motion) as problem to be solved according to the situation under consideration. The principles derived from his description will proof to apply not only to the singly body, but to systems of bodies as well and *cum grano salis* remain valid even within QM.

Euler writes the changes of position and velocity of a body of mass m being subject to the external force $\vec{F}$ during the time interval dt as[42]

$$d\vec{r} = \vec{v}dt \tag{2.7a}$$

$$d\vec{v} = \frac{1}{m}\vec{F}dt \tag{2.7b}$$

This form is an example of the general rule, that the most fundamental evolution equations are of *first* order in time.[43] In 6-vector form[44], it reads

$$d\begin{pmatrix}\vec{r}\\\vec{v}\end{pmatrix} = \begin{pmatrix}\vec{v}\\\frac{1}{m}\vec{F}\end{pmatrix}dt \tag{2.8}$$

This may be called the *Lagrange-Laplacian equation-of-state-change*, *cf* Table **2.2**.

Replacing the velocity, $\vec{v}$, with the momentum, $\vec{p} = m\vec{v}$, it becomes

$$d\begin{pmatrix}\vec{r}\\\vec{p}\end{pmatrix} = \begin{pmatrix}\frac{1}{m}\vec{p}\\\vec{F}\end{pmatrix}dt \tag{2.9}$$

Here, it proves useful to split the r.h.s. into one term containing solely the 6-vector $(\vec{r}\;\vec{p})$ and another term containing solely the external force, $\vec{F}$.

$$d\begin{pmatrix}\vec{r}\\\vec{p}\end{pmatrix} = \begin{pmatrix}\hat{0} & \frac{1}{m}\hat{1}\\\hat{0} & \hat{0}\end{pmatrix}\begin{pmatrix}\vec{r}\\\vec{p}\end{pmatrix}dt + \begin{pmatrix}\hat{0} & \hat{0}\\\hat{0} & \hat{1}\end{pmatrix}\begin{pmatrix}\vec{0}\\\vec{F}\end{pmatrix}dt \tag{2.10}$$

---

[42] Euler, *Mechanica sive motus scientia analytice exposita,* 1736; *cf* Suisky, *Euler as Physicist*, 2009, p. 154 (of course, Euler did not exploit the vector notation).
[43] *Cf* Enders, *Huygens principle as universal model of propagation*, 2009; *Physical, metaphysical and logical thoughts about the wave equation and the symmetry of space-time*, 2011.
[44] This term has been borrowed from Sommerfeld, *Elektrodynamik.*

The first term is independent of the external force and thus *internal* w.r.t. to the trajectory (but not to the body!).

$$d \begin{pmatrix} \vec{r} \\ \vec{p} \end{pmatrix}_{\text{int}} = \begin{pmatrix} \hat{0} & \frac{1}{m}\hat{1} \\ \hat{0} & \hat{0} \end{pmatrix} \begin{pmatrix} \vec{r} \\ \vec{p} \end{pmatrix} dt \equiv \widehat{U}_{\text{int}} \begin{pmatrix} \vec{r} \\ \vec{p} \end{pmatrix} dt \qquad \textbf{(2.11)}$$

The *int*ernal evolution matrix,

$$\widehat{U}_{\text{int}} = \begin{pmatrix} \hat{0} & \frac{1}{m}\hat{1} \\ \hat{0} & \hat{0} \end{pmatrix} \qquad \textbf{(2.12)}$$

transforms the '*int*ernal cause' $\vec{p}dt$ into the '*int*ernal change' of a dynamical variable, $(d\vec{r})_{\text{int}}$.

Obviously, the position, $\vec{r}$, does *not* represent a cause of change, while the momentum, $\vec{p}$, causes a change of position, $d\vec{r}$. This corroborates Newton's and Euler's state descriptions *not* to contain the position variable.

In bypassing I notice that the internal equation of state change (2.11) integrates to

$$\begin{pmatrix} \vec{r} \\ \vec{p} \end{pmatrix}(t) = \exp\{\widehat{U}_{\text{int}}t\} \begin{pmatrix} \vec{r} \\ \vec{p} \end{pmatrix}(0) \qquad \textbf{(2.13)}$$

This is a forerunner of 'motion as canonical transformation' within the Hamilton-Jacobi theory[45] and within QM.

The second term in eq. (2.10) contains the *ext*ernal force and is *ext*ernal w.r.t. to the trajectory and to the body.

$$d \begin{pmatrix} \vec{r} \\ \vec{p} \end{pmatrix}_{\text{ext}} = \begin{pmatrix} \hat{0} & \hat{0} \\ \hat{0} & \hat{1} \end{pmatrix} \begin{pmatrix} \vec{0} \\ \vec{F} \end{pmatrix} dt \equiv \widehat{U}_{\text{ext}} \begin{pmatrix} \vec{0} \\ \vec{F} \end{pmatrix} dt \qquad \textbf{(2.14)}$$

The *ext*ernal evolution matrix,

---

[45] Carl Gustav Jacob Jacobi (Jacques Simon; 1804 - 1851), *Ü ber die Reduction der Integration der partiellen Differentialgleichungen erster Ordnung zwischen irgend einer Zahl Variabeln auf die Integration eines einzigen Systemes gewöhnlicher Differentialgleichungen*, 1837; *Neue Methode zur Integration partieller Differentialgleichungen erster Ordnung zwischen irgend einer Anzahl von Veränderlichen*, 1906; *Vorlesungen über Dynamik*, 1884

$$\hat{U}_{\text{ext}} = \begin{pmatrix} \hat{0} & \hat{0} \\ \hat{0} & \hat{1} \end{pmatrix} \tag{2.15}$$

transforms the 'external cause', $\vec{F}dt$, into the '*ex*ternal change' of a dynamical variable, $d\vec{p}_{\text{ext}}$. The fact, that solely the momentum, $\vec{p}$, is changed here, distinguishes it from the position, again.

The internal and external evolution matrices do *not* commute.

$$\hat{U}_{\text{int}}\hat{U}_{\text{ext}} - \hat{U}_{\text{ext}}\hat{U}_{\text{int}} = \hat{U}_{\text{int}} \neq \hat{0} \tag{2.16}$$

This means, that the internal and external evolutions depend each of another.[46] As a consequence, eq. (2.13) cannot immediately be generalized to the case of external forces.

We thus are lead to the following Eulerian principles of change of stationary states of classical bodies.

Up to *first* order in dt,

CB1 The changes of the stationary-state quantities $(d\vec{p})$ depend – mediated through the external transform, $\hat{U}_{\text{ext}}$ – on the external causes $(\vec{F})$, but not on the non-stationary-state quantities $(\vec{r})$; in particular, $d\vec{p} = \vec{0}$, if $\vec{F} = \vec{0}$.

CB2 The changes of the stationary-state quantities $(d\vec{p})$ are independent of the state quantities $(\vec{p})$ themselves (*cf* Newton's 2nd axiom).

CB3 The changes of the non-stationary-state quantities $(d\vec{r})$ depend – mediated through the internal transform, $\hat{U}_{\text{int}}$ – on the stationary-state quantities $(\vec{p})$, but not on the external causes $(\vec{F})$.

In order to illustrate the consistency of Euler's notation, let me add the following. Accounting for $ddt = 0$ and $d\vec{F} = \vec{0}$, one obtains from eq. (2.10)

---

[46] This is contrary to the earlier conclusion in Enders & Suisky, *Quantization as selection problem*, 2005; Enders, *Von der klassischen Physik zur Quantenphysik*, 2006; *Quantization as Selection rather than Eigenvalue Problem*, 2013. Hence, the principles CB4, CS4 and QS4 presented there are obsolete. However, the other, more important principles are *not* affected by this correction. The translation to quantum systems will become even easier.

$$\mathrm{dd}\begin{pmatrix}\vec{r}\\\vec{p}\end{pmatrix} = \begin{pmatrix}\vec{0} & \frac{1}{m}\vec{1}\\\vec{0} & \vec{0}\end{pmatrix}\mathrm{d}\begin{pmatrix}\vec{r}\\\vec{p}\end{pmatrix}\mathrm{d}t = \begin{pmatrix}\frac{1}{m}\vec{F}\\\vec{0}\end{pmatrix}\mathrm{d}t^2 \tag{2.17}$$

The upper line represents Newton's equation of motion (first published – in this form – in Euler's *Mechanica*, 1736). The lower line means, that $\mathrm{dd}\vec{p}$ is not given, but to be calculated from $\mathrm{d}\vec{p}$. The same applies to all higher differentials.

$$\mathrm{d}^n\begin{pmatrix}\vec{r}\\\vec{p}\end{pmatrix} \equiv \underbrace{\mathrm{d}\cdots\mathrm{d}}_{n\ \text{times}}\begin{pmatrix}\vec{r}\\\vec{p}\end{pmatrix} = \begin{pmatrix}\vec{0}\\\vec{0}\end{pmatrix}; \ n = 3,4,\dots \tag{2.18}$$

I will return to that in the next section.

Problem 2.17. *Discuss the fact, that the internal variable velocity, $\vec{v}$, occurs on the r.h.s. of eq. (2.8)! Hint: Compare the latter one with Euler's equation-of-(stationary-)state-change, $\mathrm{d}\vec{v} = (\vec{F}/m)\mathrm{d}t$!*

Problem 2.18 ((Non-)Commutativity*). *Within CM, the addition of polar vectors like position and momentum are commutative, while that of axial or pseudo-vectors is not. Indeed, 3D pseudo-vectors like angular momentum, $\vec{L}$, are shortenings of skew-symmetrical matrices, here, $L_{jk} = -L_{kj}$: $L_i = \varepsilon_{ijk}L_{jk}$, where ε_{ijk} is the fully antisymmetric, or Levi-Civita symbol[47]. Within QM, even polar vectors are represented by matrices, which may not commute, notably, position, $\hat{r}$, and momentum, $\hat{p}$. Thus, is there any relationship between the non-commutativity of the internal and external evolution matrices, eq. (2.16), and the non-commutativity of spatial rotations around non-parallel axes (inertia!)?*

2.3.3. Bohmian Mechanics*

For methodological reasons, *viz*, illustrating the goods and odds of general reasonings, let me sidestep to the following approach to a particle dynamics[48].

The velocity, $\vec{v}$, of a particle can be taken to prescribe the change of its position, $\vec{r}$.

$$\frac{\mathrm{d}\vec{r}}{\mathrm{d}t}(t) = \vec{v}(\vec{r}, t) \tag{2.19}$$

---

[47] Gregorio Ricci-Curbastro (1853 - 1925) & Tullio Levi-Civita (1873 - 1941), *Méthodes du calcul différentiel absolu et leurs applications*, 1900
[48] *Cf* Detlef Dürr (*1951), *Bohmsche Mechanik als Grundlage der Quantenmechanik*, 2001, Sect. 8.1.

This means that we deal with a velocity *field*, $\vec{v}(\vec{r},t)$. For this field to be rotational invariant (see Problem 2.19), it is tempting to write it as the gradient of some scalar field, the velocity potential, $\psi(\vec{r},t)$.

$$\vec{v}(\vec{r},t) \sim \nabla \psi(\vec{r},t) \tag{2.20}$$

Now, on time reversal, $\vec{v}(\vec{r},t)$ and hence $\psi(\vec{r},t)$ change their signs. If $\psi(\vec{r},t)$ is real-valued, this means $\psi(\vec{r},-t) = -\psi(\vec{r},t)$. This is the case within the Hamilton-Jacobi theory[49]. Alternatively, if $\psi(\vec{r},t)$ is complex-valued[50], this means $\psi(\vec{r},-t) = \bar{\psi}(\vec{r},t)$. In the latter case, we replace eq. (2.20) with

$$\vec{v}(\vec{r},t) \sim \nabla \Im \psi(\vec{r},t) \tag{2.21}$$

Then, the boost transformation: $\vec{v} \to \vec{v}' = \vec{v} + \vec{u}$, is guaranteed, if $\psi'(\vec{r},t) = e^{i\alpha\vec{r}\cdot\vec{u}}\psi(\vec{r},t)$, where α is a real-valued constant. The final result reads

$$\vec{v}(\vec{r},t) = \frac{1}{\alpha}\Im\frac{\nabla\psi(\vec{r},t)}{\psi(\vec{r},t)} \tag{2.22}$$

With $\alpha = m/\hbar$, this is the core equation of Bohmian mechanics[51], where $\psi(\vec{r},t)$ is Schrödinger's wave function.

The simplest functional form obeying $\psi(\vec{r},-t) = \bar{\psi}(\vec{r},t)$ reads $\psi(\vec{r},t) = e^{i\omega t}\psi(\vec{r},0)$. Together with $\psi(\vec{r},t) = e^{i\alpha\vec{r}\cdot\vec{v}}\psi(\vec{0},t)$, this yields the velocity potential in the form

$$\psi(\vec{r},t) = e^{i\alpha\vec{r}\cdot\vec{v}-i\omega t}\psi(\vec{0},0) \tag{2.23}$$

It obeys the wave equation

$$\frac{i}{\omega}\frac{\partial\psi(\vec{r},t)}{\partial t} = -\frac{1}{\alpha^2}\Delta\psi(\vec{r},t) \tag{2.24}$$

---

[49] Jacobi, *Vorlesungen über Dynamik*, 1884
[50] We will see in Subsection 5.1.3, that there are complex-valued quantities of physical relevance not only within QM, but already within CM.
[51] de Broglie, *La structure atomique de la matière et du rayonnement et la Mécanique ondulatoire*, 1926; Bohm, *A Suggested Interpretation of the Quantum Theory in Terms of 'Hidden' Variables*, 1952 – notice, that the unlucky wording "Hidden Variables" is just misleading.

This is the time-dependent Schrödinger equation for a free particle of mass $m = \hbar^2 \alpha^2 / 2\omega$. External forces can be incorporated through seeking parallels to the Hamilton-Jacobi equation (*cf* Dürr, *ibid.* and eq. (2.27)).

However, the arguing above is by no means an axiomatic derivation of the time-dependent Schrödinger equation (see Problem 2.20). It is not known before the occurrence of quantum theory. More important, it provides no deeper physical meaning of the velocity potential *aka* wave function, $\psi(\vec{r}, t)$. There is a methodologically similar case in Section 1.3.

Problem 2.19. *Reason that Galileo-invariance implies rotational invariance!*

Problem 2.20. *Find the speculative elements in the reasoning above!*

2.4. MOTION

Many representations of CM begin with the description of mechanical motion. Stressing the paramount understanding of CM by Newton and Euler, I continue to adhere to their construction principles of CM, according to which the description of motion is merely a consequence of the descriptions of stationary states and their changes under the influence of external causes.

2.4.1. Euler's Derivation of Newton's Equation of Motion

> "…rational mechanics will be the science, expressed in exact propositions and demonstrations, of the motions that result from any forces whatever and of the forces that are required for any motions whatever."[52]

The relationship between the forces and the trajectories of the bodies is established by the equation of motion. The famous formulation 'force = mass times acceleration',

$$\vec{F}_{\text{ext}} = m \frac{\mathrm{d}^2 \vec{r}}{\mathrm{d}t^2} \tag{2.25}$$

---

[52] Newton, *Principia*, *Author's preface to the reader*, 1999, p. 382; see also the explanations to Definition 5.

however, results only after assuming the mass, m, to be constant. It has first been published in Euler's *Mechanica* (1736)[53].

If an external force acts upon a body at rest, it sets the body into motion[54]. The velocity, $\vec{v}$, assumed is

- proportional to strength and direction of the external force, $\vec{F}$,

- proportional to the time duration, t, of its action, and

- inversely proportional to the "steadfastness" (inertia) of the body against changes of its stationary state, measured by its mass, m.

$$\vec{v} = \frac{1}{m}\vec{F}t \tag{2.26}$$

Here, in contrast to Newton's axioms, the inertia of bodies is explicitly accounted for.

The solution to Newton's equation of motion (2.25) contains the initial values of position, $\vec{r}_0 = \vec{r}(t = 0)$, and of velocity, $\vec{v}_0 = \vec{v}(t = 0)$, as constants of integration. The assumed (!) unique dependence of the trajectory, $\vec{r}(t)$, on those initial values brought Laplace to his famous demon.

The initial values (conditions) are indispensable for the complete solution of the equation of motion. In contrast to that equation, however, they are arbitrary. This provides them with an autonomous significance. As a consequence, the old rule 'aequat causa effectum' is realized only, if they are properly accounted for, as we will see in Chapter 19.

Problem 2.21. *Kepler's 3rd law states, that the square of the orbital period of a planet is proportional to the cube of the semi-major axis of its orbit[55].*

1. Show, that – given Newton's equation of motion (2.25) – Kepler's 3rd law implies the force of gravity to be proportional to $1/r^2$!

---

[53] Newton has surely used it as he knew the concept of mechanical similarity. Together with his force of gravity, $F \sim 1/r^2$, Kepler's 3rd law follows immediately. – *Cf* also Treder, *Isaac Newton und die Begründung der mathematischen Prinzipien der Naturphilosophie*, 1977, p. 8.
[54] Notice, that in Newton's axiom *N1*, the body *itself* changes its state.
[55] Johannes Kepler (1571 - 1630), *Harmonices Mundi*, 1619, book 5, ch. 3, p. 189

2. In turn, assume Kepler's 3rd law to hold true and the force of gravity to be proportional to $1/r^2$ to derive Newton's equation of motion!

2.4.2. Newton-Eulerian Line of Reasoning

Newton's as well as Euler's representations of CM proceed along the steps

- conservation of stationary state (axioms *N1*, *N3*, *E0…E2*)

- change of stationary state (*N2*)

- change of position (equation of motion).

An equation-of-stationary-state-change describes an infinitesimal change of a stationary-state function, here, of the momentum (Newton) and velocity (Euler), respectively, while an equation of motion calculates finite changes of *non-stationary* state quantities, here, of the position of a body.

The equation of motion does *not* enter Newton's axioms[56].

2.5. APPLICATION: FROM NEWTON TO HAMILTON WITHOUT LAGRANGE

The reader may ask herself, what is this non-standard representation of CM above good for? One answer consists in that it provides a direct way from Newton's to Hamilton's equations of motion cutting short the historical process. Recall that matrix and wave mechanics are essentially non-classical Hamiltonian mechanics. Moreover, I will show the applicability of Euler's principles of stationary-state change for a single body to systems of bodies, in order to bolster their application within QM. This will provide us with a straightforward way from the stationary to the time-dependent Schrödinger equation.

In doing so I concentrate on the Hamiltonian as the most important stationary-state quantity for conservative systems and discard the other stationary-state quantities, such as the angular momentum.

---

[56] Newton himself has not stressed this all-important detail. This fact may be related to Newton's intention to avoid any discussions about his methods, wherefore he exploited a 'geometric calculus', see Tristan Needham, *Visual Complex Analysis*, 1997; *cf* also his *Newton and the Transmutation of Force*, 1993. – The thorough application of the 'analytical calculus' to mechanical tasks begins with Euler's *Mechanica*, 1736.

2.5.1. Few Historical Remarks*

The historical development from Newton's (1687) to Hamilton's (1834, 1835) representations of CM went over d'Alembert's principle (1743) and Lagrange's equations (1788). The set of (stationary-)state variables changed from conserved quantities (total momentum, total living force,...) to dynamical variables, or variables of motion (the position and velocity resp. momentum vectors of all degrees of freedom). Laplace's demon[57] calculates all future positions and velocities from the present ones. Heisenberg's 1925 matrix mechanics and Schrödinger's 1926 wave mechanics are non-classical forms of Hamiltonian mechanics. However, the amount of stationary-state variables in them points back to Newton's notion of state, see Tables **2.2** above.

Moreover, when correctly applied, Newton's and Euler's manners of thinking are still appealing. An example is Galileo's proof of the fact, that the speed of a falling body is independent of its weight, as emphasized by von Laue[58]. Another beautiful example is Stevin's most elegant proof of the law of equilibrium on an inclined plane, using the non-existence of *perpetua mobilia*, see Fig. **2.2**[59]. That kind of reasoning has been praised by Feynman[60], too.

Fig. 2.2. Detail of Stevin's tomb; the effective weight of the two balls on the right leg equals the effective weight of the four balls on the left leg.

---

[57] Pierre Simon Marquis de Laplace (1749 - 1827), *Essai Philosophique sur la Probabilité*, 1814
[58] Max Theodor Felix von Laue (1879 - 1960, Nobel Award 1914), *Zum dreihundertsten Geburtstag des ersten Lehrbuches der Physik*, 1938
[59] Simon Stevin (Stevinus, 1548 - 1620), *Beghinselen der Weegconst*, 1585; *Hypomnemata mathematica*, 1605...1608; *The Principal Works of Simon Stevin*, Vol. I, Ch II, Bk I, Th XI. The photo has been taken from en.wikipedia.org/wiki/Simon_Stevin#/media/File:Clootcrans.jpg. – See also Maurits Cornelis Escher's (1898 - 1972) 1961 picture *Waterfall*.
[60] Feynman, *Sechs physikalische Fingerübungen*, 2004, p. 134

The power of that kind of thinking will be demonstrated through deriving Hamilton's equations of motion independently of d'Alembert's principle and Lagrange's equations.

Problem 2.22 (Free fall). *Show, that the speed of a falling body is independent of its weight, without using Newton's equation of motion and Newton's force of gravity!* Hint: *Galileo,* Discorsi, *pp. 56ff.*

Problem 2.23 (Fall with friction). *Aristotle claimed the speed of a falling body to be proportional to its weight. Does this claim hold true, if the friction force is proportional to the velocity (Stokesian friction[61])? If not, how the friction force should depend on the velocity?*

Problem 2.24 (Stevin's proof). *Complete Stevin's proof given the hints in the text above!*

2.5.2. Dynamics as Extension of Statics. Fictitious Forces. D'Alembert's Principle*

"Ist das, was wir jetzt Schwungkraft oder Zentrifugalkraft nennen, etwas anderes als die Trägheit des Steines? Dürfen wir, ohne die Klarheit unserer Vorstellungen zu zerstören, die Wirkung der Trägheit doppelt in Rechnung stellen, nämlich einmal als Masse, zweitens als Kraft?"[62]

For the sake of the unity of physics, let me sidestep to the unity of dynamics and statics. From this point of view, Newtonian mechanics represents the inclusion of inertial force into the equilibrium of forces. Moreover, I will later need the notion of constraints and of generalized coordinates, velocities and momenta.

Some authors interpret axiom *N2* as a relation between cause (force) and effect (momentum change). This is an overstatement as merely the equivalence of the momentum change and a force is predicated[63]. Here, 'equivalence' means not only mathematical equivalence as given by the equality relation, but also physical

---

[61] Sir George Gabriel Stokes, 1ˢᵗ Baronet (1819 - 1903), *On the Effect of the Internal Friction of Fluids on the Motion of Pendulums*, 1851

[62] Hertz, *Die Prinzipien der Mechanik*, 1910, p. 7 – Engl.: "Is what we call centrifugal force anything else than the inertia of the stone? Can we, without destroying the clearness of our conceptions, take the effect of inertia twice into account,–firstly as mass, secondly as force?" (Engl. ed. 1899, p. 6)

[63] Ágoston Budó (1914 - 1969), *Theoretische Mechanik*, ¹²1990, Sect. 13.2, p. 67; von Borzeszkowski & Wahsner, *Newton und Voltaire,* 1980, p. 17

equivalence. Not only is a force the cause of a momentum change: in turn, a momentum change is the cause for a force.

Other authors include Newton's force of inertia, $-m\vec{a}$, into the set of fictitious, phantom or pseudo forces[64], or d'Alembertian forces[65] – like the centrifugal[66] and Coriolis forces[67].

A third approach considers axiom *N2* to be the very definition of 'force'. This point of view, however, discards the interpretation of *N2* in terms of statics: The rate of momentum change, $d\vec{p}/dt$, is just as large as being necessary to balance the external force, $\vec{F}_{\text{ext}}$ ($n = 1$).

Indeed, within statics, it is customary to separate the constraints (kinematic conditions, bonds to other bodies), $\vec{F}_{\text{constr}}$, from the external forces. Then, the equilibrium of forces reads[68]

$$\vec{F}_{\text{inert}} + \vec{F}_{\text{ext}} + \vec{F}_{\text{constr}} = \vec{0} \qquad (2.27)$$

Here (notice the minus sign!),

$$\vec{F}_{\text{inert}} = -\frac{d\vec{p}}{dt} \qquad (2.28)$$

is the force of inertia[69] according to

Definition 3 "Inherent force of matter is the power of resisting by which every body, so far as it is able, perseveres in its state either of resting or of moving uniformly straight forward". (Newton, *Principia*)

---

[64] *Eg*, Feynman, Leighton & Sands, *The Feynman Lectures on Physics*, 2006, Vol. I, section 12–5

[65] *Cf* Lanczos, *The Variational Principles of Mechanics*, 1986, p. 100; Courtney Seligman (*1944), *Fictitious Forces*, 2014.

[66] Huygens coined the term 'centrifugal force' in *De Vi Centrifuga*, 1659, and used it also in *Horologium Oscillatorium on pendulums*, 1673; after Soshichi Uchii, *Inertia*, 2001.

[67] Gustave Gaspard de Coriolis (1792 - 1843), *Sur les équations du mouvement relatif des systèmes de corps*, 1835; see also Aleksander Yul'yevitch Ishlinskii', *Preface by the editor* to Nurbei' Vladimirovitch Gulia, *Inertia*, 1982.

[68] *Cf* Sommerfeld, *Mechanik*, 1994, eq. (10.4); Budó, *Theoretische Mechanik*, 1990, §§ 12f.

[69] In the form $F_{\text{inert}} = -m\vec{a}$, it is also called "D'Alembert's force of inertia", see Budó, *Theoretische Mechanik*, 1990, Sect. 13.2, p. 66.

Unfortunately, it is often quite tedious to express the force of constraints, $\vec{F}_{\text{constr}}$, in explicit, analytical form. In order to eliminate it, one can evoke from statics the

Principle of virtual work If the constraints guide a body smoothly, their virtual work, *i.e*, their work done along the virtual displacements, $\delta\vec{r}$, that are consistent with the constraints, vanishes.[70]

$$\vec{F}_{\text{constr}} \cdot \delta\vec{r} = 0 \qquad (2.29)$$

As a consequence, the virtual work of the sum of the other forces vanishes, too. This is expressed in[71]

D'Alembert's Principle

$$(\vec{F}_{\text{ext}} + \vec{F}_{\text{inert}}) \cdot \delta\vec{r} = 0 \qquad (2.30)$$

The constraints diminish the degrees of freedom of motion. A simple example is the mathematical pendulum. Its constant length reduces the number of degrees of freedom, f, from $f = 2$ (unconstrained motion in a plane) to $f = 1$ (motion along a circle). For this, it is useful to change from Cartesian: x, y, to circular coordinates: r, θ. r becomes merely a parameter, while the motion is described solely through the coordinate $\theta(t)$[72]. Such coordinates are called *generalized coordinates* and written as $q_i | i = 1 \dots f$. The corresponding generalized velocities are $\dot{q}_i = \mathrm{d}\theta/\mathrm{d}t$. The corresponding generalized momentum, however, is defined *not* as $m\dot{q}$, as we will see in Subsection 2.5.6.

D'Alembert's principle served Lagrange (*Mécanique Analytique*, 1788) as starting point for his invention of generalized coordinates and velocities and thus for his equations. I will, however, proceed along another way, in order to stay closely to Newton's notion of (stationary) states.

Problem 2.25. *Prove the principle of virtual work for point-like particles that are bound each to another!* Hint: *Exploit axiom N3!*

---

[70] *Cf* Sommerfeld, *Mechanik*, 1994, § 8. – The term 'work' for the mechanical transfer of energy by a force acting along a path was coined only by Coriolis (after Max (Moshe) Jammer (1915 - 2010), *Concepts of Force*, 1999, p. 167, fn. 14); *cf* also Alexandre Moatti (*1963), *Gaspard-Gustave de Coriolis (1792-1843): un mathématicien, théoricien de la mécanique appliquée*, 2011, Ch. 4.
[71] Jean-Baptiste le Rond d'Alembert (1717 - 1783), *Traité de Dynamique*, 1743; *cf* Lanczos, *The Variational Principles of Mechanics*, 1986, p. 90; Sommerfeld, *Mechanik*, 1994, § 10; Szabó, *Geschichte der mechanischen Prinzipien und ihrer wichtigsten Anwendungen*, 1996, I.C, II.D.1
[72] The general method is due to Lagrange, *Mécanique Analytique*, 1788.

2.5.3. Bodies in Force Fields

Newton's Axiom *N1* can be understood such, that the stationary states of interaction-less bodies are quantitatively described by their momentum (vector), $\vec{p}$. Axiom *N3* states, that a system of two interacting bodies exhibits a new conserved quantity, *viz*, their *total momentum*, $\vec{p}_{tot} = \vec{p}_1 + \vec{p}_2$.

Now, without using the word 'field', the Definitions 5…8 in Newton's *Principa* actually describe (central) fields.[73] This means, that Newton's axioms are *not* confined to interactions *via* impacts, although some formulations in that part of the *Principia* may suggest that.

Thus, if two bodies interact each with another *via* a force field, $\vec{F}_{12}(\vec{r}_1(t), \vec{r}_2(t))$, and body 2 is fixed at its position: $\vec{r}_2(t) = \vec{r}_2(0) \equiv \vec{r}_{2,0}$, then, body 1 moves in the external force field $\vec{F}_1(\vec{r}_1(t)) = \vec{F}_{12}(\vec{r}_1(t), \vec{r}_{2,0})$. Well-known examples are the Kepler problem and the pendulum. In such cases, the momentum of the moving body, $\vec{p}$ (omitting the index 1), is different for different positions, $\vec{r}$. In any conserved quantity, H, the dependence of $\vec{p}$ on $\vec{r}$ is compensated through an appropriate counter-dependence of H on $\vec{r}$: $H = H(\vec{p}, \vec{r})$.

By virtue of $H(\vec{p}, \vec{r}) = const$, there is an equation of change of stationary states as[74]

$$\mathrm{d}H_{i_1 \ldots i_n}(\vec{p}, \vec{r}) = \sum_{k=1}^{3} \frac{\partial}{\partial p_k} H_{i_1 \ldots i_n}(\vec{p}, \vec{r}) \mathrm{d}p_k + \sum_{k=1}^{3} \frac{\partial}{\partial r_k} H_{i_1 \ldots i_n}(\vec{p}, \vec{r}) \mathrm{d}r_k = 0 \quad \textbf{(2.31)}$$

Up to the sign, its simplest solution reads[75]

$$\frac{\partial}{\partial p_k} H_{i_1 \ldots i_n}(\vec{p}, \vec{r}) = H_{i_1 \ldots i_n k}(\vec{p}, \vec{r}) \frac{\mathrm{d}r_k}{\mathrm{d}t};$$

$$\frac{\partial}{\partial r_k} H_{i_1 \ldots i_n}(\vec{p}, \vec{r}) = -H_{i_1 \ldots i_n k}(\vec{p}, \vec{r}) \frac{\mathrm{d}p_k}{\mathrm{d}t} \qquad \textbf{(2.32)}$$

This solution is compatible with $\vec{p} = m\vec{v}$ and $\mathrm{d}\vec{p}/\mathrm{d}t = \vec{F}$, if $n = 0$ ($H_{i_1 \ldots i_n} = H$ being a scalar function) and $H_k(\vec{p}, \vec{r}) = const = 1$ (any constants or functions can be absorbed into $H(\vec{p}, \vec{r})$). Then,

---

[73] Enders, *Precursors of force fields in Newton's 'Principia'*, 2010
[74] $(\vec{p}, \vec{r})$ may represent a scalar, vector or even tensor function of rank n, $H_{i_1 \cdots i_n}(\vec{p}, \ \vec{r})$.
[75] *Cf* Christian Baumgarten, *Old Game, New Rules: Rethinking The Form of Physics*, 2016.

$$\frac{d\vec{p}}{dt} = -\frac{\partial H(\vec{p},\vec{r})}{\partial \vec{r}} = \vec{F}; \ \frac{d\vec{r}}{dt} = \frac{\partial H(\vec{p},\vec{r})}{\partial \vec{p}} = \frac{\vec{p}}{m} \tag{2.33}$$

If, moreover, $\vec{F}$ is independent of $\vec{p}$, then, H is of the functional form

$$H(\vec{p},\vec{r}) = T(\vec{p}) + V(\vec{r}) \tag{2.34}$$

Here, $V(\vec{r})$ equals the "disposable work storage"[76] of the body under consideration in position $\vec{r}$. Furthermore,

$$T(\vec{p}) = \frac{p^2}{2m} \tag{2.35}$$

equals the kinetic energy of the body. This justifies the signs in eqs. (2.32).

Eqs. (2.33) contain Hamilton's equations of motion[77] for the stationary case.

$$\frac{d\vec{r}}{dt} = \frac{\partial H(\vec{p},\vec{r})}{\partial \vec{p}}; \ \frac{d\vec{p}}{dt} = -\frac{\partial H(\vec{p},\vec{r})}{\partial \vec{r}} \tag{2.36}$$

Due to that, the function $H(\vec{p},\vec{r})$ has been termed *Hamiltonian*.

In order to extend these equations to the non-stationary case, Euler's principles of stationary-state change will be generalized in the next subsection.

Eqs. (2.33) presuppose the force field, $\vec{F}(\vec{p},\vec{r})$, to be a gradient field. For more general force fields, one can – in order to stay with the simple form of Hamilton's equations of motion (2.36) – replace the Cartesian (more accurately: Newtonian) momentum, $\vec{p} = m\vec{v}$, with a *generalized momentum*,

$$\vec{p}_{gen} = m\vec{v} + \vec{b}(\vec{v},\vec{r}) \tag{2.37}$$

The application is left to the reader, see Problem 2.26. I will return to this issue in Subsection 2.5.6.

Problem 2.26. *Derive the function $\vec{b}(\vec{v},\vec{r})$ and the Hamiltonian for the Newton-Lorentz's equation of motion[78] of a charge, q, in static electric, $\vec{E}(\vec{r})$, and magnetic, $\vec{B}(\vec{r})$, fields,*

---

[76] Helmholtz, *Vorlesungen über die Dynamik discreter Massenpunkte*, 1911, § 49
[77] Hamilton, *On a General Method in Dynamics*, 1834; *Second Essay on a General Method in Dynamics*, 1835

$$m\frac{\mathrm{d}^2\vec{r}}{\mathrm{d}t^2} = q\vec{E}(\vec{r}) + q\vec{v} \times \vec{B}(\vec{r}) \qquad (2.38)$$

Problem 2.27 (*). *Recent research claims, that any Hamiltonian can locally be canonically transformed to that of a free particle.*[79] *Can this result be exploited for improving our arguing above?*

2.5.4. Eulerian Principles of Stationary-State Change for Classical Systems

As the fundamental entity of the mechanics of bodies is the free, interaction-less body, the fundamental entity of the mechanics of systems is the free, conservative system. Its stationary-state function is not the time-*in*dependent momentum, $p = p(0)$, or velocity, $v = v(0)$, but the time-*in*dependent Hamiltonian, $H = H_0(p, x)$.

If a body is subject to external forces, F_{ext}, its stationary-state function, $\vec{p}$ or $\vec{v}$, is no longer a stationary-state function, but becomes time-dependent: $p \to p(t) = p(0) + \Delta p(t)$, where $\mathrm{d}p(t) = F_{ext}\mathrm{d}t$ (Newton), or $v \to v(t) = v(0) + \Delta v(t)$, where $\mathrm{d}v(t) = (F_{ext}/m)\mathrm{d}t$ (Euler).

Analogously, if a conservative system is subject to interactions with other, external systems, its stationary-state function, H, is expected to become time-dependent:

$$H \to H_0(p, x) + \Delta H(p, x, t),$$

where $\mathrm{d}H(p, x, t) = (\partial H_{ext}(p, x, t)/\partial t)\mathrm{d}t$, $\partial H_{ext}(p, x, t)/\partial t$ being the analog to F_{ext}.

Surprisingly enough, the translation of the Eulerian principles of stationary-state change for single bodies to such ones for conservative systems is straightforward. The expressions for the stationary-state and non-stationary-state quantities are different, while their mutual relationships remain the same ones.

Thus, up to first order in $\mathrm{d}t$,

---

[78] *Cf* Lorentz, *Versuch einer Theorie der elektrischen und optischen Erscheinungen in bewegten Körpern*, 1895, §§ 9, 12.
[79] Elizabeth Galindo-Linares, Esperanza Navarro-Morale, Gilberto Silva-Ortigoza, Román Suárez-Xique, Magdalena Marciano-Melchor, Ramón Silva-Ortigoza & Edwin Román-Hernández, *Any Hamiltonian System Is Locally Equivalent to a Free Particle*, 2012

CS1 The changes of the stationary-state quantities (dH) depend solely on the external causes ($\partial H_{ext}/\partial t$), but not on the changes of the non-stationary-state quantities (dx, dp); in particular, $dH = 0$, if $\partial H_{ext}/\partial t = 0$.

CS2 The changes of the stationary-state quantities (dH) are independent of the stationary-state quantities (H) themselves.

CS3 The changes of the non-stationary-state quantities (dx, dp) depend solely on the stationary-state quantities (H).

Notice, that the change of a stationary state is governed not by the external Hamiltonian, $H_{ext}(p, x, t)$, itself, but by its time-derivative, $\partial H_{ext}/\partial t$. This means, that all time-*in*dependent parts of H belong to the system under consideration. As QM is basically non-classical Hamiltonian mechanics (see also Problem 2.31), it is not surprising, that we will meet that feature there, again, see Sections 11.3 and 13.2.

The stationary state of higher-dimensional systems is determined not only by the value of the Hamiltonian. Nevertheless, the equations of motion involve only the latter one.

2.5.5. Equation of Stationary-State Change

The principles *CS1...CS3* imply the following *equation of stationary-state change*.

$$dH = \frac{\partial H}{\partial p}\,dp + \frac{\partial H}{\partial x}\,dx + \frac{\partial H}{\partial t}\,dt \overset{!}{=} \frac{\partial H_{ext}}{\partial t}\,dt \tag{2.39}$$

By virtue of $\partial H/\partial t = \partial H_{ext}/\partial t$, this means

$$\frac{\partial H}{\partial p}\,dp + \frac{\partial H}{\partial x}\,dx = 0 \tag{2.40}$$

as in the stationary case, in which $H_{ext} \equiv 0$.

2.5.6. Hamilton's Equations of Motion for Time-Dependent External Fields

Hence, the equations of motion (2.36) remain valid for the general case, in which the Hamiltonian does *explicitly* depend on time: $H = H(p(t), x(t), t)$.

$$\frac{dx(t)}{dt} = \frac{\partial H(p(t),x(t),t)}{\partial p(t)} ; \quad \frac{dp(t)}{dt} = -\frac{\partial H(p(t),x(t),t)}{\partial x(t)} \tag{2.41}$$

The equation-of state-change corresponding to that equations reads

$$d\begin{pmatrix} x \\ p \end{pmatrix} = \begin{pmatrix} \frac{\partial H}{\partial p} \\ -\frac{\partial H}{\partial x} \end{pmatrix} dt \qquad (2.42)$$

Here, the position and momentum variables, $\{x_i, p_i; i = 1 \ldots f\}$, are treated on *equal* footing. The advantages and disadvantages of such a treatment are at the heart of physics and thus will interfere throughout this book.

Like eqs. (2.33), eqs. (2.41) presuppose the force field, $\vec{F}(\vec{p}, \vec{r}, t)$, to be a gradient field. For more general force fields, one can replace the Cartesian (more accurately: Newtonian) momentum, $\vec{p} = m\vec{v}$, with the *generalized momentum*, again.

$$\vec{p}_{gen} = m\vec{v} + \vec{b}(\vec{v}, \vec{r}, t) \qquad (2.43)$$

Now, generalizing formula (2.37), the function $\vec{b}$ may explicitly depend on time. The application is left to the reader, see Problem 2.28. For the sake of generality, I also replace the Cartesian coordinates with the generalized ones, $q(t)$, as introduced in Subsection 2.5.2. Then, eqs. (2.41) become

$$\frac{dq_i(t)}{dt} = \frac{\partial H(p_{gen}(t), q(t), t)}{\partial p_{gen,i}(t)}; \ \frac{dp_{gen,i}(t)}{dt} = -\frac{\partial H(p_{gen}(t), q(t), t)}{\partial q_i(t)}; \ i = 1 \ldots f \qquad (2.44)$$

This is Hamilton's celebrated equations of motion[80].

The first eq. (2.44) defines the generalized velocities, $\dot{q}(t)$, in terms of the generalized coordinates, $q(t)$, and momenta, $p_{gen}(t)$, and *vice versa* the generalized momenta, $p_{gen}(t)$, in terms of the generalized velocities, $\dot{q}(t)$, and coordinates, $q(t)$. I will enlarge upon these relationships in the next subsection.

Like Newton's equation of motion (2.25), they are bound to motion along trajectories. As with that, it is *im*possible to generalize them to cases of motion *not* procding along trajectories. Perhaps, this observation has brought Bohr to his "principal assumption", that "the passing of the [atomic] systems between different stationary states cannot be treated on that basis [CM]"[81]. We will see,

---

[80] Hamilton, *On a General Method in Dynamics, by which the study of motions of all free systems of attracting or repelling points is reduced to the search and differentiation of one central relation or characteristic function*, 1834; *Second Essay on a General Method in Dynamics*, 1835
[81] Bohr, *On the constitution of atoms*, 1913, p. 7

however, that the Eulerian principles of state change apply to quantum-mechanical systems as well.

The set $q_i, p_{gen,i} | i = 1 \dots f$ of (independent) dynamical variables spans the phase space. Since both position and momentum variables are largely treated on equal footing (exceptions may be, *e.g*, boundary conditions), it is tempting to collect them in *one* vector, say, $\vec{s} = (s_1, \dots, s_{2f}$. In this notation, Hamilton's equations of motion (2.44) read

$$\frac{d\vec{s}(t)}{dt} = \widehat{\Omega}_{2f} \, \nabla_{\vec{s}} H(\vec{s}(1), t); \; \widehat{\Omega}_{2f} = \begin{pmatrix} \hat{0}_f & \hat{I}_f \\ -\hat{I}_f & \hat{0}_f \end{pmatrix} \tag{2.45}$$

where $\hat{I}_f$ denotes the f-dimensional unit matrix. The matrix $\widehat{\Omega}$ indicates the simplectic structure of Hamiltonian mechanics[82], which, however, are far outside the scope of this book. Nevertheless, we will meet the matrix $\widehat{\Omega}$ later-on, notably, in Problems 9.4 and 12.2[83].

Problem 2.28 *Derive the function $\vec{b}(\vec{v}, \vec{r}, t)$ and the Hamiltonian for the Newton-Lorentz's equation of motion[84] of a charge, q, in time-dependent electric, $\vec{E}(\vec{r}, t)$, and magnetic, $\vec{B}(\vec{r}, t)$, fields!*

$$m \frac{d^2 \vec{r}(t)}{dt^2} = q\vec{E}(\vec{r}(t), t) + q\vec{v}(t) \times \vec{B}(\vec{r}(t), t) \tag{2.46}$$

2.5.7. Lagrange's Equations of Motion*

Let us now sketch the historically preceding step of Lagrange's equations of motion. At once, the relationship between the generalized coordinates, velocities and momenta will become more obvious.

For this, consider the following Legendre transform[85].

$$L(\dot{q}, q, t) = p_{gen} \frac{\partial H(p_{gen}, q, t)}{\partial p_{gen}} - H(p_{gen}, q, t)$$

---

[82] de Gosson, *Introduction to Simplectic Mechanics: Lectures I-II-III*, 2006; *Symplectic Geometry and Quantum Mechanics*, 2006

[83] See also C. Baumgarten, *Minkowski Spacetime and QED from Ontology of Time*, 2015, eq. (3).

[84] *Cf* Lorentz, *Versuch einer Theorie der elektrischen und optischen Erscheinungen in bewegten Körpern*, 1895, §§ 9, 12.

[85] Hugo Touchette, *Legendre-Fenchel transforms in a nutshell*, 2005-2007 [after Adrien-Marie Legendre (1752 - 1833) and Moritz Werner Fenchel (1905 - 1988)].

$$= p_{gen}(\dot{q}, q, t)\dot{q} - H(p_{gen}(\dot{q}, q, t), q, t) \qquad (2.47)$$

Here, the function $p_{gen}(\dot{q}, q, t)$ is implicitly given by the first eq. (2.44). The function $L(\dot{q}(t), q(t), t)$ is called Lagrangian, because it plays the central role in Lagrange's representation of CM (*Mécanique Analytique*, 1788).

From the Legendre transform (2.47), using eqs. (2.44), it is straightforward to show, that

$$\frac{\partial L}{\partial q} = -\frac{\partial H}{\partial q} = \frac{dp_{gen}}{dt}; \frac{\partial L}{\partial \dot{q}} = p_{gen} \qquad (2.48)$$

The second equation is an explicit expression of the generalized momentum in terms of the generalized coordinates and velocities. Furthermore,

$$\frac{\partial L}{\partial q} = \frac{d}{dt}\frac{\partial L}{\partial \dot{q}} \qquad (2.49)$$

This is Lagrange's equation of motion of 2nd kind, perhaps, the most important equation in Lagrange's representation of CM.

If the Hamiltonian is the sum of a kinetic energy, which is a homogeneous function of degree 2 of the generalized momenta, and a potential energy, which is independent of the latter ones:

$$H(p_{gen}, q, t) = T(p_{gen}, q, t) + V(q, t); p_{gen}\frac{\partial T}{\partial p_{gen}} = 2 \qquad (2.50)$$

then, the Lagrangian is of the form

$$L(\dot{q}, q, t) = T(\dot{q}, q, t) - V(q, t) \qquad (2.51)$$

Now, analogously, to the case of Cartesian coordinates, we can introduce generalized forces, F_{gen}, potential, V_{gen}, and kinetic, T_{gen}, energies.

$$\frac{\partial L}{\partial q} = -\frac{\partial H}{\partial q} = \frac{dp_{gen}}{dt} = F_{gen} = -\frac{\partial V_{gen}}{\partial q} \qquad (2.52a)$$

$$V_{gen} = -L(\dot{q}, q, t) + \tilde{L}(\dot{q}, t) \qquad (2.52b)$$

$$\frac{\partial L}{\partial \dot{q}}\dot{q} = p_{gen}\dot{q} = 2T_{gen} \qquad (2.52c)$$

Helmholtz has called the negative of the Lagrangian: $-L = V - T$, *kinetic potential*[86]. In the generalized potential (2.52b), the coordinate-independent terms, $\tilde{L}(\dot{q}, t)$ are eliminated from L, because they do not contribute to the generalized force (2.52a).

Problem 2.29 (Few geometry). *Discuss the relationship between generalized velocity, $\dot{q} = \partial H / \partial p_{gen}$, and generalized momentum, $p_{gen} = \partial L / \partial \dot{q}$, in view of the geometric meaning of the Legendre transform*[87]*!*

Problem 2.30 (Lagrangian mechanics). *Show, that Lagrange's representation of CM does not fit into the Newton-Eulerian scheme*

> *Stationary state – change of stationary state – motion!*

What does this mean? Hints: Is the Lagrangian, L, unique? Calculate d*L!*

Problem 2.31 (Path integral QM*). *Discuss your result of Problem 2.30 in the light of Feynman's path integral mechanics*[88] *being non-classical Lagrangian mechanics! Hint: Few features are mentioned in Section 12.2.*

2.5.8. Hamilton's Principle of Least Action*

> "Cum enim Mundi universi fabrica fit perfectissima, atque a Creatore sapientissimo absoluta, nihil omnino in mundo congingit, in quo non maximi minimive ratio quaepiam eluceat: quamobrem dubium prorsus est nullum, quin omnes Mundi effectus ex causis finalibus, ope Methodi maximorum & minimorum aeque feliciter determinari queant, atque ex ipsis causis efficientibus."[89]

> "The variational principles are firmly rooted in the soil of that great century of liberalism which starts with Descartes and ends with the French Revolution and which has witnessed the lives of Leibniz, Spinoza, Goethe, and Johann Sebastian Bach. It is the only period of

---

[86] Helmholtz, *Vorlesungen über die Dynamik discreter Massenpunkte*, 1911, § 76
[87] *Cf* Royce K. P. Zia, Edward Frederick Redish & Susan R. McKay, *Making Sense of the Legendre Transform*, 2009.
[88] Feynman, *Space-Time Approach to Non-Relativistic Quantum Mechanics*, 1948; Feynman & Hibbs, *Quantum Mechanics and Path Integrals*, 2005
[89] Euler, *De Curvis Elasticis*, 1744, p. 2 – Engl.: "For since the fabric of the universe is most perfect, and is the work of a most wise Creator, nothing whatsoever takes place in the universe in which some relation of maximum or minimum does not appear. Wherefore there is absolutely no doubt that every effect in the universe can be explained as satisfactorily from final causes, by the aid of the method of maxima and minima, as it can from the effective causes themselves." (1933 transl., pp. 5f.)

cosmic thinking in the entire history of Europe since the time of the Greeks."[90]

Lagrange's equations of motion of 2nd kind (2.49) are the Euler-Lagrange equations[91] to the variational problem

$$\delta \int_{t_1}^{t_2} L(q(t), \dot{q}(t), t)\mathrm{d}t \overset{!}{=} 0 \qquad (2.53a)$$

$$\delta t_1 = \delta t_2 = \delta q(t_1) = \delta q(t_2) = \delta \dot{q}(t_1) = \delta \dot{q}(t_2) = 0 \qquad (2.53b)$$

This statement is equivalent with

Hamilton's principle (1834) The action integral

$$S \underset{def}{=} \int_{t_1}^{t_2} L(q(t), \dot{q}(t), t)\mathrm{d}t \qquad (2.54)$$

is stationary for each part of the actual trajectory, when compared with all neighboring trajectories, that have got the same initial and endpoints at the same initial and end times.

Hamilton's principle is mostly called *principle of least action*, although the action, S, may become a local maximum.

The action integral served Hamilton as starting point for his uniform representation of mechanics and optics, which Schrödinger (1926) has exploited for the founding of wave mechanics.

Despite of the paramount power of that principle, it plays almost no role in this book. For the action is not a Newtonian stationary-state quantity, and I do not see any possibility to put Schrödinger's ingenious guesses onto an axiomatic ground. Within quantum physics, there is the quantum of action, h. It is a fundamental constant of nature[92]. For this, the action is not subject to a variational principle.

---

[90] Lanczos, *The Variational Principles of Mechanics*, 1986, p. x
[91] Euler, *Methodus inveniendi lineas curvas maximi minimive proprietate gaudentes*, 1744; Lagrange, *Abhandlungen zur Variationsrechnung*, 1894; Wolfgang Yourgrau (1908 - 1979) & Stanley Mandelstam (1928 - 2016), *Variational Principles in Dynamics and Quantum Theory*, 1979, App. 1
[92] Planck, *Zur Theorie des Gesetzes der Energieverteilung im Normalspektrum*, 1900, p. 237 – notice, that in the 1901 publication of that talk, Planck back-pedaled, h being merely a combination of known spectroscopic parameters; Einstein (1905) did not use h. Interestingly enough, the universality of the constant in the Stefan-Boltzmann law suggests the existence of a universal action constant, too, see Harry Paul (*1931), *Auf dem Weg zur Quantentheorie: Die Erfindung des Hohlraums und ihre Folgen*, 2015.

CHAPTER 3

Alternative Axiomatic: Energy Conservation

Abstract: The conservation of the total energy of closed systems belongs to the most general laws of nature. This chapter treats first the common way to the mechanical one starting with Newton's equation of motion. Then, the possibility to use the energy law as foundation of classical mechanics is explored, notably, Planck's axioms, Euler's second equation of stationary-state change, Carlson's principle of the stationarity of total energy, and Leibniz's theorem on the conservation of kinetic energy and its dual on the conservation of potential energy. The chapter concludes with considerations about the relationship between energy and extension in configuration space as known from the harmonic oscillator and the Kepler ellipses, and analogous considerations in momentum configuration space.

Keywords: Axiomatic, Configuration space, Energy conservation, Extension, Harmonic oscillator, Helmholtz, Hodograph, Kepler ellipse, Leibniz's theorem, Momentum configuration space, Nemorarius' theorem, Newton's equation of motion, Schütz, State, State function, Stationary state.

> "Kein Wesen kann zu Nichts zerfallen!
> Das Ew'ge regt sich fort in allen,
> Am Sein erhalte dich beglückt!
> Das Sein ist ewig; denn Gesetze
> Bewahren die lebend'gen Schätze,
> Aus welchen sich das All geschmückt."[1]

Within the common way to the mechanical energy law, Newton's equation of motion is multiplied by the velocity, $\vec{v}$, and integrated. This ties it to the classical orbits. In order to facilitate the transition from classical to quantum mechanics, let me expose alternative thoughts (Fig. **3.1**).

---

[1] Johann Wolfgang von Goethe (1749 - 1832), *Vermächtnis*, 1829 – Engl.: None essence can decay to nothing!/ The eternal bestirs oneself in all, / Go on to be happy with the Being! / The Being is eternal, for laws / Preserve the living treasuries, / With which the universe has adorned itself.

Fig. 3.1. Goethe Monument in Chicago by Hermann Hahn (1868 - 1945). Photo by Greg Dunham.

3.1. THE COMMON WAY TO THE MECHANICAL ENERGY CONSERVATION LAW

> "Die Naturerscheinungen sollen zurückgeführt werden auf Bewegungen von Materien mit unveränderlichen Bewegungskräften, welche nur von den räumlichen Verhältnissen abhängig sind."[2]

Usually, Newton's equation of motion,

$$m\ddot{\vec{r}}(t) = \vec{F}(\vec{r}(t)) \tag{3.1}$$

is multiplied by $\dot{\vec{r}}$ (Euler, *Anleitung*, §§ 74-76) to obtain

$$m\ddot{\vec{r}} \cdot \dot{\vec{r}} = \frac{\mathrm{d}}{\mathrm{d}t}\left(\frac{m}{2}\dot{r}^2\right) = \vec{F} \cdot \dot{\vec{r}} = \frac{\mathrm{d}}{\mathrm{d}t}\int \overrightarrow{\vec{F}(\vec{r}')} \cdot \mathrm{d}\vec{r}' \tag{3.2}$$

---

[2] Helmholtz, *Über die Erhaltung der Kraft*, 1847, pp. 4f. – Engl.: "…the phaenomena of nature are to be referred back to motions of material particles possessing unchangeable moving forces, which are dependent upon con ditions of space alone." (1953, p. 116)

Now, if there is a function, $W(\vec{r})$, such, that

$$\int_{\vec{r}_0}^{\vec{r}} \vec{F}(\vec{r}') \cdot \mathrm{d}\vec{r}' = W(\vec{r}) - W(\vec{r}_0); \quad \vec{F}(\vec{r}) = \nabla W(\vec{r}) \qquad (3.3)$$

then, eq. (3.2) yields the conservation law

$$\frac{m}{2} \dot{\vec{r}}(t)^2 - W(\vec{r}(t)) = const = E \qquad (3.4)$$

Euler called the function $W(\vec{r})$ "Wirksamkeit" (effectiveness, efficacy)[3]. Such force fields are called *conservative*.

The relationship between gradient fields and conservation can nicely be illustrated by means of the *impossible staircase*[4], which often are called Penrose stairs or Penrose steps[5] and has been artistically realized by Escher, see Figs. (**3.2** and **3.3**).

The square endless staircase models the motion in a solenoidal field. Walking upstairs from the left corner *via* the upper corner to the right corner, one gains potential energy in the earth's gravity field. On the contrary, walking downstairs from the left corner *via* the lower corner to the right corner, one looses potential energy in the earth's gravity field. Hence. the earth's gravity field is irrotational.

Helmholtz was the first to realize the universal relevance of this constant of integration, E. Nevertheless, his more fundamental approach (Helmholtz 1847, 1911) starts with the conservation of the 'living force', *i.e.*, of the kinetic energy, see Subsection 3.2.6.

Problem 3.1 (Electrical charge in a static magnetic field). *Which quantities are conserved, if a point-like electrical charge, q, moves in a static magnetic field, $\vec{B}(\vec{r})$, and thus is subject to the (Maxwell-)Lorentz force[6] created by this force field: $\vec{F} = q\vec{v} \times \vec{B}(\vec{r})$? Hint: This force also depends on $\vec{v} \equiv d\vec{r}/dt$.*

---

[3] Euler *Anleitung*, 1750, § 75; *cf* also *Harmonie entre les principes généraux de repos et de mouvement de M. de Maupertuis*, 1751.

[4] Oscar Reutesvald, 1930-ies, after http://psylux.psych.tu-dresden.de/i1/kaw/ diverses Material/www. illusionworks.com/html/art_of_reutersvard.html, see also Diana Deutsch (*1938), *Pitch Circularity*, 2010, note in Ref. 1

[5] Lionel Sharples Penrose (1898 - 1972) & Roger Penrose (*1931), *Impossible objects: A special type of visual illusion*, 1958

[6] Maxwell, *On Faraday's lines of force, Pt. II. On Faraday's electrotonic state,* 1856f., Sect. D. Electro-magnetism; Lorentz, *La théorie électromagnétique de Maxwell et son application aux corps mouvants*, 1892

Fig. 3.2. *Ascending and Descending*, art print by Maurits Cornelis Escher (1898 – 1972), 1960[7]

Fig. 3.3. *Ascending and Descending*, Lego construction by Andrew Lipson & Daniel Shiu of Maurits Cornelis Escher's 1960 art print[8]

---

[7] By Official M. C. Escher website, Fair use, https://en.wikipedia.org/w/index.php?curid=6847499
[8] A. Lipson 2002, http://andrewlipson.com/escher/ascending.html, with friendly permission by Andrew Lipson

Problem 3.2. *Show, that the efficacy, $W = \int \vec{F} \cdot d\vec{r}$, is invariant against rotations of the coordinate system, while the quantity $\int \vec{F}dt$ is not (Euler, Anleitung, § 75)!*

Problem 3.3. *Show, that a body is in an equilibrium state, if W assumes a minimum or a maximum!* [9]

3.2. THE ENERGY LAW AS BASIS OF CLASSICAL POINT MECHANICS

On the other hand, the energy law is much more general than Newton's equation of motion, $m\vec{a} = \vec{F}$, or Euler's equation of stationary-state change, $d\vec{v} = (1/m)\vec{F}dt$. Thus, for the sake of the unity of physics, it is desirable to reverse the common approach of the foregoing section and to derive the former from the latter one.

Schütz has derived Newton's 3rd axiom from the conservation of kinetic energy (*cf* Subsection 3.2.6) and Galileo's principle of relativity (*cf* Subsection 2.2.1)[10]. He considers the elastic one-dimensional impact of two (non-rotating) masses, m_1 and m_2. with initial velocities, $w_{1,2}$, and final velocities, $\omega_{1,2}$. The conservation of kinetic energy implies

$$\frac{m_1}{2}w_1^2 + \frac{m_2}{2}w_2^2 = \frac{m_1}{2}\omega_1^2 + \frac{m_2}{2}\omega_2^2$$

Now, the four velocities are arbitrary up to a common value, α (Galilean relativity). Hence, the equation

$$\frac{m_1}{2}(w_1 + \alpha)^2 + \frac{m_2}{2}(w_2 + \alpha)^2 = \frac{m_1}{2}(\omega_1 + \alpha)^2 + \frac{m_2}{2}(\omega_2 + \alpha)^2$$

is valid as well. Subtracting both equations yields

$$m_1 w_1 \alpha + m_2 w_2 \alpha = m_1 \omega_1 \alpha + m_2 \omega_2 \alpha$$

Since the value of α is arbitrary, it drops out and the conservation of total momentum remains.

For the derivation of the 1st and 2nd axioms, he has presupposed the form of the potential energy (*ibid.*, eq. (5)). For the 1st axiom, however, it is sufficient to let mass m_2 vanish.

---

[9] Pierre Louis Moreau de Maupertuis (1698 - 1759), *Accord de différentes Loix de la Nature qui avoient jusqu'ici paru incompatibles*, 1744; *cf* Euler, *loc. cit.*
[10] Ignaz (J.) Robert Schütz (1867 - 1927), *Prinzip der absoluten Erhaltung der Energie*, 1897; *cf* Mach, *Die Mechanik in ihrer Geschichte*, 1988, p. 266; Minkowski, *Raum and Zeit*, 1908, IV.

Similarly, Hamel[11] has generalized Galileo's and Leibniz's results to the energy law and deduced from it Newton's equation of motion for conservative forces.

For our purposes, Planck's treatment[12] is more appropriate. I reformulate it in terms of Newton's and Euler's line of reasoning (see Subsection 2.4.2),

Conservation of stationary states – change of stationary states – motion.

3.2.1. Planck's Axioms*

Planck (*ibid.*) starts with the kinetic energy as a stationary-state quantity.

$$T = \frac{m}{2}\left[\left(\frac{dx(t)}{dt}\right)^2 + \left(\frac{dy(t)}{dt}\right)^2 + \left(\frac{dz(t)}{dt}\right)^2\right] \tag{3.5}$$

Here, the vector

$$(x(t), y(t), z(t)) = \vec{r}(t)$$

describes the trajectory of the body and

$$\left(\frac{dx(t)}{dt}, \frac{dy(t)}{dt}, \frac{dz(t)}{dt}\right) = \vec{v}(t)$$

its velocity. The analog to Newton's 1st axiom (*N1*) is thus

Axiom P1 A body remains in its state of constant kinetic energy:

$$T = const \tag{3.6}$$

as long as no external force changes it.[13]

In particular, the state at rest is given through

$$T = T_{min} = 0 \tag{3.7}$$

The analog to Newton's 2nd axiom (*N2*) reads

Axiom P2 The change of kinetic energy, dT, equals the work, $\vec{F} \cdot d\vec{r}$, done by the external force, $\vec{F}$, on the system.

$$dT = m\left[v_x dv_x + v_y dv_y + v_z dv_z\right] = F_x dx + F_y dy + F_z dz \tag{3.8}$$

---

[11] Georg Karl Wilhelm Hamel (1877 - 1954), *Mechanik I*, 1921, § II.1
[12] Planck, *Das Prinzip von der Erhaltung der Energie*, 1908, Sect. 3.1
[13] This formulation agrees with Euler's one, while in Newton's 1st axiom (*N1*), the body itself changes its state.

Since there is an energy of the interaction itself, there is no direct analog to Newton's 3rd axiom (*N3*), but in the case of elastic impact.

Problem 3.4 (*). *Discuss the fact, that the l.h.s. of eq. (3.8) represents a total differential, while its r.h.s. does only conditionally so! Hint: See also eq. (3.10)!*

3.2.2. Euler's 2nd Equations of Stationary-State Change. Newton's Equation of Motion

Planck treats the motions along the three coordinate directions like 'different forms of energy' and applies his

Conjecture 3.1 (Planck). *The energies of 'different forms' (mechanical, electrical, heat) superpose.*

Accordingly, he separates the right and left hand sides of eq. (3.8) as[14]

$$mv_x dv_x = F_x dx \tag{3.9a}$$

$$mv_y dv_y = F_y dy \tag{3.9b}$$

$$mv_z dv_z = F_z dz \tag{3.9c}$$

This is Euler's second equations of stationary-state change, where Euler (*Anleitung*, § 74) points to the condition

$$\frac{dx}{v_x} = \frac{dy}{v_y} = \frac{dz}{v_z} = dt \tag{3.10}$$

Then,

$$mv_x \frac{dv_x}{dx} = mv_x \frac{dv_x}{dt}\frac{dt}{dx} = ma_x = F_x \; etc. \tag{3.11}$$

This sums up to Newton's equation of motion.

$$m\vec{a} = \vec{F} \tag{3.12}$$

In the representations of CM by Descartes and Newton, the momentum (vector) is considered to be the *fundamental* conserved quantity. Alternatively, Leibniz

---

[14] *Cf* the criticism in Schütz, *loc. cit.*, p. 113, fn. 1.

favored the "living force", $mv^2 = 2T(v)$, as fundamental conserved quantity. Euler's introduction of the state quantity v^2 connects both points of views.[15]

3.2.3. The Forces Doing Work are Potential Forces

The expression

$$m\left[v_x \mathrm{d}v_x + v_y \mathrm{d}v_y + v_z \mathrm{d}v_z\right] = \mathrm{d}T \qquad (3.13)$$

represents a *total* differential. Correspondingly, the r.h.s. of eq. (3.8) should be a *total* differential, too.

$$F_x \mathrm{d}x + F_y \mathrm{d}y + F_z \mathrm{d}z \overset{!}{=} \mathrm{d}W \qquad (3.14)$$

W being Euler's "efficacy", eq. (3.3) (notice, that Planck has made an additional assumption for that).

In other words, with respect to its spatial distribution, the force field, $\vec{F}$, has got a potential, $V = -W$.

$$\vec{F} = \nabla W = -\nabla V \qquad (3.15)$$

For dimensional reasons, this potential represents a *potential energy*.

Problem 3.5. *Show, that this potential, V, is time-independent! Hint: Calculate $dV(x, y, z, t)$ and compare it with the requirement (3.14)!*

Problem 3.6 (*). *How non-potential forces, in particular, the Lipschitz force (3.7) below can be incorporated into this approach? Hint: Analyze the case, that $\vec{F}(\vec{r})$ represents a static vortex field: $\vec{F} = \nabla \times \vec{G}(\vec{r})$, and account for the number of functions that are necessary for describing a gradient and a vortex field, respectively!*

3.2.4. The Total Energy and its Absolute Value

From eqs. (3.8), (3.14) and (3.15) we obtain

$$dT + dV = 0, \qquad (3.16)$$

---

[15] The equivalence of the physics of conserved quantities (Parmenides (c. 520/515 - c. 460/455 BCE) – Descartes – Leibniz) and of the physics of laws of change (Heraclitus (c. 535 - c. 475 BCE) – Galileo – Newton) has been shown first by Daniel Bernoulli (1700 - 1782), *Examen principorum mechanicae*, 1726 (after Ashot Tigranovich Grigoryan (1910 – 1997) & Boris Demyanovich Kovalev, *Daniel Bernoulli*, 1981, Ch. 5).

therefore,

$$T + V = E = const. \qquad \textbf{(3.17)}$$

The constant of integration is determined by the definition of the total energy to quantify the ability of the body to deliver work to its environment. Hence, the energy of the ground state equals zero.

3.2.5. Carlson's Principle of the Stationarity of Total Energy*

Carlson rightly remarks, that the ubiquitous relevance of energy conservation contrasts its little use in axiomatic.[16] He requires the "energy state function", $E(q_i, P_i)$ $(i = 1 \dots 3N)$, to be stationary.

$$\frac{dE}{dt} = \Sigma_i \left(\frac{\partial E}{\partial q_i} \dot{q}_i + \frac{\partial E}{\partial P_i} \dot{P}_i \right) = 0 \qquad \textbf{(3.18)}$$

Here, P_i is a component of a Cartesian momentum.[17]

For "conservative" systems, the $3N$ force components, Q_i, are the negative, $3N$-dimensional gradient of a potential function, $U(q_i, P_i)$.

$$Q_i = -\frac{\partial U}{\partial q_i} = \dot{P}_i \qquad \textbf{(3.19)}$$

Moreover, Carlson restricts himself to systems the total energy of which reads

$$E(q_i, P_i) = T(P_i) + U(q_i, P_i) \qquad \textbf{(3.20)}$$

Then, eq. (3.18) implies

$$\frac{\partial E}{\partial P_i} = v_i \qquad \textbf{(3.21)}$$

So far, this is compatible with our approach in Section 2.5. Moreover, Carlson obtains interesting results for the construction of the "energy state function", $E(q_i, P_i)$, which, however, are beyond the scope of this book.

---

[16] Shawn Carlson (*1960), *Why not energy conservation?*, 2016
[17] I do *not* adopt Carlson's notions "physical momentum" and "mechanical momentum", because the other momenta are not unphysical or non-mechanical.

Recall, that the smooth transition from CM to QM needs a representation of the energy law that is not bound to the existence of trajectories. Such a representation will be considered next.

Problem 3.7 (*). *Consider the more general energy function*

$$E(q_i, P_i) = T(q_i, P_i) + U(q_i, P_i) \tag{3.22}$$

Can it brought into the form (3.20) by means of the Hertzian line element (5.3) below?

3.2.6. Leibniz's Theorem on the Conservation of Kinetic Energy

In order to reach the height h in the gravity field g, a body of mass m needs the initial speed $v = \sqrt{2gh}$; and he gets the same when falling back. For this, the amount of work equals half of the living force: $\frac{1}{2}mv^2 = mgh$. More generally, we have

Theorem 3.1 (Leibniz). *The total kinetic energy, $T(v)$, of a system assumes the same value, when the system returns to the same configuration, C.[18]*

Conclusion 3.1 (Helmholtz). *In this case, despite of being defined in terms of velocity, the kinetic energy is a pure function of the coordinates: $T = T(x)$; the existence of such a function has to be required.[19]*

Then, $\frac{\mathrm{d}}{\mathrm{d}t}T(x) = F \cdot \dot{x}$ is the total differential of a function depending solely on the coordinates, and eq. (3.2) can be written as, say,

$$\frac{\mathrm{d}}{\mathrm{d}t}T(v) = \frac{\mathrm{d}}{\mathrm{d}t}[-V(x)], \tag{3.23}$$

provided, that $F = -\nabla V(x)$ (the minus sign will be justified below; ∇ being the gradient operator in the f-dimensional configuration space).

---

[18] Helmholtz, *Über die Erhaltung der Kraft*, 1847, p. 9; *Vorlesungen über die Dynamik discreter Massenpunkte*, 1911, § 48, pp. 193f; *cf* Gottfried Wilhelm Leibniz (1646 - 1716), *Specimen dynamicum*, 1695. – Leibniz considered *periodic* motions, and QM also began with *periodic* motions, see Bohr, *On the Spectrum of Hydrogen*, 1914.
[19] *Cf* Helmholtz, *Vorlesungen über die Dynamik discreter Massenpunkte*, 1911, pp. 194f.

The gradient form represents a condition for the spatial arrangement of the forces within the system under consideration. (*Cf ibid.*)

Upon simple integration, eq. (3.23) becomes the *energy law* of CM.

$$V(x) + T(v) = E = const \qquad (3.24)$$

The constancy of the l.h.s. is realized through the motion *along a trajectory*: $x = x(t), v(t) = \dot{x}(t)$, where the increase of one term is compensated by the decrease of the other one.

I will call the stationary state with the value E of total energy shortly the 'stationary state E'.

In eq. (3.23), Helmholtz has set $-V(x)$ rather than $+W(x)$ for the potential energy function, $V(x)$, to represent the "disposable work storage" of the system.[20]

Moreover, the potential energy, $V(x)$, expresses more directly than the efficacy, $W(x)$, the relationship between energy and spatial extension, see Subsection 3.3.1.

Leibniz's theorem is related to the reduction of the phase space, within which classical systems are moving, to the configuration space, which is sufficient to describe the motion of quantum systems.[21]

Problem 3.8. *Discuss Torricelli's flux formula[22] within the framework of the energy conservation law!*

3.2.7. Nemorarius' Theorem on the Conservation of Potential Energy*

For the sake of a more symmetrical treatment of space and momentum variables, let me make the following *dual* statements.[23]

Theorem 3.2 (Nemorarius). *The potential energy of a system, $V(x)$, is unchanged,*

---

[20] Helmholtz, *loc. cit.*, p. 215. This does not mean, that the system contains a certain 'amount of work'! The work, A, refers to the exchange of energy. It is not defined through the motion along trajectories according to the laws of motion, but through the slow displacement of a body, say, from $\vec{r}_1$ to $\vec{r}_2$, where the internal forces are balanced out by an external force. If there is a potential energy function, $V(\vec{r})$, then, $A = V(\vec{r}_2) - V(\vec{r}_1)$.

[21] *Cf* Robert Paul Geroch (*1942), *Geometrical Quantum Mechanics*, 1974, p. 38.

[22] Evangelista Torricelli (1608 - 1647), *Opera geometrica*, 1644.

[23] Analogously to its (spatial) configuration, the momentum configuration of a system refers to the relative momenta of its bodies, *e.g.*, to its total momentum. The motion of the center of gravity of a system exerts no influence on its conserved quantities.

when it returns to the same momentum configuration, P.[24]

Conclusion 3.2 (Dual Helmholtz). *Despite of being defined in terms of position, the potential energy can be written as a pure function of the momenta: $V(p)$; the existence of such a function has to be required.*

Taking for this function the expression $E - T(p)$, the energy conservation law (3.24) is reobtained, where the velocities are replaced with the momenta. Choosing, again, the minus sign [this time, at the function $T(p)$], $T(p)$ denotes the kinetic energy, that represents the 'motion storage' of a system. Here, 'motion storage' means "disposable work storage" in form of 'motional energy'. Moreover, it expresses more directly than $-T(p)$ the relationship between energy and extension in momentum configuration space, see Subsection 3.3.2.

3.3. ENERGY AND EXTENSION. I

In this section, I will explore ordered sets of total energy and (characteristic) extension in space and momentum space, respectively. This will proof helpful for the transition to quantum mechanics as well as for the relationship between classical and quantum systems.

3.3.1. Energy and Spatial Extension

The equation

$$V(\vec{r}) = c = const \tag{3.25}$$

defines an equipotential surface in space. If a static system delivers work to its environment, it moves from surface $c = c_1$ to surface $c = c_2 < c_1$[25]. The smallest value of c is assumed in the *ground state*.

This observation suggests

Conjecture 3.2 (Hemlholtz's rule). *A system changes its extension in configuration space when exchanging work with its environment.*

---

[24] Jordanus Nemorarius (Jordanus de Nemora/Giordano of Nemi, 1225 - 1260), *Elementa Jordani super demonstrationem ponderis secandum sitis*; *Liber Jordani de rationi ponderis* – after Hamel, *Mechanik I*, 1921, pp. 6ff., and Simonyi, *Kulturgeschichte der Physik*, 1990, pp. 146ff., Nemorarius was the first one to assume, that the lifting of weight G to the height h and the lifting of weigt G' to the height h' creates the same "tension", if $Gh = G'h'$.
[25] *Cf* Helmholtz, *Vorlesungen über die Dynamik discreter Massenpunkte*, §§ 50-51.

For example, the characteristic extension of a linear harmonic oscillator is related to its maximum potential energy, V_{max}, as its turning points equal $\pm\sqrt{2V_{max}/\kappa}$ (κ being its force constant). At once, $V_{max} = E$. The ground state is that with $V_{max} = E = 0$.

Kepler systems and the like do *not* exhibit a ground state; they are model systems that loose their validity at some (small) distance from the center of the force field, notably at the surface of a gravitating body.

For bound states (Kepler ellipses), the characteristic extension of the orbit (its semi-major axis, see below) is the larger, the larger the *total* energy of the orbit is, while the minimum and maximum values of the potential energy play no role here.

We thus are lead to

Conjecture 3.3 (Helmholtz's rule). *A system changes its extension in configuration space when exchanging energy with its environment.*

In its ground state, a system is not able to deliver energy to its environment – otherwise, it would be a *perpetuum mobile*. It is thus reasonable, to assign to the ground state the value $E = 0$.[26]

To illustrate these ideas by means of a less trivial case, let us consider the Kepler ellipses in more detail. The semi-major, a, and semi-minor, b, axes are determined by the angular momentum, L, and the total energy, E, as (ignoring the reduced mass, $mM/(M + m)$)

$$a = \frac{GMm}{-2E} \qquad (3.26a)$$

$$b = \frac{L}{\sqrt{-2mE}} \qquad (3.26b)$$

Notice, that, for $E < 0$ (elliptical orbits), $b \le a$ implies $L \le GMm^{3/2}/\sqrt{2|E|}$.[27]

Table **3.1** illustrates the fact, that there is a monotonous relation only between total energy, E, and semi-major axis, a, but not between E and the semi-minor axis, b. For b depends not only on E, but also on the angular momentum, L. Therefore, the characteristic extension of a Kepler ellipse is given by its semi-major axis, a.

---

[26] *Cf* Planck, *Das Prinzip von der Erhaltung der Energie*, p. 111.
[27] The analog to eq. (3.26a) for the case of a Coulomb force [Charles-Augustin de Coulomb (1736 - 1806), *Premier mémoire sur l'électricité et le magnétisme*, 1785] has been used by Bohr in 1913.

Table 3.1. Semi-major, a, and semi-minor, b, axes of a Kepler ellipse for different values of total energy, E, and angular momentum, L (arbitrary units, $G = M = m = 1$).

Example	E	L	a	b
1	-1	$\dfrac{1}{10}$	$\dfrac{1}{2}$	$\dfrac{1}{10\sqrt{2}}$
	-2	$\dfrac{1}{5}$	$\dfrac{1}{4}$	$\dfrac{1}{10}$

The same is illustrated in Fig. **3.4**. Therefore, I put forward the following

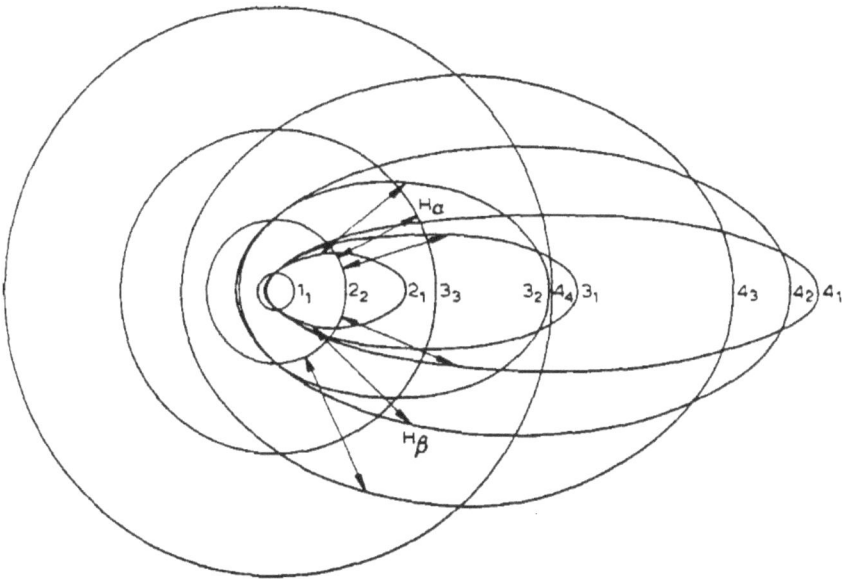

Fig. 3.4. Bohr orbitals[28]

Conjecture fig 3.4 (sets of energy and spatial extension) *For each ordered set of the total energy, $E_1 < E_2 < \cdots$, of a spatially bounded system, there is an ordered set of its spatial extension, $x_{max,1} < x_{max,2} < \cdots$.*

---

[28] Bohr, *The structure of the atom*, Nobel lecture 1922, Fig. 5; *cf* P. Kuiper, *Semi-classical electron orbits with principle quantum number n=5 in the hydrogen atom*, 2007.

Problem 3.9 (Similarity*) *Discuss the relationship between total energy and extension using the concept of mechanical similarity[29]!*

Problem 3.10 (Ground state) *In their pioneering papers, Bohr (1913) and Heisenberg (1925) have postulated the existence of a ground state. Show, that this postulate is not necessary!*

Problem 3.11 (Perpetuum mobile) *Show, that energy conservation and perpetua mobilia exclude each another! Hint: Think about bookkeeping!*

3.3.2. Energy and Extension in Momentum Space

Within momentum space, it reasonable to assume the

Dual generalized Helmholtz's rule A system changes its extension in momentum space when exchanging energy with its environment.

It holds true for the linear harmonic oscillator as

$$E = T(p_{max}) = \frac{1}{2m} p_{max}^2 \tag{3.27}$$

In contrast, the radius, v_h, of the hodograph of a Kepler ellipse[30] depends on the angular momentum, L, but *not* on the energy, E, of the orbit.

$$v_h = \frac{GMm^2}{L} \tag{3.28}$$

Here, that rule does *not* apply.

For this, a somewhat restricted hypothesis is put forward.

Conjecture 3.5 *For each ordered set of total energies, $E_1 < E_2 < \cdots$, of a system, to which the Dual generalized Helmholtz's rule applies, there is an ordered set of its extension in momentum space, $p_{max,1} < p_{max,2} < \cdots$.*

---

[29] Newton, *Principia*, Book II, Sect. 7, Prop. 32; Whittaker, *Analytical Dynamics*, § 33; Landau & Lifshitz, *Mechanics*, § 10

[30] Hamilton, *The hodograph or a new method of expressing in symbolic language the Newtonian law of attraction*, 1846; for more details, see Section 5.3 below.

CHAPTER 4

Possible and Impossible (Momentum) Configurations

Abstract: This chapter prepares a *smooth* way from *Eulerian* classical mechanics to quantum mechanics. Starting from Helmholtz's foundation of the energy conservation law using Leibniz's theorem as described in the foregoing Chapter 2, the configurations a system can (cannot) assume in a given stationary state are explored. The following constraints are taken into account: Newton-Euler's exclusion principle, d'Alembertian constraints and constraints imposed by conservation laws. The set of possible and impossible momentum configurations is also considered, using Nemorarius' theorem of the foregoing chapter.

Keywords: d'Alembertian constraints, Conservation laws, Helmholtz, Hodograph, Impossible configurations, Impossible momentum configurations, Leibniz's theorem, Momentum configuration space, Nemorarius' theorem, Newton-Euler's exclusion principle, Possible configurations, Possible momentum configurations, Schütz, State, State function, Stationary state.

The pioneers of quantum mechanics (notably, Bohr, 1913, Heisenberg, 1925, Schrödinger, 1926) have correctly stated, that there is *no smooth* way from *Newtonian* or *Hamiltonian* CM to QM. This is basically due to the fact, that Newton's 2nd axiom bounds his mechanics to motion along trajectories. This leads to constraints of the sets of (momentum) configurations a system in a given stationary state can assume. These constrains do not occur within QM. Thus, in order to prepare our *smooth* way from *Eulerian* CM to QM, they are considered in this chapter in somewhat more detail.

4.1. POSSIBLE AND IMPOSSIBLE CONFIGURATIONS

Due to the validity of the equations of motion in *whole* space and to the freedom in the choice of the initial conditions, $\vec{r}_0$ and $\vec{p}_0$, on principle, every body can move to *any other* location. Analogously, for a system of mass points, *all* configurations are possible; the phase space orbit being a sequence of possible initial conditions. Nevertheless, the motion of bodies and hence the possible configurations of a system can be limited by constraints.

4.1.1. Newton-Euler's Exclusion Principle

"Wo Du nicht bist, kann ich nicht sein"[1]

One of the constraints mentioned above is the

Newton-Euler's exclusion principle The locations of two bodies are disjunct.[2]

Problem 4.1 *Show, that this principle is a consequence of the impenetrability of the bodies (see Subsubsection 2.2.5)!*

4.1.2. D'Alembertian Constraints*

D'Alembertian constraints are constraints that lower the number of degrees of freedom of a body or system. An example is a table on that a ball rolls[3].

For the remaining degrees of freedom, the motion is unlimited, if there are no further constraints.

4.1.3. Lagrangian Constraints*

Carrying out the time derivative in Lagrange's equation of motion of 2nd kind (2.49) yields the formula (summation over $l = 1 \dots f$)

$$\frac{\partial^2 L}{\partial \dot{q}_k \, \partial \dot{q}_l} \ddot{q}_l = \frac{\partial L}{\partial q_k} - \frac{\partial^2 L}{\partial \dot{q}_k \, \partial q_l} \dot{q}_l; \ k = 1 \dots f \tag{4.1}$$

It can be resolved for the accelerations, $\ddot{q}_k$, if the matrix $W_{kl} = \partial^2 L / \partial \dot{q}_k \, \partial \dot{q}_l$ is non-singular: $\det \widehat{W} \neq 0$. If, on the contrary, $\det \widehat{W} = 0$, the generalized momenta, p_k, are not independent each of another. The dependencies between them expresses themselves as so-called Lagrangian constraints among the dynamical variables

---

[1] Fritz Löhner (Bedřich Löwy, 1883 - 1942) & Ludwig Herzer (Ludwig Herzl, 1872 - 1939), aria 'Dein ist mein ganzes Herz' in act 2 of the 1929 operetta *Das Land des Lächelns*; it is also a verse in the 1985 song *Dein ist mein ganzes Herz* by Heinz Rudolf Erich Arthur Kunze (*1956). – Engl.: Where you not are, I cannot be – the translation by Jocelyn Henry Clive 'Harry' Graham (1874 – 1936) is the negation: "And where you are, I long to be".

[2] Newton, *De Gravitatione*, pp. 1f.; Euler, *Anleitung*, § 35

[3] D'Alembert, *Traité de Dynamique*, 1743; Jorge Valenzuela José (*1949) & Eugene Jerome Saletan (*1924), *Classical Dynamics*, 1998, Sect. 2.1; see also Subsection 1.5.2.

q_k, p_k.[4] They occur rather seldom within CM, but are crucial for electromagnetism[5] and quantum field theory[6].

4.1.4. Constraints Imposed by Conservation Laws

The type of constraints, which is the most important one for this approach to QM, occurs, when the values of the stationary-state quantities are given. If the total energy, E, of a system of mass points is fixed, the initial (momentum) configurations, x_0 and p_0, are not independent each of another, but interrelated through the energy law as $T(x_0) + V(p_0) = E$. As a consequence, *not all* configurations are possible in state E.

In fact, the non-negativity of the kinetic energy: $T(p) \geq 0$, implies the crucial condition

$$V(x) \leq E \tag{4.2}$$

Physically spoken, Helmholtz's "disposable work storage" (see Subsection 2.5.3) is never larger than the total energy. As a consequence, in each *fixed* state E, the set of *possible* configurations does *not* comprise the *whole* configuration space, C^{all}, but is *limited* to the set

$$C^{poss}_{Newton} = C^{poss}_{Newton}(E) = \{x|V(x) \leq E\} \subset C^{all} = \{x\} \tag{4.3}$$

For instance, for a linear oscillator (*cf* eq. (3.27)), we have

$$C^{poss}_{Newton}(E) = \{x|x_{min}(E) \leq x \leq x_{max}(E)\} \tag{4.4}$$

In other words, by virtue of the energy law, the whole configuration space, C^{all}, is divided into two *disjunct* domains: $C^{poss}_{Newton}(E)$ and $C^{imposs}_{Newton}(E) = C^{all}\backslash C^{poss}_{Newton}(E)$. This *inherent limitation* of the motion of Newtonian classical-mechanical systems implies the

Point of generalization (1a) Under which conditions or for which mechanical systems this division into possible and impossible configurations takes *not* place, and which is the mechanics describing such systems?

---

[4] Dirac, *Lectures on Quantum Mechanics*, 1964
[5] Andrew Zangwill, *Modern Electrodynamics*, 2013, 24.5.4
[6] Steven Weinberg (*1933, Nobel Award 1979), *The Quantum Theory of Fields, Vol. I. Foundations*, 1995, § 7.6

4.2. POSSIBLE AND IMPOSSIBLE MOMENTUM CONFIGURATIONS

Due to the validity of the equations of motion in *whole* space and to the freedom in the choice of the initial conditions, $\vec{r}_0$ and $\vec{p}_0$, a body can also move to *any other* position in *momentum* space, too. Thus, generally speaking, *all* momentum configurations are possible, too.

However, if conserved quantities of a system are prescribed, similar constraints as in the foregoing section applies. In particular, if the total energy, E, of a conservative system is given, the interrelation of the initial (momentum) configurations, x_0 and p_0, through the energy law: $T(p_0) + V(x_0) = E$, limits the set of all possible momentum configurations, again.

In fact, the inequality $V(x) \geq V_{min} = 0$ implies the condition

$$T(p) \leq E \qquad (4.5)$$

Physically spoken, the 'motion storage' – being the analog to the work storage – is never larger than the total energy. As a consequence, in each stationary state E, the set of *possible* momentum configurations, P_{Newton}^{poss}, does *not* comprise the *whole* momentum configuration space, P^{all}, but is *limited* to the set

$$P_{Newton}^{poss} = P_{Newton}^{poss}(E) = \{p | T(p) \leq E\} \subset P^{all} = \{p\} \qquad (4.6)$$

For instance, for a linear oscillator, $|p(t)| \leq p_{max}(E)$, *cf* eq. (3.27). Therefore,

$$P_{Newton}^{poss}(E) = \{p | -p_{max}(E) \leq p \leq p_{max}(E)\} \qquad (4.7)$$

In other words, by virtue of the energy law, the momentum configuration space, P^{all}, is divided into two disjunct domains: $P_{Newton}^{poss}(E)$ and $P_{Newton}^{imposs}(E) = P^{all} \backslash P_{Newton}^{poss}(E)$. This *inherent limitation* of the motion of classical-mechanical systems implies the

Point of generalization (1b) Under which conditions or for which mechanical systems this division into possible and impossible momentum configurations takes *not* place, and which is the mechanics describing such systems?

Part II

Quantization as Selection Problem

"Don't just read it; fight it! Ask your own questions, look for your own examples, discover your own proofs. Is the hypothesis necessary? Is the converse true?What happens in the classical special case? What about the degenerated cases? Where does the proof use the hypothesis?"[7]

Schrödinger's 4th requirement to any quantization: The use of the classical expressions for the potential and kinetic energies should be justified (*Fourth Commun.*, p. 113), is best fulfilled through an *axiomatic* way from classical to quantum mechanics. Such a way is described in this Part.

---

[7] Paul Richard Halmos (Halmos Pál, 1916 - 2006) about doing mathematics, in: *I Want to Be a Mathematician*, 1985; quoted after https://en.m.wikipedia.org/wiki/Halmos

CHAPTER 5

Why Quantum Mechanics?

Abstract: This chapter introduces the paradigm 'quantization as selection problem' by means of few historical remarks. They include Einstein's 1907 reasoning, Schrödinger's 1926 derivation of his wave equation, and Gödel's theorem and Munchhausen's trilemma. The latter ones concern questions, which can be posed, but not be answered within CM. Such questions transcendent CM. This book poses the question, how the mechanics of oscillators *without* turning points looks like. ("oscillators without turning points" means oscillators, which may assume configurations beyond the classical turning points.)

Keywords: Axiomatic, Boundary conditions, Einstein, Gödel's theorem, Münchhausen's trilemma, Quantization as selection problem, Schrödinger's wave equation.

5.1. EINSTEIN'S REASONING FOR 'QUANTIZATION AS SELECTION PROBLEM'

"Aus dem Vorhergehenden geht klar hervor, in welchem Sinne die molekular-kinetische Theorie der Wärme modifiziert werden muß, um mit dem Verteilungsgesetz der schwarzen Strahlung in Einklang gebracht zu werden. Während man sich nämlich bisher die molekularen Bewegungen genau denselben Gesetzmäß igkeiten unterworfen dachte, welche für die Bewegungen der Körper unserer Sinneswelt gelten (wir fügen wesentlich nur das Postulat vollständiger Umkehrbarkeit hinzu), sind wir nun genötigt, für schwingungsfähige Ionen … die Annahme zu machen, daß die Mannigfaltigkeit der Zustände, welche sie anzunehmen vermögen, eine geringere sei als bei den Körpern unserer Erfahrung."[1]

---

[1] Einstein, *Die Plancksche Theorie der Strahlung und die Theorie der spezifischen Wärme*, 1907, pp. 284f. – English, shortly: It is clear, in which sense the molecular-kinetic theory of heat must be modified, in order to bring it in agreement with the distribution law of black-body radiation. The molecular motions do not obey the classical laws. We are forced to assume that the manyfold of states, which oscillating ions can assume, is smaller than that of the bodies of our everyday experience. – By the way, this paper brought him the attention of Walter Hermann Nernst (1864 - 1941, Nobel Award 1920), who was – together with Planck – crucial for drawing him to Berlin.

In his 1907 paper on the specific heat of solids just quoted, Einstein has pioneered not less than three developments of quantum theory.[2]

1. The application of quantum theory to solids, in particular, to the deviation of the experimental values of their specific heat from Dulong-Petit's rule[3], $c_v \approx 3R$, at low temperatures.

2. The calculation of the mean energy, $\langle E \rangle$, of a quantum oscillator in the still established manner.

$$\langle E \rangle_{\text{cl}} = \frac{\int_0^\infty E e^{-E/k_B T} \, \mathrm{d}E}{\int_0^\infty e^{-E/k_B T} \, \mathrm{d}E} = kT \tag{5.1a}$$

$$\langle E \rangle_{qu} = \frac{\sum_{n=1}^\infty E_n e^{-E_n/k_B T}}{\sum_{n=1}^\infty e^{-E_n/k_B T}} = \frac{\sum_{n=1}^\infty n\hbar\omega e^{-n\hbar\omega/k_B T}}{\sum_{n=1}^\infty e^{-n\hbar\omega/k_B T}} = \frac{\hbar\omega}{e^{\hbar\omega/k_B T} - 1} \tag{5.1b}$$

Solely the *discrete* subset, $\{n\hbar\omega \,|\, n = 1,2 \dots\}$, of the energy, E, leads to Planck's denominator, $[\exp(\hbar\omega/k_B T) - 1]$, and to a specific heat that decreases with decreasing temperature, see Problem 5.1.

3. Quantization is formulated as the problem to *select* the set of stationary quantum states out off the set of stationary classical states, *cf* item 2.

Einstein considers – like Planck and almost 150 years earlier Euler[4] – the *stationary* states and hence the notion of state to be *fundamental*. In later papers – which also largely influenced the development of quantum theory – Einstein returns to Planck's law and its compatibility with the energy and momentum conservation laws.[5] This way the quantum hypothesis becomes *independent* of its statistical foundation by Planck.

---

[2] *Cf* A. Douglas Stone, *Einstein and the Quantum: The Quest of the Valiant Swabian*, 2013; Enders, *State, Statistics and Quantization in Einstein's 1907 Paper 'Planck's Theory of Radiation and the Theory of Specific Heat of Solids'*, 2009.

[3] Pierre Louis Dulong (1785 - 1838) & Alexis Thérèse Petit (1791 - 1820), *Recherches sur quelque points important de la Théorie de la Chaleur*, 1819

[4] Suisky, *Euler as Physicist*, 2009, Ch. 8. *Euler's Mechanics and Schrödinger's Quantum Mechanics*, pp. 285-312

[5] Einstein, *Strahlungs-Emission and -Absorption nach der Strahlungstheorie*, 1916; *Zur Quantentheorie der Strahlung*, 1916; *Zum Quantensatz von Sommerfeld and Epstein*, 1917. De Broglie (*The wave nature of the electron*, 1929), Schrödinger (*Zweite Mitteilung*, 1926) and Born (*The statistical interpretation of quantum mechanics*, 1954) point to the close relationship of their considerations to Einstein's work. – *Cf* also Einstein, *Einleitende Bemerkungen über Grundbegriffe*, 1953, p. 4.

Problem 5.1. *Using eqs. (5.1), calculate the specific heats of a set of N (N ≫ 1) classical respectively quantum oscillators! For the former ones, you should obtain Dulong-Petit's rule mentioned in item 1 above. The latter one should agree with the classical result for sufficiently high temperature, T, and vanish as T → 0.*

5.2. ON SCHRÖDINGER'S DERIVATION OF HIS WAVE EQUATION*

"The large river of evolution that leads from classical mechanics to quantum or wave mechanics of our days is engendered by two big river gods: Hamilton and Jacobi."[6]

"Denn es zeigt sich in allen Fällen der klassischen Dynamik, die ich bisher untersucht habe, dass die Gleichung (18) [eq. (5.2) below] die Quantenbedingungen in sich trägt. Sie sondert in gewissen Fällen, und zwar in denjenigen, in denen die Erfahrung dafür spricht, selbsttätig gewisse Frequenzen und Energieniveaus als die für die stationären Vorgänge möglichen aus, ohne irgendeine weitere Zusatzannahme als die für eine physikalische Größe beinahe selbstverständliche Anforderung an die Funktion ψ: dieselbe soll im ganzen Konfigurationenraum eindeutig endlich und stetig sein."[7]

Schrödinger's view on the difference between CM and QM differs from Heisenberg's one as he expresses the "suspicion,…that our classical mechanics fails…at very small dimensions of the trajectory", so that it is necessary "to search an undulatory mechanics." (*ibid.*, pp. 496f.) This difference appears as that between a "more accurate" and a "less accurate" theories.

Rather than starting from the basic equations of CM, one has to start with a wave equation in q-space [configuration space] and to consider the manifold of the correspondingly possible processes. For processes, which depend on time only through a factor $e^{2\pi i v t}$, the ansatz for the wave equation reads (*ibid.*, pp. 509f., eq. (18)).

---

[6] Lanczos, *The Poisson bracket*, 1972; after the German translation, 1972, p. 305: "Der Strom der Evolution, der von der klassischen Mechanik zur Quanten- oder Wellenmechanik unserer Tage führt, wird von zwei großen Flußgöttern erzeugt: Hamilton und Jacobi".
[7] Schrödinger, *Quantization as eigenvalue problem. 2nd Commun.*, 1926, p. 511 – Engl.: For in all cases of classical dynamics I have investigated so far, the equation *div grad* $\psi - u^{-2} \partial^2 \psi / \partial t^2$ carries in itself the quantum conditions. It singles out certain frequencies and energy levels as being the [only] possible ones for stationary processes. There is not any additional assumption, except this one, which is almost self-understanding: within the whole configuration space, the function ψ bc unique, finite and continuous.

$$div\ grad\ \psi - \frac{1}{u^2}\ddot{\psi} = 0 \tag{5.2}$$

Here, div and $grad$ refer to the Hertzian line element, ds.

$$ds^2 = 2\bar{T}(q,\dot{q})dt^2 \tag{5.3}$$

(*ibid.*, eq. (3)), where $\bar{T}(q,\dot{q})$ is the kinetic energy in terms of the canonical coordinates and velocities.[8]

Eq. (5.2) exhibits the advantage, that it meets the "wave nature" of QM systems, which – after de Broglie – is *indispensable*. On the other hand, it contains *not any* hint to a 'quantum of action' or to any other quantity of dimension 'action'.

Such a quantity is provided by the Hamilton-Jacoby theory.[9] For this and because it connects mechanics and optics, Schrödinger combines the wave equation with that theory.

The system and state properties can enter the wave equation (5.2) solely through the phase velocity, u. Hence, u depends on the potential and total energies: $u = u(V(x), E)$. Within Hamilton-Jacobi theory,

$$u = \frac{ds}{dt} = \frac{E}{\sqrt{2(E-V)}} \tag{5.4}$$

(*2nd Commun.*, eq. (6)) is the normal velocity of the phase faces of constant action. Then, with $\ddot{\psi} = -(2\pi\nu)^2\psi$ and $E = h\nu$, eq. (5.2) becomes the stationary Schrödinger equation (*ibid.*, eq. (18'')).

$$div\ grad\ \psi + \frac{8\pi^2}{h^2}(E - V)\psi = 0 \tag{5.5}$$

Alternatively, Schrödinger obtains the equation

---

[8] Actually, Hertz's definition contains the total mass as factor on the l.h.s., so that ds exhibits the dimension 'length', see *Die Prinzipien der Mechanik in neuem Zusammenhange dargestellt*, 1910, p. 162. In contrast to Riemann and his successors, Hertz did not introduce ds *a priori*, but derived it from an Euclidian line element, see.

[9] Jacobi, *Über die Reduction der Integration der partiellen Differentialgleic hungen erster Ordnung zwischen irgend einer Zahl Variabeln auf die Integration eines einzigen Systemes gewöhnlicher Differentialgleichungen*, 1837; *Neue Methode zur Integration partieller Differentialgleichungen erster Ordnung zwischen irgend einer Anzahl von Veränderlichen*, 1906; *Vorlesungen über Dynamik*, 1884.

$$div\ grad\ \psi + \frac{8\pi^2}{h^2}(h\nu - V)\psi = 0 \tag{5.6}$$

(*ibid.*, eq. (18')). For this, he additionally *requires* to show that eq. (5.5) is the correct one, so that the energy rather than the frequency becomes quantized (*ibid.*, pp. 511, 519).

Let me return to eq. (5.2) and – using $\ddot{\psi} = -(2\pi\nu)^2\psi$ – rewrite it as

$$\frac{-1}{(2\pi\nu)^2\psi}\ div\ grad\ \psi = \frac{1}{u^2} \tag{5.7}$$

Let $\psi(s)$ to be real-valued. Because it is required "to be unique, finite and continuous over the whole configuration space" (*ibid.*, p. 511), it displays at least two turning points. For them, there are two possibilities.

1. At the turning point, the function $\psi(s)$ itself also changes its sign, while u^2 retains it.

2. Conversely, $\psi(s)$ retains its sign, while u^2 changes it.

The second case is absent within classical theory. Hence, it is obvious to connect the non-classical case $E < V$ with it, *cf* eq. (4.2); more details will follow below.

$$\frac{1}{u^2} = \frac{E - V(s)}{E^2} \tag{5.8}$$

(the denominator being suggested by dimensional reasons). The exploration of this formula is left to the reader (see also the problems at the end of this section).

Therefore, Schrödinger's ansatz (5.2) is sufficient for obtaining u^2, eq. (5.5); the recurrence to the Hamilton-Jacobi theory is dispensable.

Let me close this section with few remarks on Schrödinger's requirement, that "the function ψ be unique, finite and continuous", as quoted at the beginning of this section. Its meaning is clear for classically bounded motions. For the oscillator, it implies, that ψ vanishes at infinity, as we will see later. For the rotator, it means the periodicity of ψ on the sphere. But what does it mean for classically *un*bounded motions? If $\psi(x)$ is finite for $-\infty < x < +\infty$, Schrödinger's "weight function" (4th Commun., § 7) (more accurate, weight density), $\psi\overline{\psi} \equiv |\psi|^2$ is finite for $-\infty < x < +\infty$, too. As a consequence, the total weight, $\int_{-\infty}^{+\infty}|\psi|^2 dx$ diverges. The same problem occurs, if $|\psi|^2$ is interpreted as a probability density. For this, Schrödinger – referring to Weyl[10] and Hilb[11] – orthonormalizes not the wave functions, but the 'eigendifferentials' ('Weyl packets') (*ibid.*, § 3, eq. (28)). Then, however, $|\psi(x)|^2$ is a density not in configuration, but in momentum configuration space.[12] The approach presented here avoids such difficulties.

Problem 5.2. *Show, that the formulae*

$$\frac{1}{u^2} = \frac{E - V(s)}{E V(s)} \text{ and } \frac{1}{u^2} = \frac{E - V(s)}{V(s)^2} \tag{5.9}$$

lead to the unphysical requirement '$\psi(s) = 0$ if $V(s) = 0$'! *Hint*: Consider a potential wall of width $2a$,

$$V(s) = \begin{cases} V > 0 \\ 0 \end{cases} \text{ for } \begin{matrix} |s| \geq a \\ |s| < a \end{matrix} \tag{5.10}$$

Problem 5.3. *Show, that the formula*

$$\frac{1}{u^2} = \frac{V(s) - E}{E^2} \tag{5.11}$$

contradicts the requirement of $\psi(s)$ to be finite in the whole configuration space! *Hint*: Show, that, in this case, one would obtain

$$\lim_{|s| \to \infty} |\psi(s)| = \infty \text{ if } V(s) > E \text{ for } |s| > a > 0 \tag{5.12}$$

(for the linear oscillator, a is given by the classical turning points)!

Problem 5.4. *Perform the reasoning above within the momentum configuration space!*

Problem 5.5 (*). *The reasoning above can be performed independently in the configuration and momentum configuration spaces, while classical motion proceeds in the product space of both spaces. What does this mean and imply?*

---

[10] Weyl, *Über gewöhnliche Differentialgleichungen mit Singularitäten und die zugehörigen Entwicklungen willkürlicher Functionen*, 1910.
[11] Emil Hilb (1882 - 1929), *Über Reihenentwicklungen nach den Eigenfunktionen linearer Differentialgleichungen 2. Ordnung*, 1911.
[12] See the detailed treatment in Dmitrii Ivanovich Blokhintsev (1908 - 1979), *Principles of Quantum Mechanics*, 1976, App. III, in particular, eqs. (17), (19) and (20).

5.3. GÖDEL'S THEOREM AND MUNCHHAUSEN'S TRILEMMA

> "Kurt Godel's achievement in modern logic is singular and monumental – indeed it is more than a monument, it is a landmark which will remain visible far in space and time. ...The subject of logic has certainly completely changed its nature and possibilities with Godel's achievement."[13]

The central idea of this book is related to Gödel's first incompleteness theorem[14]. A natural language statement of it reads,

> "Any effectively generated theory capable of expressing elementary arithmetic cannot be both consistent and complete. In particular, for any consistent, effectively generated formal theory that proves certain basic arithmetic truths, there is an arithmetical statement that is true, but not provable in the theory."[15]

Generally speaking,

> "The theories we have so far are both inconsistent and incomplete."[16]

This suggests to start from a problem, which can be stated, but *not* be answered within CM.

> Such a start is underpinned by Enzensberger's poem 'Homage to Goedel'.[17]

---

[13] von Neumann, after Halmos, *The Legend of John von Neumann*, 1973, quoted in en.wikipedia.org/wiki/Kurt_G%C3%B6del.

[14] Gödel, *Über formal unentscheidbare Sätze der "Principia Mathematica" und verwandter Systeme*, 1931, Theorem VI.

[15] Stephen Cole Kleene (1909 - 1994), *Mathematical Logic*, 1967, p. 250; quoted after en.wikipedia.org/wiki/G%C3%B6del%27s_incompleteness_theorems (30.09.2013).

[16] Stephen William Hawking (*1942), *Godel and the End of the Universe*, 2002

[17] For the relation of Gödel's first incompleteness theorem to 'Munchhausen's theorem', *i.e*, to 'Munchhausen, or Agrippa's trilemma', see Hans Albert (*1921), *Traktat über kritische Vernunft*, 1991, p. 11 (cited after en.wikipedia.org/wiki/M%C3%BCnchhausen_trilemma (30.09.2013)).

HANS MAGNUS ENZENSBERGER

Hommage an Gödel[18]

Münchhausens Theorem, Pferd, Sumpf und Schopf,
ist bezaubernd, aber vergiss nicht:
Münchhausen war ein Lügner.

Gödels Theorem wirkt auf den ersten Blick
Etwas unscheinbar, doch bedenk:
Gödel hat recht.

In jedem genügend reichhaltigen System
lassen sich Sätze formulieren,
die innerhalb des Systems
weder beweis- noch widerlegbar sind,
es sei denn das System
wäre selber inkonsistent.

Du kannst deine eigene Sprache
in deiner eigenen Sprache beschreiben:
aber nicht ganz.
Du kannst dein eigenes Gehirn
mit deinem eigenen Gehirn erforschen:
aber nicht ganz.
Usw.

Um sich zu rechtfertigen
muss jedes denkbare System
sich transzendieren,
d.h. zerstoeren.
Genügend reichhaltig oder nicht:
Widerspruchsfreiheit
ist eine Mangelerscheinung
oder ein Widerspruch.

(Gewissheit = Inkonsistenz)

---

[18] Hans Magnus Enzensberger (*1929), *Die Elixiere der Wissenschaft*, 2002; quoted after http://sternenfall.de/Enzensberger–Hommage_an_G0366del.html (30.09.2013); Fig. **5.1** below has been taken from en.wikipedia.org/wiki/File:Muenchhausen_Herrfurth_7_500x789.jpg (28.09.2013).

Jeder denkbare Reiter,
also auch Münchhausen,
also auch du bist ein Subsystem
eines genügend reichhaltigen Sumpfes.Und ein Subsystem dieses Subsystems
Ist der eigene Schopf,
dieses Hebezeug
fuer Reformisten und Lügner.
In jedem genügend reichhaltigen System,
also auch in diesem Sumpf hier,
lassen sich Sätze formulieren,
die innerhalb des Systems
weder beweis- noch widerlegbar sind.

Diese Sätze nimm in die Hand
Und zieh!

Shortly, in each sufficiently rich system, theorems can be formulated, which – within that system – are neither provable, nor disprovable. Take these theorems and draw!

According to the points of generalization formulated in the foregoing two sections, the question I start with reads,

Which is the mechanics of an oscillator, the configurations it can assume in the stationary state E comprise the *whole* configuration space?

$$C^{poss} = C^{poss}_{Newton}(E) \cup C^{imposs}_{Newton}(E),$$

ie,

$$\text{BOTH } x \in C^{poss}_{Newton}(E) \text{ AND } x \in C^{imposs}_{Newton}(E)$$

are allowed.

In particular, this question is admitted as the energy law is independent of the motion along trajectories. It cannot be answered within CM, because the motion along trajectories implies $C^{poss} \subset C^{all}$. The "BOTH – AND" is related to the logical relationship between CM and QM stressed in Schrödinger's Nobel Lecture quoted in the Introduction.

Fig. 5.1 . Munchhausen (Münchhausen), picture by Oskar Herrfurth (1862 - 1934).

CHAPTER 6

A Hierarchy of Selection Problems

Abstract: This chapter brings together Einstein's 1907 vision of 'quantization as selection problem' (see Section 5.1) and our explorations of the possible and impossible (momentum) configurations in Chapter 4. The hierarchy is formulated in terms of the latter ones and consists of Newtonian systems, non-Newtonian classical systems, non-classical systems and non-mechanical systems. These explorations are assisted by an overview on selection problems in classical and quantum theory, by further reasonings for Einstein's vision and by few explanations about alternative and harmony in Euler's axioms of classical mechanics. Two fundamental conclusions for non-classical systems can be drawn already at this stage. (i), The notion of path as a point-wise relationship between the configurations and the momentum configurations looses its meaning. (ii), there are no longer coordinates describing the boundaries of a system in (momentum) like the classical turning points of an oscillator.

Keywords: Einstein, Impossible configuration, Impossible momentum configuration, Newtonian mechanics, Non-classical mechanics, Non-Newtonian classical mechanics, Possible configuration, Possible momentum configuration, Selection problem.

6.1. WHY 'SELECTION PROBLEM' FOR QUANTIZATION?

"We must always be ready to change…the axiomatic basis of physics, in order to do justice to the facts of observations in the most complete way that is logically possible."[1]

The viewpoint 'selection problem' is rather old, although seldom mentioned. Obviously, the set of physically possible motions is by far smaller than the set of geometrically possible displacements.[2] Which are the mathematical and physical criteria for selecting that set?

The most general physical criteria are the conservation laws, while there are many criteria being specific to the various branches of physics.

Within CM, the principle of least action distinguishes exactly one trajectory out off the infinite set of geometrically possible paths. Within classical wave theory, the

---

[1] Einstein, *Maxwell's Einfluss auf die Entwicklung der Auffassung des Physikalisch-Realen*, 1931; quoted is the 1996 English edition.
[2] Leibniz, *Specimen dynamicum pro admirandis naturae legibus circa corporum vires et mutuas actiones detegendis et ad suas causas revocandis*, 1695

boundary conditions exclude all waves not obeying them, even when they are dynamically possible, *i.e*, when they are compatible with the wave equation.

As quoted above, quantum systems "have got a less number of stationary states" than classical systems (Einstein, 1907). Which is the selection criterion for them (despite of compatibility with Planck's radiation formula)?

Planck (1900), Einstein (1907) and Bohr (1913) have restricted the energy spectrum to *discrete* values: $nh\nu$ (ν – frequency of an oscillator) and $\frac{n}{2}h\nu$ (ν – rotational frequency of an electron), respectively ($n = 1,2,...$).

The Bohr-Sommerfeld quantization conditions[3] – originally called *state* (!) conditions[4] – select the values nh resp. $(n + \frac{1}{2})h$ (n being entire-valued) out off the classical continuum of the action integral, $\oint pdq$[5].

Because of this discreteness and because of the fact, that its minimum is determined by the quantum of action, h, the action is no longer subject to a variational principle.[6]

Now, Einstein's 1907 selection problem consists of two parts.

1. The selection between classical and non-classical mechanical systems. I will formulate it by means of possible and impossible (momentum) configurations as used above. This avoids certain difficulties with the trajectories being absent in QM.

2. The (non-classical) selection of the (stationary) quantum states. I will formulate it by means of mathematically and physically distinguished states using, in particular, the energy conservation law. This avoids certain difficulties with the solutions to (classical!) eigenvalue problems.

This reformulation of "quantization as eigenvalue problem" (Schrödinger, 1926) to 'quantization as selection problem' (indicated by Einstein, 1907) paves the way for *deriving* the stationary and the time-dependent Schrödinger equations from CM

---

[3] Sommerfeld, *Atombau und Spektrallinien*, 1919-29
[4] Pauli, *Quantentheorie*, 1926, § 6
[5] See also Heisenberg, *Die Geschichte der Quantentheorie*, 1977; Messiah, *Quantum Mechanics*, 1999, p. 34.
[6] See also Dirac, *The Lagrangian in Quantum Mechanics*, 1933.

through a *smooth* procedure, where the motivation of generalization stems from CM itself.

In contrast to the (classical!) eigenvalue problem, the non(!)-classical selection problem can be formulated and even solved *without*

- posing (classical!) boundary conditions (which are ill defined for classically unbound systems);

- making any additional *a-priori* assumptions about the nature of quantum systems, say, about

  - the wave-particle dualism,

  - a founding wave equation, or

  - a quantized quantity like the quantum of action (the existence of such a new characteristic quantity will emerge in a most natural manner).

The *discretization* of parameter values – like that of frequency and wavelength of standing waves – is a *classical* phenomenon, as we will see in Section 22.3. The corresponding method for solving the wave equation is the eigenvalue method. Wavelength and frequency are *not* dynamical variables like position and momentum, and *not* stationary-state variables like total energy and angular momentum. The latter ones still assume *continuous* values.

In contrast, *quantization* reduces the set of classically possible values of *stationary-state variables*, such as total energy (Planck, 1900) and angular momentum (Bohr, 1913), to the "distinguished" quantum subset.[7] The problem thus consists in *how to select* that quantum subset of stationary states out of the classical continuum?

For the sake of the unity of physics as well as for pedagogical reasons, the selection should proceed along a way, where the alternatives CM and QM exclude each another and, at once, are in harmony each with another.[8] This harmony is also realized in Euler's axioms of CM as described in Subsection 2.2.5.

---

[7] Bohr, Hendrik Anthony "Hans" Kramers (1894 - 1952) & John Clarke Slater (1900 - 1976), *The quantum theory of radiation*, 1924, Sect. 1

[8] Frank Anthony Wilczek (*1951, Nobel Award 2004) & Betsy Devine (*1946), *Longing for the Harmonies*, 1988, explicate, that the whole universe, in the large as well as in the small, is in physical harmony.

6.2. NEWTONIAN *VERSUS* NON-NEWTONIAN CM

As QM is an alternative to CM, not to Newtonian CM, I begin with non-Newtonian CM as alternative to Newtonian CM. What 'non-Newtonian CM' means will become clear at once.[9]

Helmholtz's choice of the minus sign in eq. (3.23) is motivated through the relation of the function $V(x)$ to the work, A. (Remember, that this implies $V_{min} = 0$.). We have thus the

Helmholtzian conditions for Newtonian CM:

$$E \geq T(p); \; E \geq V(x) \tag{6.1}$$

Accordingly, the sets of all possible (momentum) configurations are (see Chapter 4)

$$C^{poss}_{Newton} = C^{poss}_{Newton}(E) = \{x | V(x) \leq E\} \subset C^{all} = \{x\} \tag{6.2}$$

$$P^{poss}_{Newton} = P^{poss}_{Newton}(E) = \{p | T(p) \leq E\} \subset P^{all} = \{p\} \tag{6.3}$$

For instance, the motion of a linear oscillator is bound to the closed interval between the two turning points.

$$x_{min} \leq x \leq x_{max}; \; p_{min} \leq p \leq p_{max} \tag{6.4}$$

Now, what happens, if one sets $E - T(p) = -V(x)$? Here, of course, the relationship between force and potential energy is kept: $\vec{F}(x) = -\nabla V(x)$. For, if one would at the same time set $\vec{F}(x) = +\nabla V(x)$, one would merely replace the potential energy, $V(x)$, with Euler's the efficacy, $W(x) = -V(x)$, or Gauß and Jacobi's 'potential' $\tilde{V}(x) = -V(x)$, without coming to a different physical model.

Indeed, in place of eq. (3.1) one can require the equation of motion to read

$$M\ddot{x}(t) = -F(x(t)) \tag{6.5}$$

---

[9] Non-Newtonian mechanics treating internal, notably, spinning degrees of freedom – see, *e.g*, Giovanni Salesi, *Non-Newtonian Mechanics*, 2002 – are discarded.

without coming into conflict with Newton's 1st and 3rd axioms, or with Euler's axioms. This case leads to

Helmholtzian conditions for non-Newtonian CM:

$$E \leq T(p); \; E \geq -V(x) \tag{6.6}$$

Accordingly, the sets of all possible (momentum) configurations are

$$C^{poss}_{non-Newton} = C^{poss}_{non-Newton}(E) = \{x | V(x) \leq -E\} \subset C^{all} = \{x\} \tag{6.7a}$$

$$P^{poss}_{non-Newton} = P^{poss}_{non-Newton}(E) = \{p | T(p) \geq E\} \subset P^{all} = \{p\} \tag{6.7b}$$

For instance, the motion of a linear oscillator would proceed *beyond* the turning points.

$$x \leq x_{min} \; \text{OR} \; x \geq x_{max}; \; p \leq p_{min} \; \text{OR} \; p \geq p_{max} \tag{6.8}$$

This is a classical, but *non*-Newtonian mechanics. It is free of internal logical contradictions, though not realized in our world.

Therefore, within CM, there is a *selection problem* between Newtonian and *non*-Newtonian CM.

$$\text{EITHER} \; E \geq T(p) \; \text{AND} \; E \geq V(x) \tag{6.9a}$$

$$\text{OR} \; E \leq T(p) \; \text{AND} \; E \geq -V(x) \tag{6.9b}$$

In other words,

$$\text{EITHER} \; C^{poss} = \{x | E \geq V(x)\} \; \text{AND} \; P^{poss} = \{p | E \geq T(p)\} \tag{6.10a}$$

$$\text{OR} \; C^{poss} = \{x | E \geq -V(x)\} \; \text{AND} \; P^{poss} = \{p | E \leq T(p)\} \tag{6.10b}$$

For instance, for the symmetric ($x_{min} = -x_{max}$, $p_{min} = -p_{max}$) linear oscillator, this means

$$\text{EITHER} \; |x| \leq x_{max} \; \text{AND} \; |p| \leq p_{max} \tag{6.11a}$$

$$\text{OR} \; |x| \geq x_{max} \; \text{AND} \; |p| \geq p_{max} \tag{6.11b}$$

The logical relationship (XOR) between Newtonian CM and non-Newtonian CM is the same as that between Euler's axioms *E1* and *E2* as expressed in axiom *E0*, see Subsubsection 2.2.5. These alternatives exclude each other *and*, at once, are "in harmony each with another". This harmony expresses itself in the fact, that the qualitative principles of stationary-state conservation and stationary-state change are the same (Newton's 1st and 3rd axioms, Euler's axioms), while the representations of the total energy and the equations of motion are different, though not without interrelations. These interrelations are provided through the use of the same dynamical variables, $x(t)$ and $p(t)$.

Both Newtonian and non-Newtonian CM arise by the parametrization of configuration and momentum configuration through the common parameter time. Hence, a system being able to assume configurations in the *whole* (momentum configuration spaces, C^{all} and P^{all}, does *not* move along trajectories, $x(t)$.

Problem 6.1 *Explore the two-body problem within non-Newtonian CM!*

6.3. CLASSICAL *VERSUS* NON-CLASSICAL MECHANICS

"Es ist wirklich meine Ueberzeugung, dass eine Interpretation der Rydberg-Formel[10] im Sinne von Kreis- und Ellipsenbahnen in klassischer Geometrie nicht den geringsten physikalischen Sinn hat und meine ganzen kümmerlichen Bestrebungen gehen dahin, den Begriff der Bahnen, die man doch nicht beobachten kann, restlos umzubringen und geeignet zu ersetzen."[11]

The alternative between classical (CM) and non-classical mechanics (non-CM) can be formulated in a similar, *logical*, manner. For CM, we have the

Helmholtzian condition of defining classical entities:

The difference $E - T(p)$ is semi-*definite*. In this case, the *exclusion*

---

[10] Johannes (Janne) Robert Rydberg (1854 - 1919), *Untersuchungen über die Beschaffenheit der Emissionsspektren der chemischen Elemente*, 1922

[11] Heisenberg, *Letter to Pauli,* July 9, 1925; quoted after van der Waerden, *Sources of Quantum Mechanics*, 1968, p. 27; – Engl.: "I'm really convinced, that an interpretation of the Rydberg formula in the sense of circle and elliptic orbits of classical geometry has got not the smallest meaning, and all of my meagre efforts go toward killing off and suitably replacing the concept of the orbital path which one cannot observe." (aip.org/history/heisenberg/p07.htm).

$$\text{EITHER } E - T(p) \geq 0 \text{ OR } E - T(p) \leq 0 \tag{6.12}$$

holds true for *all* possible momentum configurations, p, in stationary state E, *i.e*,

$$\text{EITHER } p|_{E \geq T(p)} \in P^{poss} \text{ OR } p|_{E \leq T(p)} \in P^{poss} \tag{6.13}$$

In other words,

$$\text{EITHER } P^{poss} = P^{poss}_{Newton} \text{ OR } P^{poss} = P^{poss}_{non-Newton} \tag{6.14}$$

In contrast, for non-CM we have the

Helmholtzian condition of defining non-classical entities:

The difference $E - T(p)$ is *in*definite. Consequently, BOTH inequalities,

$$E - T(p) \geq 0 \text{ AND } E - T(p) \leq 0 \tag{6.15}$$

are admitted for the momentum configurations, p, in state E, *i.e.*,

$$p|_{E \geq T(p)} \in P^{poss} \text{ AND } p|_{E \leq T(p)} \in P^{poss} \tag{6.16}$$

The latter means

$$P^{poss}_{non-CM} = P^{poss}_{Newton} \cup P^{poss}_{non-Newton} = P^{all} \tag{6.17}$$

As mentioned in the last section, systems with $P^{poss} = P^{all}$ do *not* move along trajectories. The corresponding formulations for the configurations are left to the reader.

Notice, that, again, the logical relationships (XOR, AND) are exactly those, which Schrödinger has stressed in his Nobel Lecture quoted in the Introduction.

6.4. MECHANICS *VERSUS* NON-MECHANICS

For completeness I mention the

Helmholtzian conditions of defining *non-mechanical* Entities:

$$\text{NEITHER } E - T(p) \geq 0 \text{ NOR } E - T(p) < 0 \tag{6.18a}$$

for *any* momentum configuration, p, in stationary state E. Also,

$$\text{NEITHER } E - V(x) \geq 0 \text{ NOR } E - V(x) < 0 \qquad \textbf{(6.18b)}$$

for *any* configuration, x, in stationary state E. Here, *no* mechanics is possible, because $C^{poss} = \emptyset$ and $P^{poss} = \emptyset$.

Again, the alternatives 'mechanics' and 'non-mechanics' exclude each another and, at once, are "in harmony each with another". The harmony is guaranteed through the energy law, while – lacking trajectories – the representations of the total energy and the equations of motion are expected to be quite different.

6.5. CONCLUSIONS FOR NON-CLASSICAL MECHANICAL SYSTEMS

Accordingly, there are two fundamental differences between classical and non-classical mechanical systems.

Fundamental Difference 1: The notion of *path* as a *point*-wise relationship between the configurations, x, and the momentum configurations, p, looses its meaning, because

- there is no longer a common parametrization as $x = x(t)$, $p = p(t)$; hence,
- there is no unique *algebraic* relation $V(x) + T(p) = const$ between them.

Fundamental Difference 2: There are no longer coordinates like the classical turning points of an oscillator, $x_{min/max}$ ($p_{min/max}$), describing the *boundaries* of a system in (momentum) space, because the expressions $E - V(x)$ and $E - T(p)$ are no longer (semi-)definite.

For lacking the *algebraic* relation $V(x) + T(p) = E$, both fundamental differences are interrelated in that

1. *All* (momentum) configurations should be considered *together*, and *all* configurations are related to *all* momentum configurations, and *vice versa*; there are no initial (momentum) configurations that can serve as *single* representatives through determining the total energy *via* $E_{cl} = V(x_0) + T(p_0)$ (*cf* Schrödinger, 4th *Commun.*, § 7).

2. *Novel* entities are necessary for describing the extension of a system in (momentum) configuration space.

As a consequence, another, *non*-classical representation of the energy is necessary. We will find such one in the following chapter.

CHAPTER 7

Non-Classical Representations of Potential, Kinetic and Total Energies

Abstract: This chapter presents the first concrete consequences of the foregoing one. The possibility of configurations, x, for which the classical expression for the potential energy, $V(x)$, is larger than the total energy, E, implies, that this expression, $V(x)$, does no longer represent the contribution of the configuration x to the work storage of the system. For this, a 'limiting function', $F(x)$, is introduced such, that the non-classical contribution of the configuration x to the work storage, $V_{\mathrm{ncl}}(x) = F(x)V(x)$, is smaller than the total energy. The same is done for the momentum configurations and the kinetic energy. Moreover, since there are no trajectories, the non-classical representation of the energies become integral expressions, in agreement with Schrödinger's vision. Then, the general properties of the limiting functions are deduced. Limiting amplitudes (dimensionless wave functions) are introduced, in order to find an integral relationship between the motions in space and in momentum space, as envisaged by Schrödinger, again.

Keywords: Characteristic length, Fourier transform, Limiting function, Non-classical kinetic energy, Non-classical potential energy, Non-classical total energy, Normalization, Ordered sets, Weight amplitude, Weight function.

7.1. NON-CLASSICAL CONTRIBUTIONS OF THE (MOMENTUM) CONFIGURATIONS TO THE POTENTIAL (KINETIC) ENERGY. SCHRÖDINGERIAN WEIGHT FUNCTIONS

"$\psi\bar{\psi}$ ist eine Art *Gewichtsfunktion* im Konfigurationenraum des Systems. Die *wellenmechanische* Konfiguration des Systems ist eine *Superposition* vieler, streng genommen *aller*, kinematisch möglichen punktmechanischen Konfigurationen. Dabei steuert jede punktmechanische Konfiguration mit einem gewissen *Gewicht* zur wahren wellenmechanischen Konfiguration bei, welches Gewicht eben durch $\psi\bar{\psi}$ gegeben ist."[1]

Thus, a non-classical system can assume (momentum) configurations, for which $E \leq V(x)$ and $E \leq T(p)$, respectively (recall, that we have $V(x) \geq 0$ and $T(p) \geq$

---

[1] Schrödinger, *4th Commun.*, § 7, p. 135 – Engl.: $\psi\bar{\psi}$ is is a kind of *weight function* in the configuration space of the system. The *wave-mechanical* configuration of the system is a *superposition* of many, strictly speaking, of *all* kinematically possible point-mechanical configurations. Here, every point-mechanical configuration contributes with a certain *weight* to the true wave-mechanical configuration, which is just given by $\psi\bar{\psi}$.

0). The total energy is no longer the sum of the potential and kinetic energies: $E \neq V(x) + T(p)$. A novel, non-classical representation of the total energy as a function of the (momentum) configurations is required!

On the other hand, the total energy of non-classical, but still *mechanical* systems is expected to be composed of contributions of both x- and p-dependent terms, too. For otherwise, the number of degrees of freedom would be lowered. These terms will be called non-classical potential energy, $V_{ncl}(x)$, and non-classical kinetic energy, $T_{ncl}(p)$, respectively. They are expected to be of the following forms.

$$V_{ncl}(x) \underset{def}{=} F_E(x)V(x) \equiv V_E(x) \leq E; \; x \in C^{all} \tag{7.1a}$$

$$T_{ncl}(p) \underset{def}{=} G_E(p)T(p) \equiv T_E(p) \leq E; \; p \in P^{all} \tag{7.1b}$$

Here, $F_E(x)$ and $G_E(p)$ represent dimensionless 'limiting factors' to be determined in what follows.

This is the simplest modification of the classical expressions, $V(x)$ and $T(p)$. Two other ansatzes are proposed to be explored by the reader in Problems 7.1 and 7.2. The harmony between the classical and non-classical functions is realized in that the latter ones 'contain' the former ones.

Problem 7.1 (Another $V_{ncl}(x)$*). *Another ansatz for the non-classical potential energy is $V_{ncl}(x) = V(x) + \Delta V(x)$. Discuss the advantages and disadvantages of this ansatz! Hint: It suggests that there are two contributions, a classical and a non-classical ones.*

Problem 7.2 (Yet another $V_{ncl}(x)$*). *Yet another ansatz for the non-classical potential energy is $V_{ncl}^{(E)}(x) \sim V(x)^v$ Discuss the advantages and disadvantages of this ansatz! Hint: It needs a factor of variable dimension and thus unclear physical meaning.*

Problem 7.3. *Assume the potential energy function to be confining as in the case of the harmonic oscillator: $\lim_{|x| \to \infty} V_E(x) = \infty$. Show that, in this case, $\lim_{|x| \to \infty} F_E(x) = 0$!*

Problem 7.4. *Assume the potential energy function to represent an infinite wall: $V_E(x) = \infty$ for $x \geq a < \infty$. Show that, in this case, $F_E(x) = 0$ for $x \geq a$!*

7.2. NON-CLASSICAL REPRESENTATION OF TOTAL ENERGY

As the motion is *not* along trajectories, $(x(t), p(t))$, there is *not* a *point*-wise relationship between configurations, $x(t)$, and momentum configurations, $p(t)$. For this, the classical *algebraic* formula for the total energy,

$$E_{\text{cl}} = H_{\text{cl}}(x, p) = V_{\text{cl}}(x) + T_{\text{cl}}(p) \tag{7.2}$$

can*not* be replaced with a formula like (*cf* the definitions (7.1))

$$E_{\text{ncl}} = H_{\text{ncl}}(x, p) = V_{\text{ncl}}(x) + T_{\text{ncl}}(p) \equiv F_{E_{\text{ncl}}}(x)V_{\text{cl}}(x) + G_{E_{\text{ncl}}}(p)T_{\text{cl}}(p) \tag{7.3}$$

The alternative consists in that *all* configurations and *all* momentum configurations of a system contribute to its total energy. Then, the representation of the total energy reads (omitting the index 'ncl')

$$E \;\; = \frac{\int_{C\text{all}} V_E(x)\mathrm{d}x}{\int_{C\text{all}} F_E(x)\mathrm{d}x} + \frac{\int_{p\text{all}} T_E(p)\mathrm{d}p}{\int_{p\text{all}} G_E(p)\mathrm{d}p} = \frac{\int_{C\text{all}} F_E(x)V(x)\mathrm{d}x}{\int_{C\text{all}} F_E(x)\mathrm{d}x} + \frac{\int_{p\text{all}} G_E(p)T(p)\mathrm{d}p}{\int_{p\text{all}} G_E(p)\mathrm{d}p} \tag{7.4}$$

(the denominators being added for dimensional reasons).

In contrast to the *explicit* equation (7.2) of CM, this is an *implicit* equation for E_{ncl}. Like Bohr's 1913 atom model, it contains *no external* parameters. (For more details about external and external system parameters, see Subsection 9.1.1.) In this sense, E_{ncl} becomes an *internal* parameter like the frequency of a harmonic oscillator (Suisky & Enders, 2003; Enders & Suisky, 2005; Enders, 2006). On the other hand, anyway, the actual energy of a system depends on its initial state, which is still set from *out*side. In this sense, E_{ncl} remains to be an *external* parameter (Enders, 2013).

7.3. PROPERTIES OF THE LIMITING FACTORS, $F_E(x)$ AND $G_E(p)$

7.3.1. Dimensionlessnes. Characteristic Extensions I

By their very definition, the novel functions, $F_E(x)$ and $G_E(p)$, are dimensionless. Hence, there are (possibly, state-, *i.e*, energy-dependent) 'reference' values, x_E and p_E, such, that actually $F_E = F_E(x/x_E)$ and $G_E = G_E(p/p_E)$. This means, that even *unbounded* non-classical systems exhibit such a 'characteristic' length in both configuration and momentum configuration spaces.

For free quantum particles, this complies with the *de Broglie wavelength*[2], $\lambda = h/p$.

7.3.2. Normalization. Characteristic Extensions II

$F_E(x/x_E)$ and $G_E(p/p_E)$ occur as *weights* for the contributions of the (momentum) configurations to the total energy. Therefore, without loss of generality, I set

$$\int_{C\text{all}} F_E\left(\frac{x}{x_E}\right)\frac{\mathrm{d}x}{x_E} = \int_{p\text{all}} G_E\left(\frac{p}{p_E}\right)\frac{\mathrm{d}p}{p_E} = 1 \qquad (7.5)$$

Then, the characteristic extensions are given as

$$x_E = \int_{C\text{all}} F_E\left(\frac{x}{x_E}\right)\mathrm{d}x; \quad p_E = \int_{p\text{all}} G_E\left(\frac{p}{p_E}\right)\mathrm{d}p \qquad (7.6)$$

I will deepen upon that later.

7.3.3. Non-Negativity

$V_E(x)$ represents a partial work storage, therefore, $V_E(x) \geq 0$ ($V_{\min} = 0$, as above). $T_E(p)$ represents a partial motion storage, therefore, $T_E(p) \geq 0$, too. Since $V(x) \geq 0$ and $T(p) \geq 0$, the limiting factors are *non-negative*.

$$F_E(x/x_E) \geq 0; \quad G_E(p/p_E) \geq 0 \qquad (7.7)$$

7.3.4. Classical Limit. I*

Formally, the classical eq. (3.24) is obtained through setting $F_E(x/x_E) = x_E\delta(x - x(t))$ and $G_E(p/p_E) = p_E\delta(p - p(t))$. But this would make the functions $F_E(x/x_E)$ and $G_E(p/p_E)$ time-dependent, in contradiction to their definitions (7.1). For this, I postpone this issue to the part Quantum Dynamics, where the time-dependence will be dealt with.

---

[2] de Broglie, *Recherche sur la théorie des quanta*, 1924. De Broglie argued as follows. Planck's and Einstein's relationship $E = h\nu = \hbar\omega$ concerns the temporal component of the special-relativistic 4-vectors ($c = 1$) $p = (\vec{p}, E)$ and $k = (\vec{k}, \omega)$ of a body. Relativistic invariance implies the same relationship to hold true between the spatial components of p and k, i.e, $\vec{p} = \hbar\vec{k}$.

7.3.5. No integral Transform Between $F_E(x)$ and $G_E(p)$!

"Von einem Wort läßt sich kein Jota rauben."[3]

Fig. 7.1. Goethe's Faust, illustration by Eugène Delacroix (1798 - 1863)

As stressed above, there is no relationship between *single* configurations, x, and momentum configurations, p. Hence, *all* configurations are connected with *all* momentum configurations. This suggests the existence of an *integral* transform between $F_E(x/x_E)$ and $G_E(p/p_E)$.

$$G_E(\eta) = \int K_E(\eta, \xi) F_E(\xi) \mathrm{d}\xi; \; \xi \equiv \frac{x}{x_E}, \eta \equiv \frac{p}{p_E} \qquad (7.8)$$

Now, for an interaction-less particle, all configurations are equivalent, while its momentum equals $p = p_E = \sqrt{2mE}$. In this case, we have

$$F_E(\xi) = F_E^{\text{free}}(\xi) = \iota^2; \; G_E(\eta) = G_E^{\text{free}}(\eta) = \delta(\eta - 1) \qquad (7.9)$$

where ι is an infinitesimal number such, that $\int_{-\infty}^{+\infty} \iota^2 \mathrm{d}\xi = 1$. (We will see later on, how such a number can be circumvented by means of wave packets.) Inserting these functions into relation (7.9) yields

$$\delta(\eta - 1) = \int K_E(\eta, \xi) \iota^2 \mathrm{d}\xi \qquad (7.10)$$

---

[3] Goethe, *Faust 1*, verse 2000 (Mephistopheles), http://www.gutzitiert.de/zitat_autor_johann_wolfgang_von_goethe_thema_verantwortung_zitat_21089.html – Engl.: "No word suffers a jot from thieving.", www.poetryintranslation.com/PITBR/German/Fausthome.htm – Fig. **7.1** has been taken from http://www.poetryintranslation.com/PITBR/German/Fausthome.htm.

Therefore, $K_E(\eta, \xi) = \delta(\eta - 1)$. Since this kernel is independent of ξ, the integral transform (7.8) is not meaningful; see also Problem 7.5.

Problem 7.5. *Show, that the kernel $K_E(\eta, \xi)$ cannot be made meaningful by means of a potential-depending term, say,*

$$K_E(\eta, \xi) = \delta(\eta - 1) + V(\eta)k_E(\eta, \xi) \tag{7.11}$$

7.4. WEIGHTING AMPLITUDES, $f_E(x)$ AND $g_E(p)$. INTEGRAL TRANSFORM BETWEEN THEM

The fact, that both $F_E(\xi)$ and $G_E(\eta)$ are non-negative and normalized, suggests to set

$$F_E(\xi) = |f_E(\xi)|^2; \; G_E(\eta) = |g_E(\eta)|^2; \; \xi \equiv \frac{x}{x_E}, \eta \equiv \frac{p}{p_E} \tag{7.12}$$

and to seek a norm-conserving (unitary) integral transform between $f_E(\xi)$ and $g_E(\eta)$ (see also Problem 7.6). $f_E(\xi)$ and $g_E(\eta)$ will be called *weighting amplitudes*. In contrast to the standard approaches,[4] their normalizability (or that of the wave functions) has not to be required, but is a natural property of them, see eqs. (7.5).

According to formulae (7.9), the weighting amplitudes for a free, interaction-less particle equal (up to irrelevant constant phase factors)

$$f_E^{\text{free}}(\xi) = \iota e^{i\kappa_E\xi}; \; g_E^{\text{free}}(\eta) = \sqrt{\delta(\eta - 1)} \tag{7.13}$$

Here, the parameter κ_E can be absorbed into the parameter x_E. Then, these two functions are just the Fourier transforms each of another, see Problem 7.7.

$$f_E^{\text{free}}(\xi) = \sqrt{\frac{|b_E|}{2\pi}} \int_{-\infty}^{+\infty} e^{-ib_E\xi\eta} g_E^{\text{free}}(\eta)d\eta; \tag{7.14a}$$

$$g_E^{\text{free}}(\eta) = \sqrt{\frac{|b_E|}{2\pi}} \int_{-\infty}^{+\infty} e^{ib_E\xi\eta} f_E^{\text{free}}(\xi)d\xi \tag{7.14b}$$

---

[4] See, *e.g*, Bohm, *Quantum Theory*, 1989, § 13.5; Hermann Haken (*1927), *Quantenfeldtheorie des Festkörpers*, 1973, § 3.

Here, b_E is a free, real-valued, possibly state-dependent parameter.[5] Eq. (7.5) implies the symmetric form, *cf* Plancherel's theorem[6].

Now, one could assume, that the integral transform between the functions $f_E(\xi)$ and $g_E(\eta)$ depends on the potential function, $V(x)$. Then, however, the requirement of norm-conservation would make it to become *non*-linear. As a consequence, the resulting equation of motion would become non-linear, too, in contradiction to the interference experiments, notably, with single (!) electrons[7] (superposition principle).

In terms of these functions, the non-classical representation (7.4) of the energy becomes

$$E = \int \ |f_E(\xi)|^2 V(x_E\xi)\mathrm{d}\xi + \int \ |g_E(\eta)|^2 T(p_E\eta)\mathrm{d}\eta \qquad (7.15)$$

Problem 7.6. *Assume, that instead of formulae (7.12), we would set*

$$F_E(\xi) = f_E(\xi)^2; \ G_E(\eta) = g_E(\eta)^2$$

Show, that there would be no integral transform between $f_E(\xi)$ and $g_E(\eta)$!

Problem 7.7 *(i) Show, that the free-particle amplitudes (7.13) obey the Fourier transform (7.14a)! Hint: Approximate $g_E(\eta)$ as a Dirac sequence of a Gaussian (heat kernel)![8] (ii) Show, that the limit $N \to \infty$ of the function*

$$\delta_N^{(1/2)}(\xi) \equiv \tfrac{1}{\sqrt{N}} \Sigma_{\nu=0}^{N-1} \exp(-i\xi 2\pi\nu) \qquad (7.16)$$

is not suitable here, because it is not the square root of the function $\delta(\xi)$. Instead,[9]

$$\lim_{N\to\infty} |\delta_N^{(1/2)}(\xi)|^2 = \Sigma_{j=-\infty}^{\infty} \ \delta(j - \xi) \qquad (7.17)$$

---

[5] *Cf* Weisstein, *Fourier Transform*, 2012.
[6] Michel Plancherel (1885 - 1967) & Magnus Gösta Mittag-Leffler (1846 - 1927), *Contribution a l'etude de la representation d'une fonction arbitraire par les integrales définies*, 2010
[7] Roger Bach, Damian Pope, Sy-Hwang Liou & Herman Batelaan, *Controlled double-slit electron diffraction*, 2013
[8] For an animated picture, see, *e.g*, https://en.wikipedia.org/wiki/Dirac_delta_function.
[9] Wolfgang P. Schleich (*1957), *Quantum Optics in Phase Space*, 2001, App. P

CHAPTER 8

Conservation of Stationary Quantum States Stationary Schrödinger Equation

Abstract: In this chapter, the stationary Schrödinger equations in configuration and momentum configuration spaces are derived from the non-classical expression of the total energy (7.15) found in the foregoing chapter. The new parameter in the Fourier transform is identified as $\hbar \equiv h/2\pi$. The roles of boundary conditions, probability concepts and independent dynamical variables are shortly discussed.

Keywords: Boundary conditions, Independent dynamical variables, Momentum representation, Probability, Stationary Schrödinger equation, Wave function.

8.1. EQUATION FOR $f_E(\xi)$

The Fourier transform (7.14a) enables us to eliminate the function $g_E(\eta)$ from eq. (3.15).

$$E = \int_{-\infty}^{+\infty} |f_E(\xi)|^2 V(x_E\xi)\mathrm{d}\xi \ (b_E = 1)$$

$$+ \int_{-\infty}^{+\infty} \frac{1}{2\pi} \int_{-\infty}^{+\infty} e^{-i\xi\eta}\bar{f}_E(\xi)\mathrm{d}\xi \int_{-\infty}^{+\infty} e^{i\xi'\eta} f_E(\xi')\mathrm{d}\xi' T(p_E\eta)\mathrm{d}\eta$$

$$= \int_{-\infty}^{+\infty} |f_E(\xi)|^2 V(x_E\xi)\mathrm{d}\xi$$

$$+ \int_{-\infty}^{+\infty} \frac{1}{2\pi} \int_{-\infty}^{+\infty} e^{-i\xi\eta}\bar{f}_E(\xi)\mathrm{d}\xi \int_{-\infty}^{+\infty} T\left(-ip_E \frac{\partial}{\partial\xi'}\right) e^{i\xi'\eta} f_E(\xi')\mathrm{d}\xi'\mathrm{d}\eta$$

$$= \int_{-\infty}^{+\infty} |f_E(\xi)|^2 V(x_E\xi)\mathrm{d}\xi + \int_{-\infty}^{+\infty} \bar{f}_E(\xi)T\left(-ip_E \frac{\partial}{\partial\xi}\right) f_E(\xi)\mathrm{d}\xi \qquad \textbf{(8.1)}$$

($\bar{f}$ denoting the complex-conjugated value of f), or

$$E = \int_{-\infty}^{+\infty} \bar{f}_E(\xi) \left[V(\xi x_E) + T\left(-ip_E \frac{\partial}{\partial\xi}\right)\right] f_E(\xi)\mathrm{d}\xi \qquad \textbf{(8.2)}$$

A sufficient condition for this equation to hold true reads

$$V(x_E\xi)f_E(\xi) + T\left(-ip_E \frac{\partial}{\partial\xi}\right) f_E(\xi) = E f_E(\xi) \qquad \textbf{(8.3)}$$

Since $T(p)$ is a quadratic function in (the components of) p, this equation is even a necessary condition for the minimum value of the r.h.s. of eq. (8.2), *i.e*, for the ground state.

Problem 8.1 (Ground state). *Verify the last statement! Hint: Explore the corresponding variational problem and account for $\int_{-\infty}^{+\infty} \bar{f}_E(\xi) f_E(\xi) d\xi = 1$!*

8.2. EQUATION FOR $g_E(\eta)$

Reversely, $f_E(\xi)$ can be eliminated from eq. (7.15) to obtain

$$E = \int_{-\infty}^{+\infty} \bar{g}_E(\eta) \left[V\left(ix_E \frac{\partial}{\partial \eta}\right) + T(p_E \eta) \right] g_E(\eta) d\eta \tag{8.4}$$

or

$$\int_{-\infty}^{+\infty} \bar{g}_E(\eta) \left[V\left(ix_E \frac{\partial}{\partial \eta}\right) + T(p_E \eta) - E \right] g_E(\eta) d\eta = 0 \tag{8.5}$$

A sufficient condition for this equation to hold true is

$$\left[V\left(ix_E \frac{\partial}{\partial \eta}\right) + T(p_E \eta) - E \right] g_E(\eta) = 0 \tag{8.6}$$

The range of validity of this equation is the same as that of eq. (8.3).

8.3. EINSTEIN OSCILLATOR: $x_E p_E = \hbar$

According to Einstein[1], Planck's discrete spectrum: $E_n = nh\nu$, applies to quantum harmonic oscillators as well. (The ground-state energy, $h\nu/2$, has been added later, and this for both systems.) Speeding ahead of the next chapter, the solution of eqs. (8.3) and (8.6) for the linear harmonic oscillator, for which we have

$$T(p) = \frac{1}{2m} p^2; \ V(x) = \frac{m}{2} \omega^2 x^2 \tag{8.7}$$

yields the energy spectrum

$$E = \left(n + \frac{1}{2}\right) x_E p_E \omega \tag{8.8}$$

Therefore, $x_E p_E = \hbar \equiv h/2\pi$.

---

[1] Einstein, *Die Plancksche Theorie der Strahlung und die Theorie der spezifischen Wärme*, 1907

The same result: $x_E p_E = \hbar$, is obtained when comparing the solution to the Coulomb problem with Bohr's 1913 model of hydrogen. This suggests $x_E p_E$ to be state- and system-independent and to universally equal $\hbar$.

8.4. SCHRÖDINGER'S WAVE FUNCTIONS AND EQUATIONS

8.4.1. Position Representation

Schrödinger's spatial wave functions, $\psi_E(x)$, are not dimensionless weight amplitudes, but dimensional weight *density* amplitudes.

$$\int |\psi_E(x)|^2 \mathrm{d}x = 1 \tag{8.9}$$

For this, I set

$$\psi_E(x) = \frac{1}{\sqrt{x_E}} f_E\left(\frac{x}{x_E}\right) \tag{8.10}$$

Then, with $T(p) = p^2/2m$, eq. (8.3) becomes the stationary Schrödinger equation in configuration representation.

$$-\frac{\hbar^2}{2m}\frac{\partial^2 \psi_E(x)}{\partial x^2} + V(x)\psi_E(x) = E\psi_E(x) \tag{8.11}$$

It is noteworthy, that this equation does *not* necessarily imply, that

$$E|\psi_E(x)|^2 = -\bar{\psi}_E(x)\frac{\hbar^2}{2m}\frac{\partial^2 \psi_E(x)}{\partial x^2} + V(x)|\psi_E(x)|^2 \tag{8.12}$$

equals the energy density, see Section 13.2.

Notice, that Schrödinger has already considered the more general case, that the mass – and thus the kinetic energy – is a function of x.[2] This leads to the more general form

$$H\left(p = -i\hbar\frac{\partial}{\partial x}, x\right)\psi_E(x) = E\psi_E(x) \tag{8.13}$$

For E being real-valued, terms, in which both x and $p = -i\hbar\,\partial/\partial x$ occur, are usually arranged in Hermitian[3] form.

---

[2] Such a situation occurs, *e.g*, in inhomogeneous semiconductors, see Enders, *Schrödinger Equation and Wave-Function Matching Conditions for Spatially Varying Effective Mass*, 1987.

[3] Charles Hermite (1822 - 1901), *Sur la Théorie des Formes Quadratiques*, 1853 – see also Problem 8.4!

Problem 8.2 (*). *Explore the fact, that the expression $\bar{\psi}_E(x)\partial^2\psi_E(x)/\partial x^2$ is real-valued!*

Problem 8.3 (*). *Explore the use of Sturm-Liouville[4] rather than Hermitian operators! Which matrices correspond to them?[5]*

Problem 8.4 (Quantization of $x^m p^n$. I*). *Explore the quantization of classical expressions proportional to $x^m p^n$![6] I will return to this issue in Problem 17.9.*

Problem 8.5 (*). *Explore the relationship between the Schrödinger equation (8.13) and the Hertzian line element (5.3)!*

8.4.2. Boundary Conditions

To obtain a unique solution to the stationary Schrödinger equation (8.11), one usually prescribes boundary conditions for $\psi_E(x)$. Arguing, that the electron in a bound state of the *H*-atom is not at infinity, Schrödinger (1926, *First Commun.*) has posed the boundary conditions

$$\lim_{|\vec{r}|\to\infty} \psi_E(\vec{r}) = 0 \qquad (8.14)$$

On the other hand, he wrote, that the "quantum equation" should "carry the quantum conditions in itself." (*Second Commun.*, pp. 511f. – in Section 1.4 we have considered this 'first requirement' in more detail). Strictly speaking, the boundary conditions are part of the differential *operator*, to which the differential equation belongs, not of the latter one. Even if Schrödinger himself had not in mind this subtle difference, in the next Chapter, I will show, that the stationary Schrödinger equation can uniquely be solved *without* posing boundary conditions. This feature gives it "maximum strength" in the sense of Einstein[7].

---

[4] Jacques Charles François Sturm (1803 - 1855) & Joseph Liouville (1809 - 1882), *EXTRAIT D'un Mémoire sur le développement des fonctions en séries dont les différents termes sont assujettis à satisfaire à une même équation diff érentielle linéaire, contenant un paramètre variable,* 1837; Sturm, *Cours d'analyse de l'Ecole polytechnique,* 1877

[5] For the relationship between Hermitian wave and matrix mechanics, see, *e.g,* Schrödinger, *Über das Verhältnis der Born -Heisenberg -Jordanschen Quantenmechanik zu der meinen,* 1926.

[6] *Cf* de Gosson, *The Angular Momentum Dilemma and Born-Jordan Quantization,* 2015, and refs. herein; *Reconsidering the Schrödinger Picture of Quantum Mechanics,* 2015.

[7] Einstein, *Grundzüge der Relativitätstheorie,* 2014, Appendix II

8.4.3. Probability*

The fact, that $|\psi_E(x)|^2$ is a *weight density*, suggests $|\psi_E(x)|^2 dx$ to represent the probability of a quantum system in state E to assume a configuration, x, in the interval $[x, x + dx]$. Based on the novel probability concepts introduced in 1916 by Einstein (*Zur Quantentheorie der Strahlung*) and in 1924 by Bohr, Kramers & Slater (*The Quantum Theory of Radiation*), this probability interpretation has been mentioned by Schrödinger (1926, *ibid.*) and brought essentially by Born (*Quantenmechanik der Stoßvorgänge*, 1926) into the nowadays widely accepted form.

After Heisenberg (*Die Geschichte der Quantentheorie*, 2000, pp. 17f.), the 'probability wave' contains more than a statement about our knowledge. It means something like a tendency towards a certain process, the quantitative version of Aristotle's notion '$\delta\upsilon\nu\alpha\mu\iota$' or 'potentia'. It introduces a strange kind of physical reality that stands about in the middle between possibility and reality.

This probability concept is compatible with the quantum-logical approach to QM, where the Hilbert space of wave functions is constructed using the probabilistic outcome of quantum measurements. QM is thus the theory of the measurement of quantum-mechanical systems by means of quantum methods.[8] However, since this approach yields the concept of wave functions, but not the concrete form of the operators acting upon them, its axiomatic power appears to be limited.

8.4.4. Momentum Representation

The momentum representation is quite analogous to the position (or direct) representation. The wave functions, $\phi_E(p)$, are not dimensionless weight amplitudes, but dimensional weight *densities*, here, in momentum (configuration) space.

$$\int |\phi_E(p)|^2 dp = 1 \tag{8.15}$$

Hence, we have

$$\phi_E(p) = \frac{1}{\sqrt{p_E}} g_E\left(\frac{p}{p_E}\right) \tag{8.16}$$

As a consequence, $\phi_E(p)$ is the Fourier transform of $\psi_E(x)$.

---

[8] Notice, that classical-mechanical systems are usually measured *not* by means of classical methods – like the impact of bodies –, but by means of quantum methods, notably, light.

The stationary Schrödinger equation in momentum space reads

$$\frac{p^2}{2m}\phi_E(p) + V\left(i\hbar\frac{\partial}{\partial p}\right)\phi_E(p) = E\phi_E(p) \tag{8.17}$$

More generally,

$$H\left(p, x = i\hbar\frac{\partial}{\partial p}\right)\phi_E(p) = E\phi_E(p) \tag{8.18}$$

For E being real-valued, terms in which both p and $x = i\hbar\frac{\partial}{\partial p}$ occur have to be suitably ordered, again (see Problem 8.4).

8.4.5. Independent Dynamical Variables*

The number of degrees of freedom of a mechanical system equals the number of independent dynamical variables describing it. Within CM, position, $x(t)$, and velocity, $v(t)$, or momentum, $p(t)$, are such ones, because their initial values can be chosen independently each of another.

In contrast, $\psi_E(x)$ and $\phi_E(p)$ are *not* independent dynamical variables, because the one is uniquely given by the other one through the Fourier transform. How then the number of independent dynamical variables is conserved during the transition from CM to QM, as claimed above?

Generally speaking, this conservation is guaranteed through the complex-valuedness of $\psi_E(x)$ and $\phi_E(p)$.

Problem 8.6 (Two-component wave functions*). *Explore the Schrödinger equation emerging from the weight function*

$$F_E(x) = \left(f_E^{(1)}(x)\right)^2 + \left(f_E^{(2)}(x)\right)^2 \tag{8.19}$$

This means, that the weight amplitudes and subsequently the wave functions are two-component vectors, the Hamiltonian a 2×2 matrix operator. Compare the action of the non-diagonal matrix elements with that of the diagonal ones! Can they be exploited for the description of novel effects? Hint: See also Problems 9.19 and 9.18!

Non-Classical Selection of the Physical Solutions

Abstract: The energy parameter in the stationary Schrödinger equation is primarily continuous. Schrödinger has selected the discrete quantum states by means of boundary conditions. This, however, is mathematics for classical systems like strings and pipes. And, more important, it hides the *intrinsic* discrete structure of that equation. The latter is represented by the recursion relations between solutions to different values of the energy parameter. These follows from Whittaker's integral expressions of the solutions and hold true for *all* solutions, not only for Schrödinger's eigensolutions. The physically relevant solutions are distinguished by certain mathematical properties as well as physical criteria, in particular, by their compliance with the absence of *perpetua mobilia*. This non-classical approach is exemplified by means of the model system harmonic oscillator. For this, the chapter starts with a sketch of the classical harmonic oscillator, including quite general topics such as the separation of internal and external system parameters and Huygens' principle, which will reappear in the quantum realm.

Keywords: Boundary conditions, Eigensolutions, Energy parameter, External system parameter, Harmonic oscillator, Huygens' principle, Internal system parameter, Perpetuum mobile, Stationary Schrödinger equation, Selection problem, Wave function, Weber's equation, Whittaker's integral expressions.

> "...nicht alle Lösungen jener [der klassischen] Differentialgleichungen scheinen nämlich wirklich realisierbaren Vorgängen zu entsprechen, sondern die Natur trifft unter den unendlich vielen Lösungen eine gewisse ganz spezielle Auswahl ..."[1]

On principle, when accounting for the boundary conditions (8.14), eq. (8.3) can be solved as a classical eigenvalue problem. However, this would discard the non-classical content of the wave function, *cf* Schrödinger's 1st and 2nd requirements to any quantization quoted in the Preface. For this, let us look for a solution method, which disclosures the discreteness *without* using boundary conditions like (8.14). This way, we will realize, that the stationary Schrödinger equation exhibits an *immanent* discreteness, which is *in*dependent of the boundary conditions.

---

[1] Planck, *Akademie-Ansprache vom 29. Juni 1922 (Leibniztag)*, pp. 45f. – Engl.: ...not all solutions of those [the classical] differential equations namely seem to correspond to the actually realizable processes, but the nature makes a certain special selection among the infinitely many solutions...

The novel selection method – discovered by Dieter Suisky[2] – is best demonstrated using the widest applicable model of physics, the linear harmonic oscillator[3].

Although I consider quantization to be *not* an eigenvalue problem, the stationary Schrödinger equation is an eigenvalue equation.[4] And in our derivation of it, the Fourier transform plays a major role. This suggests the question, what about the eigenfunctions of the Fourier transform? In the function space $L^2(\mathbb{R})$, these are the Hermite functions[5]. As we will see below, the Hermite functions obey a differential equation that is isomorphic with the stationary Schrödinger equation for the linear harmonic oscillator. – Conversely, the Fourier transform can be defined by means of that property of the Hermite functions.[6]

In general, there is a mathematical part to be solved first and a physical part yielding the final solution. For the linear harmonic oscillator, the physical criterion of compliance with the energy conservation law will turn out to be sufficient for the selection. For methodical reasons, however, the selecting mathematical properties and other physical criteria will be described, too.

9.1. CLASSICAL LINEAR HARMONIC OSCILLATOR

Being, perhaps, the simplest non-trivial system, the classical linear harmonic oscillator enables us to describe some important concepts in a straightforward manner.

9.1.1. Internal and External System Parameters

Internal system parameters are those, the values of which can*not* be altered by external influences, in particular, not by initial and boundary conditions. Within CM, best known examples are the mass of the bodies and other material parameters, such as the force constant of a spring. Moreover, the extreme values of the functions $V(x)$ und $T(p)$ are internal parameters. According to Euler, these extreme values can be calculated by means of the calculus, and that independently

---

[2] Suisky & Enders, *On the derivation and solution of the Schrödinger equation. Quantization as selection problem*, 2003; Enders & Suisky, *Quantization as selection problem*, 2005

[3] See, *e.g.*, Sylvan C. Bloch, *Introduction to Classical and Quantum Harmonic Oscillators*, 2013.

[4] This is not a contradiction, but refers to the distinction between differential *equation* and differential *operator*, the latter one includes the boundary conditions.

[5] Helmut Fischer (*1936) & Helmut Kaul (*1936), *Mathematik für Physiker*, Vol. 2, 2004, § 12 Sect. 4.2, pp. 300f.

[6] Norbert Wiener (1894 - 1964), after Javier Duoandikoetxea, *Fourier Analysis*, 2001

of the motion of the system under consideration. In particular, the minima of the function $V(x)$ determine its stable equilibrium states.[7]

In contrast, *external* system parameters are those, the values of which are set by external influences. Best known are the initial conditions, $x_0 \equiv x(t = 0)$ and $p_0 \equiv p(t = 0)$. All functions of them are external parameters, too, in particular, the total energy, $E = V(x_0) + T(p_0)$, and the turning points of an oscillator, $x_{min/max}$ and $p_{min/max}$, as

$$V(x_{min/max}) = E; \ T(p_{min/max}) = E \qquad (9.1)$$

In agreement with the statements about internal system parameters above, $V(x_{min/max})$ and $T(p_{min/max})$ are not extreme values of the functions $V(x)$ and $T(p)$, respectively.

Generally speaking, the actual state of a system is determined through both its internal and external parameters.

An important exception is the state at rest. It is determined by the *simultaneous* conditions $V = V_{min}$ AND $T = T_{min}$, therefore, *solely* by *internal* parameters.

This suggests the following generalizations.

Conjecture 9.1 (internal parameters). *Parameters, the values of which are determined by the condition $V = V_{min}$ AND $T = T_{min}$ are* internal *parameters.*

Conjecture 9.2 (internal states). *States, in which the extension in (momentum) configuration space belongs to extreme values of appropriate functions (e.g, $V(x)$ and $T(p)$), are* internal *states.*

In contrast to the Kepler orbits, the atomic shells are structurally stable[8]. Consequently, their energy spectra are not continuous functions of continuous external variables like the classical initial conditions. In fact, the representations of the energy of a quantum system in Section 7.2 contain initial wave functions, which cannot be chosen arbitrarily.

---

[7] de Maupertuis, *Accord de différentes Loix de la Nature qui avoient jusqu'ici paru incompatibles*, 1744

[8] *Cf* Franck & Hertz, *Über Zusammenstöße zwischen langsamen Elektronen und den Molekülen des Quecksilberdampfes und die Ionisierungsspannung derselben*, 1914; *Über die Erregung der 2536-Å-Quecksilberresonanzlinie durch Elektronenstöße und die Ionisierungsspannung derselben*, 1914; Franck & Jordan, *Anregung von Quantensprüngen durch Stöße*, 1926.

Problem 9.1. *Formulate the statement about the state at rest for the case, that the function $V(x)$ has got more than one local minimum!*

Problem 9.2. *Show, that there is no state at rest in a Kepler system!*

9.1.2. Separation of External and Internal System Parameters. I

> "Il apparut que, entre deux vérités du domaine réel, le chemin le plus facile et le plus court passe bien souvent par le domaine complexe."[9]

Here is a point, again, where the recursion to Eulerian thoughts supports the advancement of the unity of physics and of the axiomatic of quantum theory. Euler's care for external and internal system parameters suggests a description of the motion of the classical linear harmonic oscillator, which will, (i), exemplify Huygens' construction (see next subsection) and, (ii), yield a factorization of the classical Hamiltonian that will proof useful in the quantum case.

The Hamiltonian equals

$$H(p, x) = \frac{1}{2m}p^2 + \frac{m}{2}\omega^2 x^2 \tag{9.2}$$

Solving Hamilton's equation of motion (2.36), one obtains

$$\begin{pmatrix} x(t) \\ p(t) \end{pmatrix} = \begin{pmatrix} \cos(\omega t) & \frac{1}{m\omega}\sin(\omega t) \\ -m\omega\sin(\omega t) & \cos(\omega t) \end{pmatrix} \begin{pmatrix} x(0) \\ p(0) \end{pmatrix} \underset{def}{=} \widehat{D}(t) \begin{pmatrix} x(0) \\ p(0) \end{pmatrix} \tag{9.3}$$

Here, the external parameters: the initial conditions, $x(0)$ and $p(0)$, are already separated from the internal parameters: m and ω, that enter the matrix[10]

$$\widehat{D}(t) \underset{def}{=} \begin{pmatrix} \cos(\omega t) & \frac{1}{m\omega}\sin(\omega t) \\ -m\omega\sin(\omega t) & \cos(\omega t) \end{pmatrix}; \ |\widehat{D}(t)| = 1 \tag{9.4}$$

---

[9]   Paul Painlevé (1863 - 1933), *Analyse des travaux scientifiques*, 1900 (quoted after homepage.math.uiowa.edu/jorgen/hadamardquotesource.html, retrieved 01.04.2014) – Engl.: It came to appear that, between two truths of the real domain, the easiest and shortest path quite often passes through the complex domain. – Schrödinger (*4th Commun.*, 1926, p. 139) hesitated to introduce his time-dependent equation, since it is essentially complex-valued.

[10] The matrix (9.4) is simplectic. Such matrices play a crucial role within Hamiltonian mechanics, see, notably, de Gosson, *Introduction to Simplectic Mechanics: Lectures I-II-II*, 2006, and also Problem 9.4..

Problem 9.3. *Rewrite formula (9.3) in the form[11]*

$$\begin{pmatrix} x(t) \\ p(t) \end{pmatrix} = \exp(\hat{A}t) \begin{pmatrix} x(0) \\ p(0) \end{pmatrix} \tag{9.5}$$

Problem 9.4. *Find the variables* $X(t) = X(x(t), p(t))$ *and* $Y(t) = Y(x(t), p(t)))$*, in which the equations of motion read as*

$$\frac{\mathrm{d}}{\mathrm{d}t} \begin{pmatrix} X(t) \\ Y(t) \end{pmatrix} = \omega \widehat{\Omega}_2 \begin{pmatrix} X(t) \\ Y(t) \end{pmatrix} \tag{9.6}$$

where the matrix $\widehat{\Omega}_2$ *is given in eq. (2.45)!*

9.1.3. Huygens' Principle. I

The matrix $\widehat{D}(t)$ describes rotations in phase space. Being a symplectic matrix (see the foregoing subsection), it is a member of a symplectic group.

$$\widehat{D}(t) = \widehat{D}(t - t') \cdot \widehat{D}(t'); \ 0 \le t' \le t \tag{9.7}$$

This group multiplication rule represents a variant of the Chapman-Kolmogorov equation[12] and thus a point-mechanical analog of Huygens' construction[13].

Problem 9.5. *Show that the group property (9.7) is obvious in view of* $\widehat{D}(t) = \exp(\hat{A}t)$*, see eq. (9.5)!*

9.1.4. Separation of External and Internal System Parameters. II

Since the harmonic temporal behavior is essentially related to the angular frequency, ω, it is tempting to separate the mass, m, from it. Moreover, it is expected that the diagonalization of $\widehat{D}(t)$ leads to simplifications.

---

[11] *Cf Hint*: Henz & Langhake, *Pfade durch die Theoretische Mechanik 1*, 2016, Section 4.4.4!

[12] Sydney Chapman (1888 - 1970) & Thomas G. Cowling, *The Mathematical Theory of Non-Uniform Gases. An account of the kinetic theory of viscosity, thermal conduction, and diffusion in gases*, 1939; Andrej Nikolajewitsch Kolmogoroff (1903 - 1987), *Über die analytischen Methoden der Wahrscheinlichkeitsrechnung*, 1931; *Grundbegriffe der Wahrscheinlichkeitsrechnung*, 1933

[13] Originally (*Horologium oscillatorium*, 1673), it is the superposition of the increments of velocity during the subsequent influences of external forces, see Simonyi, *Kulturgeschichte der Physik*, 1990, pp. 241f.

That diagonalization yields the two eigenvalues of $\hat{D}(t)$, $\exp(\pm i\omega t)$. The eigenvector with the 'main axes' of $\hat{D}(t)$, $\tilde{x}(t)$ and $\tilde{p}(t)$, reads

$$\begin{pmatrix}\tilde{x}(t)\\\tilde{p}(t)\end{pmatrix} = \begin{pmatrix}\sqrt{\frac{m\omega}{2}}\,x(t) + \frac{i}{\sqrt{2m\omega}}p(t)\\\sqrt{\frac{m\omega}{2}}\,x(t) - \frac{i}{\sqrt{2m\omega}}p(t)\end{pmatrix} = \begin{pmatrix}\tilde{x}(0)e^{i\omega t}\\\tilde{p}(0)e^{-i\omega t}\end{pmatrix} \quad (9.8)$$

Here, I have distributed the factor $\sqrt{m\omega}$ such, that the new dynamical variables: $\tilde{x}(t)$ and $\tilde{p}(t)$,

- are complex-conjugated each to another:

$$\tilde{p}(t) = \tilde{x}(t)^* \quad (9.9)$$

- factorize the Hamiltonian:

$$\tilde{x}(t)\tilde{p}(t) = \frac{m}{2}\omega^2 x^2(t) + \frac{1}{2m}p^2(t) = H = \tilde{x}(0)\tilde{p}(0) \quad (9.10)$$

- are intimately related to two independent integrals of motion:

$$I_1 = e^{-i\omega t}\tilde{x}(t) = \tilde{x}(0) = const; \; I_2 = e^{i\omega t}\tilde{p}(t) = \tilde{p}(0) = const \quad (9.11)$$

Since there are no further independent integrals of motion, the total energy, E, depends on $\tilde{x}(t)$ and $\tilde{p}(t)$, or on $x(t)$ and $p(t)$, solely *via* I_1 and I_2 (9.11). Indeed, $E = I_1 I_2 = \tilde{x}(0)\tilde{p}(0) = const$, in agreement with eq. (9.10).

The new variables, $\tilde{x}(t)$ and $\tilde{p}(t)$, will acquire a novel meaning within QM, see Section 10.2. They are also used for exploring geometric phases[14]. We will meet such a phase in the Ehrenberg-Siday-Aharonov-Bohm effect in Subsection 17.3.2.

Problem 9.6 (*). *Can the Hertzian line element, ds, in eq. (5.3) simplify the arguing above? If yes, how? If not, (i), why, and, (ii), is there another, appropriate line element?*[15]

---

[14] Shivaramakrishnan Pancharatnam (1934 - 1969), *Generalized Theory of Interference, and Its Applications. Part I. Coherent Pencils*, 1956; Sir Michael Victor Berry (*1941), *Quantal Phase Factors Accompanying Adiabatic Changes*, 1984. For a review and more original texts, see Alfred D. Shapere & Wilczek (Eds.), *Geometric Phases in Physics*, 1989. For newer applications, see en.wikipedia.org/wiki/Geometric_phase.

9.2. THE MATHEMATICALLY DISTINGUISHED SOLUTIONS

9.2.1. Rationalization of the Schrödinger Equation to Weber's Equation, the Energy Parameter Becoming the Only Free Parameter

For the Hamiltonian (9.2), the Schrödinger equation reads ($\alpha_E \equiv x_E p_E$)

$$\frac{m}{2}\omega^2 x^2 \psi_E(x) - \frac{\alpha_E^2}{2m}\frac{\partial^2}{\partial x^2}\psi_E(x) = E\psi_E(x) \tag{9.12}$$

One standard form of this equation is Weber's equation[16], one of the equations of the parabolic cylinder.[17]

$$\frac{d^2 y_a(\xi)}{d\xi^2} - \left(\frac{1}{4}\xi^2 + a\right) y_a(\xi) = 0 \tag{9.13}$$

I thus set

$$x = x_E\xi; \; x_E = \sqrt{\frac{\alpha_E}{2m\omega}}; \; y_a(\xi) = \sqrt{x_E}\psi_E(x_E\xi) \tag{9.14a}$$

$$E = -a\varepsilon; \; \varepsilon = \frac{\alpha_E^2}{2mx_E^2} = \alpha_E\omega \tag{9.14b}$$

When compared with eq. (9.12), the reduction of the number of relevant parameters is obvious; solely the 'energy parameter', a, remains. In particular, the frequency, ν, does not occur as a self-standing parameter; therefore, the quantization is not affecting it. This way, Schrödinger's 3rd requirement (see Section 1.4) is already fulfilled (see also Problem 9.8!). This feature demonstrates the power of Newton's and Euler's methodology to begin with the stationary states, again.

Nevertheless, a is primarily a *mathematical* parameter of the differential equation under consideration; it becomes a *physical* parameter only after having passed

---

[15] See also Schrödinger, 1926, *1st Commun.*, *Note added in proof 28. II. 1926*, p. 376.

[16] Heinrich Friedrich Weber (1843 - 1912), *Über die Integration der partiellen Differentialgleichung* $\partial u/\partial x^2 + \partial u/\partial y^2 + k^2 u = 0$, 1869

[17] I largely follow Milton Abramowitz (1915 - 1958) & Irene Anne Stegun (1919 - 2008), *Handbook of Mathematical Functions*, 1964, in the selection by Michael Danos (1922 - 1999) & Johann Rafelski (* 1950), *Pocketbook of mathematical functions*, 1984, Ch. 19.

suitable criteria (*cf* Schrödinger, 1926, *1ˢᵗ Commun.*, pp. 363f.). These criteria will be developed next.

Problem 9.7 (*). *Explore the*

> **Conjecture 9.3.** *The limiting (weight) functions, $F_E(x/x_E)$ and $G_E(p/p_E)$, depend solely on those (internal) parameter combinations, on which E and x_E ($p_E = \hbar/x_E$) depend.*

Problem 9.8. *Show, that replacing in eq. (9.12) E with hν leads to a contradiction!*

9.2.2. The Total Energy is Strictly Positive

Multiplying Weber's equation (9.13) with $y_a(\xi)$ and integrating over ξ yields

$$a \int_{-\infty}^{+\infty} y_a^2(\xi)d\xi = \int_{-\infty}^{+\infty} y_a(\xi) \frac{d^2 y_a(\xi)}{d\xi^2} d\xi - \int_{-\infty}^{+\infty} \frac{1}{4}\xi^2 y_a^2(\xi)d\xi$$

$$= y_a(\xi)\frac{dy_a(\xi)}{d\xi}\Big|_{-\infty}^{+\infty} - \int_{-\infty}^{+\infty} \left(\frac{dy_a(\xi)}{d\xi}\right)^2 d\xi - \int_{-\infty}^{+\infty} \frac{1}{4}\xi^2 y_a^2(\xi)d\xi < 0 \quad \textbf{(9.15)}$$

Here, I have used the implication

$$\lim_{r\to\infty} V(\vec{r}) = \infty \;\Rightarrow\; \lim_{r\to\infty} f_E(\vec{r}) = 0, \lim_{r\to\infty} \nabla f_E(\vec{r}) = \vec{0} \quad \textbf{(9.16)}$$

Hence, $a < 0$, or the integrals do not exists. The latter case is excluded by virtue of the role they are playing in the fundamental representation (7.15) of the energy. Thus, the energy, $E = -a\hbar^2/2mx_E^2$, is strictly positive.

9.2.3. Factorization of Weber's Equation

The factorization (9.10) of the classical Hamiltonian suggests to factorize Weber's equation (9.13), too.

In fact, for and only for the values $a = \mp 1/2$, the l.h.s. of eq. (9.13) factorizes.

$$\left(\frac{d}{d\xi} - \frac{1}{2}\xi\right)\left(\frac{d}{d\xi} + \frac{1}{2}\xi\right) y_{-1/2}(\xi) = 0 \quad \textbf{(9.17a)}$$

$$\left(\frac{\mathrm{d}}{\mathrm{d}\xi} + \frac{1}{2}\xi\right)\left(\frac{\mathrm{d}}{\mathrm{d}\xi} - \frac{1}{2}\xi\right) y_{+1/2}(\xi) = 0 \qquad\qquad \textbf{(9.17b)}$$

Therefore, the values $a = -1/2$ and $a = +1/2$ are *mathematically distinguished* when compared with all other a-values.

Their connection to the factorization (9.10) of the classical Hamiltonian suggests, that they have got a physical relevance, too. However, while the classical variables $\tilde{x}(t)$ and $\tilde{p}(t)$ commute: $\tilde{x}(t)\tilde{p}(t) = \tilde{p}(t)\tilde{x}(t)$, their quantum counterparts do *not* commute. This indicates the difference between eqs. (9.17a) and (9.17b) to need further exploration.

Indeed, $y_{-1/2}(\xi)$ and $y_{+1/2}(\xi)$ are mathematically equivalent, but physically different. $y_{-1/2}(\xi) = y_{-1/2}(0)\exp(-1/4\,\xi^2)$ is a limiting amplitude, while $y_{+1/2}(\xi) = y_{+1/2}(0)\exp(+1/4\,\xi^2)$ is *not*. This fact physically distinguishes the value $a = -1/2$ over the value $a = +1/2$.

This result is compatible with our conclusion $a < 0$ in the foregoing subsection.

If there would be no other physically distinguished a-value, there would be only *one* state: $a = -1/2$. However, a system having got just *one* state cannot exchange energy with its environment. This implies the existence of further stationary states.

9.2.4. Recurrence Relations

"…die wahren Gesetze der Quantenmechanik bestünden nicht in bestimmten Vorschriften für die einzelne Bahn, sondern in diesen wahren Gesetzen seien die Elemente der ganzen Bahnenmannigfaltigkeit eines Systems durch Gleichungen verbunden, so daß scheinbar eine gewisse Wechselwirkung zwischen den verschiedenen Bahnen bestehe."[18]

---

[18] Schrödinger, *2nd Commun.*, p. 508 – Engl.: …the true laws of quantum mechanics would not consist in certain prescriptions for the *single orbit*, but that in these true laws, the elements of the whole manifold of orbits of a system would be connected through equations, so that apparently a certain interaction between the orbits exists.

Schrödinger's surmise is realized by the recurrence relations[19]. The for our purpose relevant ones are

$$\left(\frac{d}{d\xi} + \frac{\xi}{2}\right) U(a, \xi) + \left(a + \frac{1}{2}\right) U(a + 1, \xi) = 0 \qquad \textbf{(9.18a)}$$

$$\left(\frac{d}{d\xi} - \frac{\xi}{2}\right) V(a, \xi) - \left(a - \frac{1}{2}\right) V(a - 1, \xi) = 0 \qquad \textbf{(9.18b)}$$

Here, $U(a, \xi)$ and $V(a, \xi)$ are the standard solution of Weber's equation (9.13). In Whittaker's notation[20],

$$U(a, \xi) = D_{-a-\frac{1}{2}}(\xi) \qquad \textbf{(9.19a)}$$

$$V(a, \xi) = \frac{1}{\pi} \Gamma\left(\frac{1}{2} + a\right) \left\{ \sin(\pi a) D_{-a-\frac{1}{2}}(\xi) + D_{-a-\frac{1}{2}}(-\xi) \right\} \qquad \textbf{(9.19b)}$$

These recurrence relations base on Whittaker's representation of the solutions as contour integrals[21]. They have been obtained well before the advance of QM and are not related to any boundary conditions. This makes them capable to become part of the solution methods that account for the *non*-classical character of the quantization problem (*cf* Schrödinger's 2nd requirement in Section 1.4).

The recurrence relations (9.18)

- hold true for *all* a-values: $-\infty < a < +\infty$, not only for the common (Schrödinger's eigen) solutions, hence,

- reflect the *genuine discrete structure* saught for, *in*dependent of boundary conditions;

- do *not* occur within the common solution methods;

- interrelate solution functions with *finite* differences between their a-values, *viz*, $\Delta a = \pm 1$;

---

[19] Abramowitz & Stegun, *loc. cit.*, 19.3, 19.6.1, 19.6.5 – this examination is due to Dieter Suisky; though the following deviates from Suisky & Enders, *On the derivation and solution of the Schrödinger equation. Quantization as selection problem*, 2003, and Enders & Suisky, *Quantization as selection problem*, 2005.
[20] Whittaker & George Neville Watson (1886 - 1965), *A Course of Modern Analysis*, 1927, § 16.5
[21] Whittaker, *On the Functions associated with the Parabolic Cylinder in Harmonic Analysis*, 1902; Whittaker & Watson, *loc. cit.*, § 16.6.

- allow for most elegant calculations of general properties of the solution functions, *without* knowing them explicitly, see next section.

9.2.5. Three Classes of Values of the Energy Parameter

As a consequence, the recurrence relations (9.18) divide the continuous set $\{a|-\infty < a < +\infty\}$ of the energy parameter, a, into 3 classes.

Class A:

$$a = \cdots, -\frac{5}{2}, -\frac{3}{2}, -\frac{1}{2} \tag{9.20}$$

The recurrence relations (9.18a) break at $a = -1/2$; this is the mathematically *and* physically distinguished value found in eq. (9.17a).

Class B:

$$a = \cdots, \frac{5}{2}, \frac{3}{2}, \frac{1}{2} \tag{9.21}$$

The recurrence relations (9.18b) break at $a = +1/2$; this is the mathematically, but *not* physically distinguished value found in eq. (9.17b).

Class C:

$$a = \cdots, -\frac{3}{2} + \varsigma, -\frac{1}{2} + \varsigma, +\frac{1}{2} + \varsigma, +\frac{3}{2} + \varsigma, \cdots; \ 0 < \varsigma < 1 \tag{9.22}$$

For these values of a, there is *no* break in the recurrence relations (9.18). Hence, this class is *not* mathematically – and thus not physically – distinguished.

9.2.6. The Representing Energy-Parameter Interval

The smallest interval representing *all* solutions to eq. (9.13) is the closed interval $a = [-1/2, +1/2]$, as all other solutions being related to it through the recursion formulae (9.18). Being its boundary values, $a = -1/2$ (class A) and $a = +1/2$ (class B), are mathematically distinguished, again. All inner points: $-1/2 < a < +1/2$ (class C), are mathematically equivalent each among another and, consequently, *not* distinguished mathematically, again.

9.3. THE PHYSICALLY DISTINGUISHED SOLUTIONS

9.3.1. Class *A versus* Class *B*

As mentioned above, the two mathematically distinguished classes,

$A.\ a = -1/2, -3/2, -5/2, \cdots$, and

$B.\ a = +1/2, +3/2, +5/2, \cdots$

are mathematically, but *not* physically equivalent, because class A contains the physically distinguished value $a = -1/2$. An even more striking argument in favor of class A emerges from the energy conservation law in form of the absence of *perpetua mobilia*.

9.3.2. Absence of *perpetua mobilia*

> "Im Gegensatz zu der seit dem Alterthum bekannten Thatsache, dass Bewegung nur als <u>relative</u> Bewegung wahrnehmbar ist, war die Physik auf den Begriff der <u>absoluten</u> Bewegung gegründet. … Sie [die] setzte voraus, dass es physikalisch bevorzugte in der Natur nicht gebe und fragte nach den Folgerungen, welche aus dieser Voraussetzung bezüglich der Naturgesetze gezogen werden können. Die Methode der Relativitätstheorie ist derjenigen der Thermodynamik weitgehend analog; denn diese letztere Wissenschaft ist nichts weiter als die systematische Beantwortung der Frage: Wie müssen die Naturgesetze beschaffen sein, damit es unmöglich sei ein *perpetuum mobile* zu konstruieren?"[22]

---

[22] Einstein, letter to Maurice Solovine (1975 - 1958) dated 20. IV 20, in: Einstein, *Lettres à Maurice Solovine / Briefe an Maurice Solovine*, 1960, p. 18 – French: "En opposition avec le fait connu depuis l'antiquité que le mouvement n'est perceptible que comme mouvement *relatif*, la physique était basée sur la la notion de mouvement *absolut*. … Elle [la Théorie de la relativité] supposait qu'il n'y a pas dans la nature d'états de mouvement physiques privilégiés, et se demandait quelles sont les conséquences qui peuvent être tirées de

In his pioneering 1847 talk *On the Conservation of Force*, Helmholtz begins the treatment of energy conservation with the presupposition, that there are no *perpetua mobilia*. As a matter of fact, conservation implies bookkeeping – bookkeeping, however, is meaningless, if not impossible, for quantities having got *un*limited resources.

In other words, each conservative system has got a stationary state with minimum total energy content, in which is it not able to deliver energy to its environment. This state has been called "permanent state" (Bohr, 1913), "ground state" (*ibid.*), or "normal state" (Heisenberg, 1925, Sect. 3 – "ground state" in the 1967 English translation, after eq. (16)).

Such a (stationary) state exists *solely* in class A with the energies

$$\cdots, E_{-\frac{5}{2}} = \frac{5}{2}\alpha_E\omega,$$

$$E_{-\frac{3}{2}} = \frac{3}{2}\alpha_E\omega, \qquad\qquad (9.23)$$

$$E_{-\frac{1}{2}} = \frac{1}{2}\alpha_E\omega$$

This compatibility with the energy conservation law physically distinguishes class A when compared with classes B and C, again.

9.3.3. The Physically Relevant Solutions. $\alpha_E = \hbar$

Thus, the physical solutions are those with $-a - 1/2 = n = 0,1,2,\cdots$, *i.e*, $E = 1/2\alpha_E\omega, 3/2\alpha_E\omega, \cdots$. This energy spectrum is in agreement with Schrödinger's result, if $\alpha_E = \hbar$.

---

cette supposition concernant les lois de la nature. La méthode de la Théorie de la relativité est très analoque à celle de la Thermodynamique; car cette dernière science n'est autre chose que la réponse systématique à la question : Comment les lois de la nature doivent-elles être constituées pour qu'il soit impossible de construire un perpetuum mobile ?" (*ibid.*, p. (19)) – Engl.: Contrary to the fact – known since the ancient world – that motion is perceptible solely as <u>relative</u> motion, physics was built on the notion of <u>absolute</u> motion. …It [the theory of relativity] presumed that there are no physically preferred states of motion in nature and asked for the consequences for the laws of nature. The method of the theory of relativity is largely analogous to that of thermodynamics, for the latter science is nothing else than the systematic answer to the question, how the laws of nature must be designed to prevent the construction of a *perpetuum mobile*.

Notice, this result has been obtained through analyzing *solely* the stationary Schrödinger equation, *without* resorting to boundary or finiteness conditions and any solution methods for eigenvalue problems, such as series ansatzes like $\psi(x) = \sum_i \alpha_i x^i$.

Setting $a = -1/2$ in eq. (9.18a) yields the equation for the ground state.

$$\left(\frac{\mathrm{d}}{\mathrm{d}\xi} + \frac{\xi}{2}\right) U\left(-\frac{1}{2}, \xi\right) = 0 \tag{9.24}$$

Its solution reads

$$U\left(-\frac{1}{2}, \xi\right) = y_{-\frac{1}{2}}(\xi) = y_{-\frac{1}{2}}(0) e^{-\frac{1}{4}\xi^2} \tag{9.25}$$

The free parameter, $y_{-1/2}(0)$, does not represent a boundary value, but a normalization constant. This confirms, that this solution method is *not* a classical (eigenvalue-)boundary-value method.

All other solutions can be calculated by means of the recursion formula (9.18a) as

$$U\left(-n - \frac{1}{2}, \xi\right) = D_n(\xi) = e^{-\frac{1}{4}\xi^2} He_n(\xi) = \frac{1}{\sqrt{2^n}} e^{-\frac{1}{4}\xi^2} H_n\left(\frac{\xi}{\sqrt{2}}\right); \; n = 0,1,2, \dots \tag{9.26}$$

where He_n and H_n denotes the two variants of the n-th Hermite polynomial[23].

$$He_n(x) = (-1)^n e^{\frac{1}{2}x^2} \frac{\mathrm{d}^n}{\mathrm{d}x^n} e^{-\frac{1}{2}x^2}; n = 0,1,2, \dots \tag{9.27a}$$

$$H_n(x) = (-1)^n e^{x^2} \frac{\mathrm{d}^n}{\mathrm{d}x^n} e^{-x^2}; n = 0,1,2, \dots \tag{9.27b}$$

(Rodrigues' formulae[24])

The set of functions

---

[23] Hermite, *Sur un nouveau développement en série de fonctions*, 1864; see also de Laplace, *Mémoire sur les intégrales définies et leur applications aux probabilités, et spécialement à la recherche du milieux qu'il faut choisir entre les résultats des observations*, 1811; Pafnuty Lvovich Chebyshev (1821 - 1894), *Sur le développement des fonctions à une seule variable*, 1859. – The formulae are taken from Abramowitz & Stegun, *loc. cit.*, 19.13.1, 22.2, 22.11. and Whittaker & Watson, *loc. cit.*, 1927, § 16.7.

[24] Olinde Rodrigues (1795 - 1851), *De l'attraction des sphérodes*, 1816, has invented this type of representation for the Legendre polynomials, see https://en.wikipedia.org/wiki/Rodrigues'_formula. For $He_n(x)$, it is due to Hermite (*loc. cit.*; after Whittaker & Watson, *loc. cit.*, p. 350 ($D_n(x) \equiv He_n(x)$)).

$$y_{-n-\frac{1}{2}}(\xi) = \frac{1}{\sqrt{\sqrt{2\pi}n!}} e^{-\frac{1}{4}\xi^2} He_n(\xi)$$

$$= \frac{1}{\sqrt{\sqrt{2\pi}2^n n!}} e^{-\frac{1}{4}\xi^2} H_n(\xi/\sqrt{2}); \; n = 0,1,2,\dots \quad \textbf{(9.28)}$$

forms a complete orthonormal system in the space of all square-integrable functions on the real axis, $L^2(\mathbb{R})$, see the two problems at the end of this subsection. Due to the factor $\exp\{-1/4\,\xi^2\}$, Schrödinger's condition: $y_a(\xi) \to 0$ to be finite in the whole configuration space ($\xi \to \pm\infty$), is fulfilled *automatically*.

Hence, this solution method accounts for the quantum nature of the object under consideration, what Schrödinger had asked for (*cf 2ⁿᵈ Commun.*, pp. 511ff.).

Problem 9.9 (Uniqueness). *To each value on n, there are two linearly independent solutions. Show, that any solution, $z_{-n-\frac{1}{2}}(\xi)$, that is linearly independent of $y_{-n-\frac{1}{2}}(\xi)$, is not a limiting function! Hint: Consider the Wronskian, $W = y'z - z'y = const$[25]!*

Problem 9.10 (Orthogonality). *Exploit the differential equation (9.13) to show, that the set of functions $D_n(\xi)$ (9.26) represents an orthogonal system! Hint: Whittaker & Watson, loc. cit., § 16.7.*

Problem 9.11 (Normalization). *Exploit the recurrence relations (9.18a) to show, that*

$$\int_{-\infty}^{+\infty} D_n^2(\xi)\mathrm{d}\xi = \sqrt{2\pi}n! \quad \textbf{(9.29)}$$

Hint: Find the relationship between $\int_{-\infty}^{+\infty} D_n^2(\xi)\mathrm{d}\xi$ and $\int_{-\infty}^{+\infty} D_{n+1}^2(\xi)\mathrm{d}\xi$ (Whittaker & Whatson, *ibid.*)!

Problem 9.12 (*). *Discuss the fact, that $x_E = \sqrt{\hbar/(2m\omega)}$ is independent of E! Notice: The classical characteristic extension: the oscillation amplitude, $x_{max} - x_{min}$, does depend on E.*

Problem 9.13 (Momentum configuration space*). *Show, that the corresponding Schrödinger equation in momentum configuration space can be solved quite*

---

[25] Faraggi & Matone, *The Equivalence Postulate of Quantum Mechanics*, 1999, eq. (14.18)

analogously and calculate $y_a(\eta)$*! Then, calculate the weight amplitude,* $g_E(p)$*, for the case of vanishing potential and compare your result with* $g_E(p) = \sqrt{\delta(p/p_E - 1)}$ *for a free particle in eq. (7.13)!*

Problem 9.14 (Factorizing the 1D Hamiltonian*). *Formally, many one-dimensional time-independent quantum Hamiltonians of the form* $\hat{H} = p^2 + U(x)$ *(to bring the Hamiltonian into this form is part of the problem!) can be factorized as*[26]

$$\hat{H} = \hat{A}^+\hat{A} \tag{9.30}$$

where

$$\hat{A} = \frac{d}{dx} - w(x); \ \hat{A}^+ = -\frac{d}{dx} - w(x); \ U(x) = w(x)^2 + \frac{dw(x)}{dx} \tag{9.31}$$

Explore the similarities of this representation with as well as its differences to the recurrence equation approach above! How this approach is related to supersymmetric QM[27]?

Problem 9.15 (Factorizing the nD Hamiltonian*) *Generalize the results of Problem 9.14 to n dimensions!*

$$\hat{H} = \hat{A}^+\hat{A} \tag{9.32}$$

where

$$\hat{A} = \frac{d}{dx} - \frac{dw(x)}{dx}; \quad \hat{A}^+ = -\frac{d}{dx} - \frac{dw(x)}{dx}; \quad U(x) = \left(\frac{dw(x)}{dx}\right)^2 + \frac{\partial^2 w(x)}{\partial x^2} \tag{9.33}$$

Explore the similarities of this representation with as well as its differences to the recurrence equation approach above!

9.4. SCHRÖDINGER'S WAVE FUNCTIONS AND ENERGIES

Finally, we have

$$\psi_E(x) \equiv \psi_n(x) = \frac{1}{\sqrt{x_E}} y_{-n-\frac{1}{2}}\left(\frac{x}{x_E}\right) \tag{9.34a}$$

$$= \sqrt[4]{\frac{m\omega}{\pi\hbar}} \frac{1}{\sqrt{n!}} \exp\left(-\frac{m\omega}{2\hbar}x^2\right) He_n\left(\sqrt{\frac{2m\omega}{\hbar}}x\right) \tag{9.34b}$$

$$= \sqrt[4]{\frac{m\omega}{\pi\hbar}} \frac{1}{\sqrt{2^n n!}} \exp\left(-\frac{m\omega}{2\hbar}x^2\right) H_n\left(\sqrt{\frac{m\omega}{\hbar}}x\right) \tag{9.34c}$$

[26] Schrödinger, *A Method of Determining Quantum-Mechanical Eigenvalues and Eigenfunctions*, 1940; *Further Studies on Solving Eigenvalue Problems by Factorization*, 1940; Leopold Infeld (1898 - 1968) & T. E. Hull, *The Factorization Method*, 1951; Daniel Zwillinger (*1957), *Handbook of Differential Equations*, 1998

[27] R. de Lima Rodrigues, *The Quantum Mechanics SUSY Algebra: An Introductory Review*, 2001

$$E = E_n = (n + 1/2)\hbar\omega; \; n = 0,1,2,\ldots \qquad \textbf{(9.34d)}$$

Of course, this agrees with the common result (Schrödinger, 1926, *2nd Commun.*, Sect. 3.1, pp. 514ff.).

This success of our approach to the harmonic oscillator suggests to apply it to other situations as well. The answer to that challenge is left to the reader, who is encouraged to solve the following problems.

Problem 9.16 (Coulomb potential*). *Show that the stationary Schrödinger equation for the Coulomb potential can be solved in the same manner! What can be said about the unbound states? Moreover, check the inequality* $V_{ncl} \le E$!

Hint 1: The corresponding recurrence relations can be found in Abramowitz & Stegun, 1964, 14.2.

Hint 2: The radial part of the equation can be transformed to that of the harmonic oscillator (Schrödinger, 1926, *2nd Commun.*, p. 518, fn. 1).

Hint 3: The angular part can be tackled by means of the angular momentum operator, $\hat{\vec{L}}$. The operators $\hat{L}^2$ and, say, $\hat{L}_z$ have got common eigenfunctions as

$$\hat{L}^2\psi_{lm}(\theta,\phi) = l^2\hbar^2\psi_{lm}(\theta,\phi); \; \hat{L}_z\psi_{lm}(\theta,\phi) = m\hbar\psi_{lm}(\theta,\phi) \qquad \textbf{(9.35)}$$

The eigenvalues are $(l\hbar)^2$ and $m\hbar$, respectively. Notice, that these relationships are not considered as an classical eigenvalue problem, *i.e.*, boundary values are *not* taken into account! There are recurrence relations between the functions $\psi_{lm}(\theta,\phi)$. They yield the sets l and m, *without* calculating the eigenfunctions[28].

Problem 9.17 (Generalization to orthogonal polynomials*). *Generalizing the calculations for the harmonic oscillator and in Problem 9.16, show, that the bound states of any stationary Schrödinger equation can be calculated in the same manner, if its potential energy is of such a form, that it can be brought into the differential equation for the orthogonal polynomials (cf Abramowitz & Stegun, loc. cit., 22.1;* $f' \equiv df/dx$),

$$g_2(x)f_n''(x) + g_1(x)f_n'(x) + a_nf_n(x) = 0; \; n = 0,1,\ldots \qquad \textbf{(9.36)}$$

Here, 'orthogonal' means

---

[28] Munir Al-Hashimi, *Accidental Symmetry in Quantum Physics*, 2008, pp. 18-19

$$\int_a^b w(x)f_m(x)f_n(x)\mathrm{d}x = |f_n|^2 \delta_{mn}; \ m,n = 0,1,\dots \tag{9.37}$$

where $|f_n|^2$ is the norm of $f_n(x)$ w.r.t. the weight function, $w(x)$. *Hint*: Derive recurrence formulae between neighboring states,

$$h_2^{\pm}(x)f_n'(x) + h_1^{\pm}(x)f_n(x) = h_0^{\pm}(x)f_{n\pm 1}(x) \tag{9.38}$$

using the general Rodrigues' formula,

$$f_n(x) = \frac{1}{e_n w(x)} \frac{\mathrm{d}^n}{\mathrm{d}x^n}\left[w(x)r(x)^n\right] \tag{9.39}$$

(e_n – numerical factor, $r(x)$ – polynomial in x).[29] What can be said about the unbound states? Moreover, check the inequality $V_{\mathrm{ncl}} \le E$!

Problem 9.18 (Generalization to multi-indexed orthogonal polynomials*). *There are recurrence relations not only between the classical orthogonal polynomials dealt with in the foregoing problem, but also between their generalization to the so-called multi-indexed orthogonal polynomials.[30] Exploit the reasoning of this chapter to find the physically relevant solutions to the corresponding differential equations! Hint: See also the following problem and Problem 8.6!*

Problem 9.19 (Generalization to rational extensions of the harmonic oscillator potential*). *There are so-called rational extensions of the harmonic oscillator potential, for which recurrence relations exist, too.[31] Exploit the reasoning of this chapter to find the physically relevant solutions to the corresponding differential equations! Hint: See also the foregoing problem and Problem 8.6!*

---

[29] For more special cases, see the list on https://en.wikipedia.org/wiki/List_of_quantum-mechanical_potentials.

[30] Satoru Odake, *Recurrence relations of the multi-indexed orthogonal polynomials*, 2013, 2015; Odake & Ryu Sasaki, *Discrete Quantum Mechanics*

[31] David Gómez-Ullate, Yves Grandati & Robert Milson, *Rational extensions of the quantum harmonic oscillator and exceptional Hermite polynomials*, 2014

CHAPTER 10

Additional Results

"Gar Manches rechnet Erwin schon
Mit seiner Wellenfunktion.
Nur wissen möcht' man gern[e] wohl,
Was man sich dabei vorstell'n soll."[1]

Abstract: This chapter contains additional topics of time-independent QM: Critically examinations of the 'tunnel effect' and 'quantum leaps', the occupation number representation (because of its relationship to the factorization of the classical Hamiltonian), energy *versus* (effective) extension, and the absence of 'quantum states at rest'. I hope this helps the reader to grasp something more of the meaning of QM.

Keywords: Energy, Extension, Occupation number representation, Rest, Tunnel effect.

10.1. ABOUT THE 'TUNNEL EFFECT'. I. STATIONARY CASE

10.1.1. The myth: $V(x) > E$, but actually $V_{ncl}(x) < E$

The observation of quantum particles, that are crossing spatial domains, $\{x\}$, in which the total energy is *lower* than the (classical!) potential energy: $E < V(x)$, has led to the notion 'tunnel effect'.[2] This wording is a nice illustration – though it masks the fact, that the *effective* potential energy is *not* the classical expression, $V(x)$, but the *non*-classical expression, $V_{ncl}(x) = F(x)V(x)$, which, by its very definition (7.1a), is nowhere larger than E. From this point of view, there is *no* tunneling at all.

---

[1] Erich Armand Arthur Joseph Hückel (1896 - 1980), 1925 (quoted after http://www.physik.tu-berlin.de/dschm/publi/schroedinger_lektuere/werk.htm) – Engl.: "Erwin with his psi can do / Calculations quite a few. / But one thing has not [yet] been seen: / Just what does psi really mean?" (transl. by Felix Bloch (1905 - 1983, Nobel Award 1952), after https://en.wikipedia.org/wiki/Erich_H%C3%BCckel).
[2] Friedrich Hund (1896 - 1997), *Zur Deutung der Molekelspektren. I*, 1927; *III. Bemerkungen über das Schwingungs- und Rotationsspektrum bei Molekeln mit mehr als zwei Kernen*, 1927; Nordheim, *Zur Theorie der thermischen Emission und der Reflexion von Elektronen an Metallen*, 1928; George Gamow (1904 - 1968), *Zur Quantentheorie des Atomkernes*, 1928; the term 'tunnel effect' has been coined by Yakov Il'ich Frenkel (1894 - 1952), *Wave Mechanics*, 1932.

For the linear harmonic oscillator, the inequality $V_{ncl}(x) = F_E(x)V(x) \leq E$ (7.1a) can be proven as follows. In the dimensionless quantities of eq. (9.13) and its solution (9.28), it reads

$$v_n(\xi) \equiv y_n^2(\xi)\frac{1}{4}\xi^2 \leq n + \frac{1}{2}; \quad -\infty < \xi < +\infty, n = 0,1,\cdots \qquad \textbf{(10.1)}$$

The verification is left to the reader, see Problem 6.2. An illustration is given in Fig. **(10.1)**.

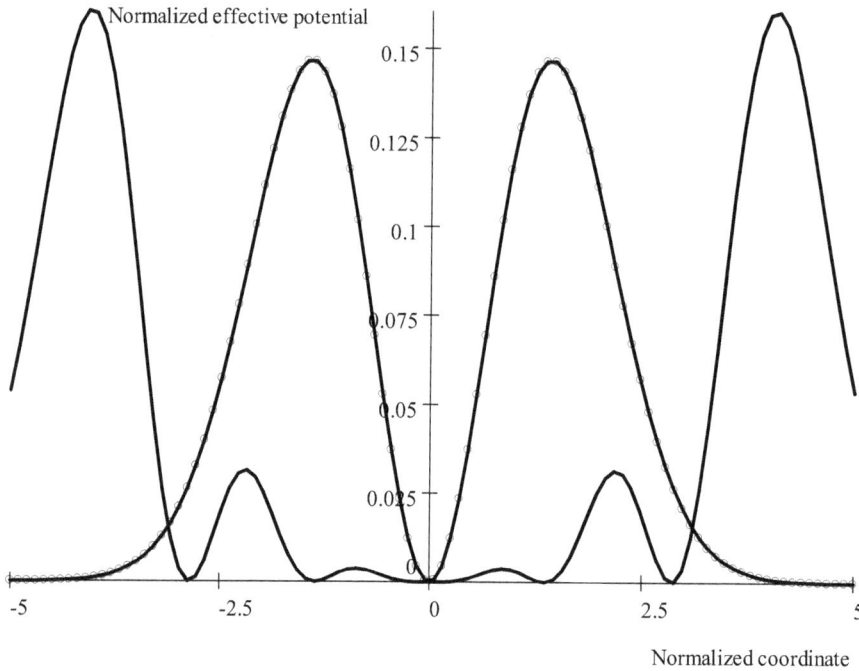

Fig. 10.1. 'Tunneling': The *non*-classical potential energy, $V_E(x)$, is everywhere *smaller* than the total energy, E: $V_E(x)/E = v_n(\xi)/(n + 1/2) < 1$, *cf* inequ. (10.1). Line with circles: $n = 0$, line without circles: $n = 5, -5 \leq \xi \leq +5$.

Hence, the quantum oscillator does *not* 'tunnel into the potential hills' that ly beyond the classical turning points, but it 'crawls' over the effective potential hills, $V_{ncl}(x) = F_E(x)V(x) \leq E$ for configurations with $V(x) > E$.

Problem 10.1. *Discuss the accuracy of the notion 'quantum tunneling' in view of the fact, that the classical tunneling of a body is not accompanied by any change of its attributes!*

Problem 10.2 *Proofineq. (10.1)! Hints: Use the recurrence formula*[3]

$$H_{n+1}(x) = 2xH_n(x) - 2nH_{n-1}(x) \tag{10.2}$$

and the inequality (ibid., 22.14.17)

$$|H_n(x)| < e^{x^2/4}\sqrt{n!}\, k; \quad k \approx 1.086435 \tag{10.3}$$

You should arrive at

$$v_n(\xi) \equiv y_n^2(\xi)\frac{1}{4}\xi^2 \leq \frac{k^2}{\sqrt{\pi}}\left(n + \frac{1}{2}\right) \approx 0.666n + 0.499;$$

$$-\infty < \xi < +\infty, n = 0,1,\cdots \tag{10.4}$$

Problem 10.3 (Rectangular 'barrier'). *Show the validity of the inequality* $V_{ncl}(x) \leq E$ *for the one-dimensional rectangular potential step of width* $2a$ *and height* V_0! *Hint: Solve the Schrödinger equation for* $u(\xi) = \psi(a\xi)$; *normalize* $u(\xi)$ *to* $\int_{-\infty}^{+\infty} u(\xi)^2 d\xi = 1$; *show, that* $u(\pm 1)^2 < E/V_0$ *holds true. This means, (i),* $\cos(q)^2 + \sin(2q)/2q > 0$ *for the even solutions and, (ii),* $\sin(q)^2 - \sin(2q)/2q > 0$ *for the odd solutions, where* $q \equiv \sqrt{2ma^2E/\hbar^2}$. *Hint:* Cf *Schwabl, Quantenmechanik, 1993, § 3.4.*

Problem 10.4 (sin⁻² 'barrier'*). *Explore the quantum-mechanical problem for the classical potential*

$$V(x) = \frac{V_0}{\sin^2(x/a)}; \quad -\frac{\pi}{2} \leq \frac{x}{a} \leq +\frac{\pi}{2} \tag{10.5}$$

Hint: *Set*

$$\frac{2ma^2V_0}{\hbar^2} = \beta; \quad \frac{x}{a} = \xi; \quad \frac{2ma^2E}{\hbar^2} = \varepsilon \tag{10.6}$$

There are solutions to the stationary Schrödinger equation in terms of the ultraspherical or Gegenbauer polynomials, $C_n^\alpha(\cos\xi)$. *For which values of* α *this solution is most convenient?*

---

[3] Abramowitz & Stegun, *Handbook of Mathematical Functions*, 1964, 22.7.13

Problem 10.5 (Eckart potential). *Explore the quantum-mechanical problem for the symmetric Eckart potential[4],*

$$V(x) = \frac{V_0}{\cosh^2(x/a)}; \quad -\infty < \frac{x}{a} < +\infty \tag{10.7}$$

10.1.2. 'Tunneling' in Momentum Space

It is tempting to exploit the stationary Schrödinger equation in momentum representation, because $p^2/2m \geq 0$. Indeed, it implies

$$\frac{1}{g_E(p)} V(\hat{p}) g_E(p) = E - \frac{p^2}{2m} < E \tag{10.8}$$

This suggests $g_E^{-1}(p) V(\hat{p}) g_E(p)$ to represent the contribution of the momentum configuration, p, to the potential energy of a quantum system. However, this suggestion is abolished by the fact, that $p^2/2m = T(p)$, the *classical* kinetic energy, does *not* represent the contribution of the momentum configuration, p, to the kinetic energy of a quantum system. For there are no relations like $V_{\mathrm{ncl}}(x) + T_{\mathrm{ncl}}(x) = E$ or $V_{\mathrm{ncl}}(p) + T_{\mathrm{ncl}}(p) = E$.

10.1.3. Analog to Undersized Waveguides

For classically bounded systems, the behavior of the wave function, in configuration space, $\psi(x)$, changes at the classical turning points as these are inflection points. Between the classical turning points, the wave function oscillates. Beyond the classical turning points, the wave function decreases exponentially. The electromagnetic analog of the latter is the attenuation of waves propagating into undersized wave guides. For this, the 'tunnel effect' still represents an attractive object.[5] Last but not least, this interest has been kept alive due to the observation of superluminal signal propagation through similar spatial arrangements[6].

Therefore, the 'tunneling' electromagnetic wave 'crawls' *over* a *quantum* hill rather than 'travels *through* a *classical* hill'.

---

[4] Carl Henry Eckart, *The penetration of a potential barrier by electrons*, 1930

[5] See, *e.g.*, Mathew Tomes, Kerry J. Vahala & Tal Carmon, *Direct imaging of tunneling from a potential well*, 2009.

[6] Andreas Enders & Günter Nimtz (*1936), *Evanescent-mode propagation and quantum tunneling*, 1993; Daniela Mugnai, Anedio Ranfagni & L. Ronchi, *The question of tunneling time duration: A new experimental test at microwave scale*, 1998

Problem 10.6. *Compare the quantum-mechanical 'tunnel effect' with the electromagnetic 'tunnel effect'[7]! Hint: Exploit the relation between refraction index and energy known already to Johann Bernoulli[8]!*

10.2. OCCUPATION NUMBER REPRESENTATION FROM RECURSION RELATIONS

The recurrence relations (9.18) provide a natural way to the occupation number representation, which has been applied most successfully not only in quantum mechanics, but also in quantum field theory.

The equality of the energy levels of the harmonic quantum oscillator, $E_n = \left(n + \frac{1}{2}\right)\hbar\omega$, with those of Planck's "vibrating resonators" suggests n to count the number of certain quanta 'being on', or 'occupying' the oscillator. Are there operators, which describe, (i), that occupation and, (ii), the transitions between different occupations? The recurrence relations (9.18) lend themselves to the latter ones.

According to eqs. (9.18), the solutions (9.34c) obey the recurrence relation

$$\left(\sqrt{\frac{\hbar}{2m\omega}}\frac{\mathrm{d}}{\mathrm{d}x} + \sqrt{\frac{m\omega}{2\hbar}}\,x\right)\psi_n(x) = \sqrt{n}\,\psi_{n-1}(x) \qquad (10.9)$$

(the factor $\sqrt{n}$ is left on the r.h.s., in order to avoid difficulties in the case $n = 0$). Writing this equation in the form[9]

$$\hat{b}|n\rangle = \sqrt{n}|n-1\rangle \qquad (10.10)$$

it becomes obvious, that the operator

$$\hat{b} \equiv \sqrt{\frac{\hbar}{2m\omega}}\frac{\mathrm{d}}{\mathrm{d}x} + \sqrt{\frac{m\omega}{2\hbar}}\,x \qquad (10.11)$$

lowers the occupation number, n, by one, or annihilates one quantum on the oscillator. For this, it has been termed *annihilation operator*.

---

[7] Peter Reinecker & Michael Schulz, *Theoretische Physik. Elektrodynamik*, 2006, § 9.5.6
[8] After Lanczos, *The Variational Principles of Mechanics*, 1970
[9] Dirac, *The quantum theory of emission and absorption of radiation*, 1927; Wladimir Alexandrowitsch Fock (Fok, 1898 - 1975), *Konfigurationsraum und zweite Quantelung*, 1932

For obtaining the operator that *increases* the number of quanta, I do not exploit relation (9.18b), because it belongs to the *non*-physical functions $V(a, \xi)$. Instead, I invert eq. (10.9) through multiplying it with

$$\hat{b}^+ \equiv -\sqrt{\frac{\hbar}{2m\omega}} \frac{\mathrm{d}}{\mathrm{d}x} + \sqrt{\frac{m\omega}{2\hbar}} \, x \qquad (10.12)$$

as suggested by the factorizations (9.17).

$$\left(-\sqrt{\frac{\hbar}{2m\omega}} \frac{\mathrm{d}}{\mathrm{d}x} + \sqrt{\frac{m\omega}{2\hbar}} x\right) \sqrt{n}\psi_{n-1}(x)$$

$$= \left(-\sqrt{\frac{\hbar}{2m\omega}} \frac{\mathrm{d}}{\mathrm{d}x} + \sqrt{\frac{m\omega}{2\hbar}} x\right)\left(\sqrt{\frac{\hbar}{2m\omega}} \frac{\mathrm{d}}{\mathrm{d}x} + \sqrt{\frac{m\omega}{2\hbar}} x\right) \psi_n(x)$$

$$= \left(-\frac{\hbar}{2m\omega} \frac{\mathrm{d}^2}{\mathrm{d}x^2} - \frac{1}{2} + \frac{m\omega}{2\hbar} x^2\right) \psi_n(x)$$

$$= \left(\frac{1}{\hbar\omega} \hat{H} - \frac{1}{2}\right) \psi_n(x) = n\psi_n(x) \qquad (10.13)$$

that is,

$$\hat{b}^+|n-1\rangle = \sqrt{n}|n\rangle \qquad (10.14)$$

Hence, the operator $\hat{b}^+$ (10.12) rises the number of quanta by one, or creates one quantum on the oscillator. For this, it has been termed *creation operator*. $\hat{b}^+$ is Hermitian conjugated to b, since, by partial integration,

$$\int_{-\infty}^{+\infty} \psi_m(x) \frac{\mathrm{d}}{\mathrm{d}x} \psi_n(x)\mathrm{d}x = -\int_{-\infty}^{+\infty} \psi_n(x) \frac{\mathrm{d}}{\mathrm{d}x} \psi_m(x)\mathrm{d}x$$

Furthermore, combining eqs. (10.10) and (10.14), one obtains

$$\hat{b}^+\hat{b}|n\rangle = n|n\rangle \qquad (10.15)$$

yielding the number of quanta, n, in state $|n\rangle$. Consequently, $\hat{n} \equiv \hat{b}^+b$ has been termed 'occupation number operator'.[10]

---

[10] *Cf* also Schrödinger, *The Factorization of the Hypergeometric Equation*, 1941.

Problem 10.7. *Verify the recurrence relation (10.9) through deriving $\psi_E(x)$ w.r.t. x! Hint: $\partial He_n(\xi)/\partial\xi = nHe_{n-1}(\xi)$ (Abramowitz & Stegun, loc. cit., 22.8.8)*

Problem 10.8. *Give a simple physical reasoning for $\hat{b}\hat{b}^+ \neq \hat{b}^+\hat{b}$! Hint: Consider the state $|0>$.*

Problem 10.9 (*). *Explore the analogs between the operators $\hat{b}$ and $\hat{b}^+$ and the operators $\hat{L}_\pm$ in Problem 5.16!*

Problem 10.10 (*). *Discuss the results of this section in terms of the dynamical variables $\tilde{x}(t)$ and $\tilde{p}(t)$ that factorize the classical Hamiltonian, eq. (9.10)! Which consequences could this have for the foundation of the occupation number representation, or even for the axiomatic of QM itself?*

10.3. ENERGY AND EXTENSION. II*

As required in the reasoning for $F(x) \geq 0$, eq. (7.7), the Helmholtzian monotony between energy and extension applies to the quantum oscillator. Moreover, I will treat a more general potential, in order to illustrate the application of the quantum virial theorem.

10.3.1. Example Harmonic Oscillator

The values, $\xi_{max,n}$, where the non-classical contribution of the configuration ξ to the total energy E_n, $v_n(\xi) = y^2_{-n-1/2}(\xi)\frac{1}{4}\xi^2$, assumes its maximum value, increase with n: $\xi_{max,0} < \xi_{max,1} < \cdots$, cf Fig. (**10.1**).

By virtue of the symmetry between ξ and η, the dual result applies as well. The values, $\eta_{max,n}$, where the non-classical contribution of the momentum configuration η to the total energy E_n, $t_n(\eta) = y^2_{-n-1/2}(\eta)\frac{1}{4}\eta^2$, assumes its maximum value, increase with n: $\eta_{max,0} < \eta_{max,1} < \cdots$.

Hence, the ordering relations between energy and effective extensions in configuration and momentum configuration spaces are realized as in the classical case.

10.3.2. Generalization to Potentials $V(\vec{r}) = V_x|x|^{n_x} + V_y|y|^{n_y} + V_z|z|^{n_z}$

In order to tackle more general potentials, I invoke the quantum-mechanical virial theorem[11].

$$2 \int \bar{\psi}_E(\vec{r}) T(-i\hbar\nabla) \psi_E(\vec{r}) \mathrm{d}^3\vec{r} = \int \bar{\psi}_E(\vec{r})(\vec{r} \cdot \nabla V(\vec{r})) \psi_E(\vec{r}) \mathrm{d}^3\vec{r} \qquad (10.16)$$

For homogeneous potentials,

$$V(\vec{r}) = V_x|x|^{n_x} + V_y|y|^{n_y} + V_z|z|^{n_z}; \; V_{x,y,z} \geq 0, n > 0 \qquad (10.17)$$

we have[12]

$$\vec{r} \cdot \nabla V(\vec{r}) = n_x V_x|x|^{n_x} + n_y V_y|y|^{n_y} + n_z V_z|z|^{n_z} \qquad (10.18)$$

Inserting the r.h.s. into eq. (6.16) and using the separation

$$\psi_E(\vec{r}) = \psi_{E_x}(x)\psi_{E_y}(y)\psi_{E_z}(z) \qquad (10.19)$$

one obtains for the total energy

$$E = \int \bar{\psi}_E(\vec{r}) T(-i\hbar\nabla) \psi_E(\vec{r}) \mathrm{d}^3\vec{r} + \int \bar{\psi}_E(\vec{r}) V(\vec{r}) \psi_E(\vec{r}) \mathrm{d}^3\vec{r}$$

$$= \int \bar{\psi}_E(\vec{r}) \left(\frac{\vec{r}}{2} \cdot \nabla V(\vec{r}) + V(\vec{r}) \right) \psi_E(\vec{r}) \mathrm{d}^3\vec{r}$$

$$= \left(\frac{n_x}{2} + 1 \right) \int V_x|x|^{n_x} \psi_{E_x}(x)^2 \mathrm{d}x + (x \to y) + (x \to z) \qquad (10.20a)$$

$$= E_x + E_y + E_z \qquad (10.20b)$$

The remainder of the arguing is left to the reader.

---

[11] *Cf* Arthur Edward Ruark (1899 - 1979), *The Zeeman effect and Stark effect of hydrogen in wave mechanics: the force equation and the virial theorem in wave mechanics*, 1928. – Originally, the theorem stems from mechanical heat theory. Rudolf Clausius (1822 - 1888) termed the quantity $\frac{1}{2}\Sigma_a \, \overline{\vec{r}_a(t) \cdot \vec{K}_a(t)}$ *virial*, after the Latin word *vir* for force; the sum runs over all bodies, *a*, of the system, the bar denotes the temporal average (*Über einen auf die Wärme anwendbaren mechanischen Satz*, 1870). It is presupposed, that the motion and the forces are bounded; see also Landau & Lifshitz, *Mechanics*, § 10.

[12] *Cf* Euler's theorem for homogeneous functions, http://mathworld.wolfram.com/EulersHomogeneous FunctionTheorem.html (07.10.2013); encyclopediaofmath.org/index.php/Homogeneous_function (07.10. 2013); R. J. Tykodi, *On Euler's theorem for homogeneous functions and proofs thereof*, 1982.

Problem 10.11. *Prove eq. (10.16)! Hint: Resolve the commutator in the identity*
$\int \bar{\psi}_E(\vec{r})[\hat{H}(\vec{r}), \vec{r} \cdot \hat{\vec{p}}]\psi_E(\vec{r})d^3\vec{r} \equiv 0!$[13]

Problem 10.12. *Show, that with increasing energy, $E_{x(y,z)}$, the maxima of the weight functions, $F_{E_{x(y,z)}} = \psi^2_{E_{x(y,z)}}(x)$, are located at increasing values of $|x|$ ($|y|$, $|z|$)! Hint: Exploit general Sturm-Liouville theory*[14]*!*

Problem 10.13 (*). *Explore the extension in (momentum) configuration spaces for the hydrogen atom! Is there a quantum analogue to the hodograph (Section 19.3)?*

10.3.3. Ordered Sets of Energies and Extensions. General Case

For classically bounded systems, the relationship between energy and extension through ordered sets can be retained, although their non-classical counterparts are, on principle, unbounded. Because of the latter, however, 'extension' means 'characteristic extension'. It can defined as follows.

For the total energy, E, in eq. (7.4) to be finite, we have the condition $\lim_{x \to \pm\infty} V_E(x) = 0$. For this, the (non-negative) function $V_E(x)$ exhibits EITHER one global maximum and no additional local maxima, e.g, $V_E(x) = V_E(0)\exp(-x^2/2x_E^2)$. In this case, the characteristic extension is the standard deviation, x_E, of that Gaussian function. OR, the function $V_E(x)$ exhibits at least two local maxima. Here, the characteristic extension, x_E, equals the distance between the configurations of the two outermost local maxima.

Now, it is expected, that – as in the classical case – there are pairs of ordered sets of energies and extensions:

$$E_1 < E_2 < \cdots; \; x_{E_1} < x_{E_2} < \cdots \qquad (10.21)$$

Conversely, for every value of E, there may be a length parameter, x_E, such, that

$$x_{E'} > x_{E''} \text{ if } E' < E'' \qquad (10.22)$$

---

[13] *Cf* Franz Schwabl (1938 - 2009), *Quantenmechanik*, 1993, p. 206.
[14] Sturm & Liouville, *EXTRAIT D'un Mémoire sur le développement des fonctions en séries dont les différents termes sont assujettis à satisfaire à une même équation différentielle linéaire, contenant un paramètre variable*, 1837; Sturm, *Cours d'analyse de l'Ecole polytechnique*, 1877

Then, x_E represents a characteristic extension in configuration space in state E, too. An example is the de Broglie wavelength mentioned above, $x_E = h/p = h/\sqrt{2mE}$.

The characteristic extension in momentum configuration space can be defined analogously, though this left to the reader, see Problem 10.14.

Problem 10.14. *Define characteristic extensions in momentum configuration space, including free particles!*

10.3.4. Effective Occupation of (Momentum, Phase) Space

While the stationary-state functions $|\psi_E(x)|^2$ and $|\phi_E(p)|^2$ are distributions in (momentum) configuration space, Wigner's quasi-probability distribution[15],

$$W_E(p,x) = \frac{1}{2\pi\hbar} \int_{-\infty}^{+\infty} \exp\left(-\frac{i}{\hbar}p\xi\right) \psi_E^*\left(x - \frac{1}{2}\xi\right) \psi_E\left(x + \frac{1}{2}\xi\right) d\xi \qquad (10.23)$$

is a distribution in phase space. Among the many remarkable properties of the function $W_E(p,x)$, the quantity

$$A_E = \frac{1}{\iint_{-\infty} W_E(p,x)^2 dxdp} = h \qquad (10.24)$$

equals the area any quantum-mechanical system in any state E assumes in phase space. This suggests the quantities

$$A_E^{(x)} = \frac{1}{\int_{-\infty}^{+\infty} |\psi_E(x)|^4 dx} \qquad (10.25a)$$

$$A_E^{(p)} = \frac{1}{\int_{-\infty}^{+\infty} |\phi_E(p)|^4 dx} \qquad (10.25b)$$

to be measures of the areas assumed by the system under consideration in state E in (momentum) configuration space. The exploration of this idea is left to Problem 10.15.

Problem 10.15 ($A_E^{(x)}$ for the oscillator*). *Check the suggestion, that $A_E^{(x)}$ (10.25a) is a measure for space occupation, for the linear harmonic oscillator!*

---

[15] Eugene Paul Wigner (Wigner Jenö Pál, 1902 - 1995, Nobel Award 1963), *On the quantum correction for thermodynamic equilibrium*, 1932; I largely follow Schleich, *Quantum Optics in Phase Space*, 2001, Ch. 3, fore pure states.

- Try to calculate $A_E^{(x)}$, or compute it for few values of E!

- Show, that $A_E^{(x)}$ is a decreasing function of E! *Hint:*[16]

$$\int_{-\infty}^{+\infty} \exp(-c^2 x^2) H_n^4(cx) \mathrm{d}x = \frac{\sqrt{\pi}}{c} 2^{2n} (n!)^2 \sum_{k=0}^{n} \binom{m}{k}^2 \binom{2k}{k} \qquad \textbf{(10.26)}$$

- Argue for a generalization of your result to all classically bound systems!

Problem 10.16 (Bohmian distribution*). *Within Bohmian (quantum) mechanics, a single quantum particle of mass* m *exhibits the 'quantum trajectory' (see Sections 25.10 and 26.1)*

$$\vec{r}(t) = \vec{r}(0) + \int_0^t \vec{v}(t') \mathrm{d}t' \qquad \textbf{(10.27a)}$$

$$\vec{v}(t) = \frac{\hbar}{m} \Im \left(\frac{\nabla \psi(\vec{r},t)}{\psi(\vec{r},t)} \right)_{\vec{r}=\vec{r}(t)} \qquad \textbf{(10.27b)}$$

This allows for the classical-like Bohmian distribution $(\vec{p}(t) = m\vec{v}(t))$[17]

$$B(\vec{r},\vec{p}) = \lim_{N \to \infty} \frac{1}{N} \sum_{n=1}^{N} \delta(\vec{r} - \vec{r}_n(t)) \delta(\vec{p} - \vec{p}_n(t)) \qquad \textbf{(10.28)}$$

There are N independent particles or repeated motions of one particle with different initial conditions. Is this distribution suitable for describing the occupation of phase space?

Problem 10.17 (Limiting function for general Hamiltonians*). *Suppose, that the Hamiltonian,* $H(p,x)$, *is not the sum of kinetic,* $T(p)$, *and potential,* $V(x)$, *energies, and introduce the limiting function in phase space,* $\Phi_E(p,x)$, *as*

$$H_{\mathrm{ncl}}(p,x) = \Phi_E(p,x) H(p,x) \leq E \qquad \textbf{(10.29)}$$

- *Try to relate the function* $\Phi_E(p,x)$ *to Wigner's quasi-probability distribution (10.23)! Hint: Explore the bound* $|W(p,x)| \leq 1/\pi\hbar$[18]*!*

---

[16] Anatolii' Platonovich Prudnikov (*1927), Yurii' Aleksandrovitch Brychkov (*1944) & Oleg Igorevitch Marichev (*1945), *Integrals and Series. Special Functions*, 1983, 2.20.17.6

[17] *Cf* Colomés, Zhen Zhan & Oriols, *Comparing Wigner, Husimi and Bohmian distributions: Which one is a true probability distribution in phase space?*, 2015, eq. (19).

[18] Schleich, *Quantum Optics in Phase Space*, 2001, p. 73

- *Try to relate the function $\Phi_E(p, x)$ to Husimi[19] and Bohmian, eq. (10.28) distributions in phase space[20]!*

10.4. ABSENCE OF 'QUANTUM STATE AT REST'*

The minimum extension of quantum systems in their ground state is always larger than zero: $x_{max,0} > 0$, because

$$\int_{-\infty}^{+\infty} x^2 |\psi_0(x)|^2 dx > 0 \tag{10.30}$$

Accordingly, even the ground state exhibits a *non*-vanishing momentum in the sense, that

$$\int_{-\infty}^{+\infty} p^2 |\phi_0(p)|^2 dp > 0 \tag{10.31}$$

Both facts imply, that a quantum system does *not* exhibit a state at rest in the classical sense.

10.5. TRANSITIONS BETWEEN STATIONARY STATES: WHAT ARE 'QUANTUM LEAPS'

"I. That an atomic system can, and can only, exist permanently in a certain series of states corresponding to a discontinuous series of values for its energy, and that consequently any change of the energy of the system, including emission and absorption of electromagnetic radiation, must take place by a complete transition between two such states."[21]

"Dabei liegt jedoch der Wellenmechanik überhaupt nicht die Vorstellung eines plötzlichen Überganges aus dem einen in den anderen Schwingungszustand zugrunde, sondern das betreffende Partialmoment – wie ich es kurz nennen will – entspringt nach ihr aus dem *Zusammenbestehen* der beiden Eigenschwingungen und dauert solange an, als beide gleichzeitig angeregt sind."[22]

---

[19] Kôdi Husimi (1909 - 2008), *Some formal properties of the density matrix*, 1940

[20] Colomés, Zhen Zhan & Oriols, *Comparing Wigner, Husimi and Bohmian distributions: Which one is a true probability distribution in phase space?*, 2015

[21] Bohr, *On the Quantum Theory of Line-Spectra*, 1918, Pt. I, Sect. 1; 1967, p. 97

[22] Schrödinger, *3rd Commun.*, 1926, p. 465 – Engl.: And yet the imagination of a sudden transition from the one to the other oscillation state is not at the base of wave mechanics. On the contrary, the corresponding, say,

Bohr, the Godfather of the Copenhagen school, tried very hard to convince Schrödinger of the necessity of quantum jumps[23]. But Schrödinger insisted on the "leap" in a 'quantum leap' to concern only the *finite* difference between the initial and final values of, say, the energy, while the transition itself proceeds *continuously*.

What does the experiment say? Consider a two-level atom in a multi-mode cavity. The initial state be that the atom is in the excited state and that there is no radiation field. The atom spontaneously emits radiation. If this radiation returns to the atom before it has reached the ground state, the transition is reversed.[24] This experimental finding supports Schrödinger's rather than Bohr's view.

Figuratively, the excitation from a lower to a higher stationary quantum state looks like the climbing of staircases, where the second feet takes off the lower stage only after the first feet has trodden the higher stage.

This arguing brings us directly to the next big task, the axiomatic foundation of the motion of quantum systems.

---

partial moment emanates from the *common existence* of both eigen oscillations and continue as long as both are simultaneously excited. (accentuation as in the original text) – For more details, see Schrödinger, *Energieaustausch nach der Wellenmechanik*, 1927.

[23] After Klein, *Introduction, loc. cit.*, and Straumann, *Schrödingers Entdeckung der Wellenmechanik, loc. cit.*, pp. 22f.

[24] M. Ligare & R. Olivieri, *The calculated photon: Visualization of a quantum field*, 2002, Fig. **5**. See also M. Komma, *Moderne Physik mit Maple* 1996/1998, in particular, the animated figures on mikomma. de/fh/embuch.html.

Part III

Quantum Dynamics

Remarkably enough, all the results above follow alone from the most general principles of stationary-state description according to Newton, Leibniz, Euler and Helmholtz as well as from Schrödinger's requirements, *without* making any additional assumptions like Planck's quantum of action, de Broglie's matter waves, or assuming particular boundary conditions. This part will show that – contrary to Bohr's 1913 claim quoted above – the power of those principles extends even to the dynamics of quantum systems. The time-dependent Schrödinger equation will be *derived*, and this by means of *Eulerian* principles of stationary-state change for quantum systems.

CHAPTER 11

Conservation and Change of Stationary States

Abstract: The time-independent Schrödinger equation represents a stationary-state equation. The stationary wave functions obtained so far are time independent. Their time-dependence is obtained by means of rather general arguments. Then, stationary-state functions are found. The next step in Newton's and Euler's representations of classical mechanics is (to derive) the equation of change of stationary-states. Here, Euler's principles of stationary-state change are generalized to quantum-mechanical systems. This enables us to derive the quantum-mechanical equation of change of stationary-states. The time-independent Schrödinger equation, *i.e*, the equation of motion will follow in the next chapter.

Keywords: Equation of change of stationary-states, Euler's principles of stationary-state change, Stationary-state functions, Stationary wave function, Time-independent Schrödinger equation.

While Heisenberg in 1925 and Schrödinger in 1926 started from a time-dependent classical equation, I have worked so far with the set of all possible (momentum) configurations of a system in its stationary states, where time plays no role. In order to incorporate time, I will proceed along the Newton-Eulerian steps that proved to be so successful in the classical case:

Conservation of stationary states – changes of stationary states – motion.

This will lead us – in the next chapter – to the time-dependent Schrödinger equation as equation of motion in an *axiomatic* manner, again.

11.1. TIME-DEPENDENCE OF STATIONARY WAVE FUNCTIONS

"What, then, is time? If no one ask of me, I know; if I wish to explain to him who asks, I know not. Yet I say with confidence, that I know that if nothing passed away, there would not be past time; and if nothing were coming, there would not be future time; and if nothing were, there would not be present time. Those two times, therefore, past and future, how are they, when even the past now is not; and the future is not as yet? But should the present be always present, and should it not pass into time past, time truly it could not be, but eternity. If, then, time present – if it be time – only comes into existence because it passes into time past, how do we say that even this is, whose cause of being is that it shall not be – namely, so that we cannot truly say that time is, unless because it tends not to be?"[1]

11.1.1. General Considerations

If there is any time-dependence in the stationary states, it is hidden in the stationary-state equation (7.15). E, $V(x)$ and $T(x)$ are time independent, hence, $F_E(x)$ and $G_E(p)$ are, too. Therefore, at most $f_E(x)$ and $g_E(p)$ can be generalized to time-dependent functions: $f_E(x,t)$ and $g_E(p,t)$, where $|f_E(x,t)| = \sqrt{F_E(x)}$ and $|g_E(p,t)| = \sqrt{G_E(p)}$ are time independent. The time dependence of $f_E(x,t)$ and $g_E(p,t)$ is hence described by a multiplicative factor with unit modulus.

Furthermore, the Fourier transform between $f_E(x)$ and $g_E(p)$ applies to $f_E(x,t)$ and $g_E(p,t)$ as well. Hence, that time-dependent factor is the same for both functions.

$$f_E(x,t) = f_E(x)\theta_E(t); \quad g_E(p,t) = g_E(p)\theta_E(t) \tag{11.1}$$

$$\theta_E(t) = \exp\{i\chi_E(t/t_E)\} \tag{11.2}$$

χ_E is a phase to be determined next. Since it is dimensionless, there is a reference time, t_E, which is expected to depend on energy, E.

This time dependence is not related to a new quantization problem, so that there is no reason for a new quantum constant, say, h'. Then, $t_E = \varsigma h/E$ ($\varsigma = const$) is the only system-independent form of a reference time (see also Problem 11.1 on p. 144).

11.1.2. Free Particles

For free particles, we have from eqs. (7.13)

$$g_E(p,t) = \sqrt{\delta\left(\frac{p}{p_E} - 1\right)}\, e^{-i\omega_E t}; \quad \omega_E \equiv -\frac{1}{t}\chi_E\left(\frac{\varsigma E}{h}t\right) \tag{11.3a}$$

$$f_E(x,t) = \iota e^{ik_E x - i\omega_E t}; \quad k_E \equiv p_E/\hbar \tag{11.3b}$$

---

[1] Augustinus Aurelius (354 - 430), *Confessiones*, XI, 14

The group velocity, v_g, equals the (time-independent) particle velocity, $v_p = p_E/m = \hbar k_E/m$.

$$v_g = \frac{d\omega_E}{dk_E} = -\frac{1}{t}\chi'\frac{\varsigma t}{h}\frac{dE}{dk_E} = -\chi'\frac{\varsigma}{h}\frac{\hbar^2 k_E}{m} = v_p = \frac{\hbar k_E}{m} \tag{11.4}$$

(χ' being the derivative of χ). Hence, $\chi = -Et/\hbar$, and

$$f_E(x,t) = f_E(x)e^{-iEt/\hbar}; \quad g_E(p,t) = g_E(p)e^{-iEt/\hbar} \tag{11.5}$$

Problem 11.1 (ς). *Reason, that ς is system independent!*

Problem 11.2 (Classical equals quantum temporal phase factor*). *Show, that the phase factor, $exp\{-iEt/\hbar\}$, agrees with that in the variable $\tilde{x}$ in eq. (9.8)! How can this fact be used to simplify the arguing above?*

Problem 11.3 (Quantum spatial and temporal behavior*). *Comment on the fact, that one and the same universal constant, $\hbar$, governs both the non-classical extension in phase space and the non-classical temporal behavior!*

11.1.3. Time-dependent Schrödinger Equation. I

As a consequence, the (still stationary) functions $f_E(x,t)$ and $g_E(p,t)$ obey the time-dependent Schrödinger equation in position and momentum representation, respectively (see Section 12.1).

$$i\hbar\frac{\partial f_E(x,t)}{\partial t} = \hat{H}_0(x)f_E(x,t) \tag{11.6a}$$

$$i\hbar\frac{\partial g_E(p,t)}{\partial t} = \hat{H}_0(p)g_E(p,t) \tag{11.6b}$$

where $\hat{H}_0(x)f_E(x,t) = Ef_E(x,t)$ and $\hat{H}_0(p)g_E(p,t) = Eg_E(p,t)$

I thus identify,

$$\psi_E(x,t) = \frac{1}{\sqrt{x_E}}f_E\left(\frac{x}{x_E},t\right) = \psi_E(x)e^{-iEt/\hbar} \tag{11.7a}$$

$$\phi_E(p,t) = \frac{1}{\sqrt{p_E}}g_E\left(\frac{p}{p_E},t\right) = \phi_E(p)e^{-iEt/\hbar} \tag{11.7b}$$

as Schrödinger's stationary time-dependent wave functions. The obey the time-dependent Schrödinger equations, too.

$$i\hbar \frac{\partial \psi_E(x,t)}{\partial t} = \hat{H}_0(x)\psi_E(x,t) \qquad \text{(11.8a)}$$

$$i\hbar \frac{\partial \phi_E(p,t)}{\partial t} = \hat{H}_0(p)\phi_E(p,t) \qquad \text{(11.8b)}$$

For later use in Section 12.1, I write down the operator form of this time-dependence as

$$\psi_E(x,t_2) = \hat{U}_0(x;t_2,t_1)\psi_E(x,t_1); \ \hat{U}_0(x;t_2,t_1) = e^{-(i/\hbar)\hat{H}_0(x)(t_2-t_1)} \qquad \text{(11.9)}$$

$\hat{U}_0$ is a (unitary) time-development operator.[2] This operator obeys the time-dependent Schrödinger equation, too.

$$i\hbar \frac{\partial \hat{U}_0(x;t_2,t_1)}{\partial t_2} = \hat{H}_0(x)\hat{U}_0(x;t_2,t_1) \qquad \text{(11.10)}$$

I finish this preparation of the non-stationary case in Section 12.1 with the formula

$$d\hat{U}_0(x;t,t) = \frac{1}{i\hbar}\hat{H}_0(x)\hat{U}_0(x;t,t)dt = \frac{1}{i\hbar}\hat{H}_0(x)dt \qquad \text{(11.11)}$$

by virtue of formula (11.9) for $\hat{U}_0(x;t,t)$

However, at this stage, the time-dependent Schrödinger equations represent merely kinematic, not yet dynamical relations, not yet general equations of motion. This means, that it would not be justified to simply carry over them to the non-stationary case, in which the Hamiltonian explicitly depends on time. A dynamical foundation of them will be prepared in the following sections and presented in the next chapter.

Problem 11.4 (Evolution equation). *Reason, that the time-dependent stationary wave functions obey a differential equation of first order in time! Can this fact be used to simplify the arguing above?*

11.1.4. Free Wave Packets

By virtue of the time-dependent Schrödinger equations (11.8), the wave functions (11.7) become

---

[2] For an exact treatment of operators like $\exp(-i\hat{H}t/\hbar)$, see Stone's theorem: Marshall Harvey Stone (1903 - 1989), *Linear Transformations in Hilbert Space. III. Operational Methods and Group Theory*, 1930; *On one-parameter unitary groups in Hilbert Space*, 1932; J. v. Neumann, *Über einen Satz von Herrn M. H. Stone*, 1932 (cited after en.wikipedia.org/wiki/Stone%27s_theorem_on_one-parameter_unitary_groups).

$$\phi_k(p,t) = \frac{1}{\sqrt{p_k}}\sqrt{\delta\left(\frac{p}{p_k}-1\right)}\,e^{-i\omega_k t} \tag{11.12a}$$

$$\psi_k(x,t) = \frac{\iota}{\sqrt{x_k}}\,e^{ikx-i\omega_k t};\quad \hbar\omega_k = \frac{\hbar^2 k^2}{2m} = E \tag{11.12b}$$

where ι is the infinitesimal number introduced in formula (7.9). x_k and p_k are the analogues to x_E and p_E, respectively, with $x_k p_k = \hbar$.

Now, in contrast to the stationary Schrödinger equation, the time-dependent Schrödinger equation (11.8a) is linear in $\psi_E(x,t)$, and any linear combination of its solutions is again a solution. For vanishing potential: $V(x) \equiv 0$, this allows to superpose wave functions of the form (11.12b) to *wave packets* as

$$\psi_{\text{wp}}(x,t) = \frac{1}{\sqrt{2\pi}}\int_0^{+\infty} \varpi\left(\frac{k}{\Delta k}\right)\frac{1}{\sqrt{x_k}}\,e^{ikx-i\omega_k t}\,\frac{dk}{\Delta k} \tag{11.13}$$

Here, the dimensionless function $\varpi(k/\Delta k)$ describes the profile of the wave packet, while Δk is a measure of its width.

$\psi_{\text{wp}}(x,t)$ is normalized over the whole real axis, if

$$\int_{-\infty}^{+\infty}\left|\psi_{\text{wp}}(x,t)\right|^2 dx = \int_0^{+\infty}\frac{1}{x_k\Delta k}\left|\varpi\left(\frac{k}{\Delta k}\right)\right|^2\frac{dk}{\Delta k} = 1 \tag{11.14}$$

An example is the Gaussian profile ($x_k = 1/k$)

$$\varpi_{\text{Gaussian}}\left(\frac{k}{\Delta k}\right) = \frac{1}{\sqrt{N}}\exp\left(-\frac{(k-k_0)^2}{2(\Delta k)^2}\right) \tag{11.15a}$$

$$N = \frac{1}{2}\exp\left(-\frac{k_0^2}{(\Delta k)^2}\right) + \frac{\sqrt{\pi}}{2}\frac{k_0}{\Delta k}\,erfc\left(-\frac{k_0}{\Delta k}\right) \tag{11.15b}$$

For wave packets of finite width: $\Delta k > 0$, the infinitesimal number ι does no longer occur.

Problem 11.5 (Normalization of a wave packet). *Prove formulae (11.14) and (11.15)!*

Problem 11.6 (Energy of a wave packet*). *In his seminal paper about time-dependent tunneling (see Section 14.1), Hartman has chosen for the incident wave the not-normalized Gaussian wave packet*[3]

$$\varphi_1(x,t) = \frac{1}{\Delta k\sqrt{2\pi}} \int_0^{+\infty} \exp\left[-\frac{(k-k_0)^2}{2(\Delta k)^2}\right] \exp\left[ik(x+x_0) - i\frac{\hbar k^2}{2m}t\right] dk \quad \textbf{(11.16)}$$

He called $E(k) = \hbar^2 k^2/2m$ "the total energy" (*loc. cit.*, p. 3429, immediately after eq. (6); his k_1 is our k). Is this justified?

11.2. STATIONARY-STATE FUNCTIONS

11.2.1. $< \psi|\hat{H}|\psi >$ as Quantum Stationary-State Function

Since the Hamiltonian is a suitable classical (stationary-)state function, it is certainly reasonable to exploit the non-classical representation (7.15) of the total energy, $E = < \hat{H} >$, as quantum stationary-state function. As in the classical case, the 'external causes' be given through $\partial V_{\text{ext}}(x,t)/\partial t$ and $\partial T_{\text{ext}}(p,t)/\partial t$; if V_{ext} (T_{ext}) is time-independent, it should be absorbed into $V(x)$ ($T(p)$), again. The corresponding non-classical stationary-state function reads

$$Z_{\langle \hat{H} \rangle}(t) = \frac{\langle \psi(x,t)|V(x)+V_{\text{ext}}(x,t)|\psi(x,t)\rangle}{\langle \psi(x,t)|\psi(x,t)\rangle} + \frac{\langle \phi(p,t)|T(p)+T_{\text{ext}}(p,t)|\phi(p,t)\rangle}{\langle \phi(p,t)|\phi(p,t)\rangle} \quad \textbf{(11.17a)}$$

$$= \frac{\langle \psi(x,t)|\hat{H}(x,t)|\psi(x,t)\rangle}{\langle \psi(x,t)|\psi(x,t)\rangle}; \quad \hat{H}(x,t) = \hat{H}_0(x) + V_{\text{ext}}(x,t) + \hat{T}_{\text{ext}}(x,t) \quad \textbf{(11.17b)}$$

$$= \frac{\langle \phi(p,t)|\hat{H}(p,t)|\phi(p,t)\rangle}{\langle \phi(p,t)|\phi(p,t)\rangle}; \quad \hat{H}(p,t) = \hat{H}_0(p) + \hat{V}_{\text{ext}}(p,t) + T_{\text{ext}}(p,t) \quad \textbf{(11.17c)}$$

Here, $\psi(x,t)$ is the *general* time-dependent wave function in configuration space, $\{x\}$, while $\phi(p,t)$ is the same in momentum configuration space, $\{p\}$. The energy, E, is no longer a characteristic parameter of the system and hence not indicated. For continuity reasons, $\psi(x,t)$ and $\phi(p,t)$ are Fourier transforms each of another in the same manner as in the stationary case, in which $\psi(x,t) = \psi_E(x,t)$, $\phi(p,t) = \phi_E(p,t)$, and $Z_{\langle \hat{H} \rangle}(t) = E = const.$ At this stage, the expressions $\langle \psi(x,t)|\psi(x,t)\rangle$ and $\langle \phi(p,t)|\phi(p,t)\rangle$ cannot yet be set equal to one, however.

---

[3] Thomas E. Hartman, *Tunneling of a Wave Packet*, 1962, eqs. (4), (6), (7)

11.2.2. $|\psi(x,t)|^2$ and $|\phi(p,t)|^2$ as Quantum Stationary-State Functions

Moreover, the modulus-squared stationary wave functions, $|\psi_E(x,t)|^2$ and $|\phi_E(p,t)|^2$ are actually time-*in*dependent. This suggests

$$Z_\psi(t) = |\psi(x,t)|^2 \text{ and } Z_\phi(t) = |\phi(x,t)|^2 \qquad (11.18)$$

to be non-classical stationary-state functions, too.

This role fortifies the relevance of the weight functions, $F_E(x) = x_E^2 |\psi_E(x)|^2$ and $G_E(p) = p_E^2 |\phi_E(p)|^2$. Moreover, it suggest to write the wave function in configuration space in the polar form[4]

$$\psi(x,t) = R(x,t)\exp\left\{\frac{i}{\hbar}S(x,t)\right\} \qquad (11.19)$$

Here, $R(x,t)$ and $S(x,t)$ are real-valued functions we will meet later on, again. Here, I only notice, that the quantity

$$Z_\psi(t) = |\psi(x,t)|^2 = R(x,t)^2 \qquad (11.20)$$

is independent of the function $S(x,t)$.

11.3. EULERIAN PRINCIPLES OF STATIONARY-STATE CHANGE FOR QUANTUM SYSTEMS

> "The problem of the action of time-dependent external forces can be regarded as a limiting case of the interaction between two systems in which the influence of the interaction on one of the two systems…is so small that the action upon the other system…remains unaffected by this influence."[5]

It is straightforward to translate the Eulerian principles of stationary-state change for classical systems, CS1…CS3, in Subsection 2.5.4 to the *non*-classical stationary-state functions, $Z_{\hat{H}}$ and Z_ψ, Z_ϕ, and non-stationary-state functions, ψ, $\bar{\psi}$ and ϕ, $\bar{\phi}$.

Up to first order in dt,

---

[4] Erwin Madelung (1881 - 1972), *Eine anschauliche Deutung der Gleichung von Schrödinger*, 1926; *Quantentheorie in hydrodynamischer Form*, 1927
[5] Born, Heisenberg & Jordan, *Über Quantenmechanik II*, 1926 ('Three-men paper'), Sect. 5

QS1 the changes of the stationary-state quantities $(Z_{\hat{H}}, Z_\psi, Z_\phi)$ depend solely on external causes $(\partial \hat{H}_{ext} / \partial t)$, but not on internal causes $(\hat{H}_0)$;

QS2 the changes of the stationary-state quantities $(dZ_{\hat{H}}, dZ_\psi, dZ_\phi)$ are independent of the stationary-state quantities $(Z_{\hat{H}}, Z_\psi, Z_\phi)$ themselves;

QS3 the changes of non-stationary-state quantities $(d\psi, d\bar{\psi})$ depend directly solely on stationary-state quantities $(Z_{\hat{H}}, Z_\psi, Z_\phi)$; the external causes $(\partial \hat{H}_{ext} / \partial t)$ affect non-stationary-state quantities only indirectly, *viz*, *via* changes of stationary-state quantities.

Problem 11.7. *Show that the statement by Born, Heisenberg & Jordan quoted at the beginning of this section applies to both classical and quantum systems!*

11.4. EQUATIONS-OF-STATIONARY-STATE-CHANGE

The stationary-state functions (11.18) change as

$$dZ_\psi = 2\Re(\overline{\psi}d\psi); \quad dZ_\phi = 2\Re(\overline{\phi}d\phi) \tag{11.21}$$

According to Eulerian principle QS3, both changes should vanish to first order.

Furthermore, the stationary-state function (11.17b) changes as

$$dZ_{<\hat{H}>} \equiv d\frac{<\psi|\hat{H}|\psi>}{<\psi|\psi>} \tag{11.22a}$$

$$= \frac{<d\psi|\hat{H}|\psi>+<\psi|d\hat{H}|\psi>+<\psi|\hat{H}|d\psi>}{<\psi|\psi>} \tag{11.22b}$$

$$- \frac{<d\psi|\psi>+<\psi|d\psi>}{<\psi|\psi>^2} <\psi|\hat{H}|\psi> \tag{11.22c}$$

$$\overset{!}{=} \frac{<\psi|d\hat{H}|\psi>}{<\psi|\psi>} = \frac{<\psi|\frac{\partial}{\partial t}\hat{H}|\psi>}{<\psi|\psi>}dt \tag{11.22d}$$

The fourth line is requested by the Eulerian principles QS1 and QS2. It holds true, if

$$(<d\psi|\hat{H}|\psi>+<\psi|\hat{H}|d\psi>)<\psi|\psi>$$

$$= (<d\psi|\psi>+<\psi|d\psi>)<\psi|\hat{H}|\psi> \tag{11.23}$$

Now, before the external, time-dependent influence ('perturbation') sets in, the system be in a stationary state, say, E. This means, that wave function and Hamiltonian change as

$$\psi = \psi_E + d\psi;\ \hat{H} = \hat{H}_0 + d\hat{H} \tag{11.24}$$

with

$$\hat{H}_0 \psi_E = E\psi_E \tag{11.25}$$

Consequently, the condition (11.23) is identically fulfilled as

$$(< d\psi|E\psi_E > +< \psi_E|Ed\psi >) < \psi_E|\psi_E >$$

$$\equiv (< d\psi|\psi_E > +< \psi_E|d\psi >) < \psi_E|E\psi_E > \tag{11.26}$$

In the next chapter, this limitation to $\hat{H}(x, 0) = \hat{H}_0(x)$ will be lifted. Instead of infinitesimal changes: $d\psi = \psi - \psi_E$ and $d\hat{H} = \hat{H} - \hat{H}_0$, general changes of the wave function, $\psi(x, t)$, for general Hamiltonians, $\hat{H}(x, t)$, will be considered.

11.5. EXAMPLE: HELLMANN-FEYNMAN THEOREM

Let us assume, that a time-independent Hamiltonian and thus stationary wave function depends on some parameter λ: $\hat{H}_0 = \hat{H}_0(\lambda)$, $\psi_E = \psi_E(\lambda)$. In this case, the Eulerian principles of stationary-state change for quantum systems, QS1…QS3 in Section 11.3, imply that

$$d\langle\psi_E(\lambda)|\hat{H}_0(\lambda)|\psi_E(\lambda)\rangle = \langle\psi_E(\lambda)|d\hat{H}_0(\lambda)|\psi_E(\lambda)\rangle \tag{11.27}$$

Hence,

$$\frac{dE(\lambda)}{d\lambda} = \left\langle\psi_E(\lambda)\left|\frac{d\hat{H}(\lambda)}{d\lambda}\right|\psi_E(\lambda)\right\rangle \tag{11.28}$$

This is the Hellmann-Feynman theorem[6] being most useful within quantum chemistry. It thus corroborates our generalization of the Eulerian principles of stationary-state changes for classical bodies to quantum systems. I will return to it in Subsection 16.6.6.

---

[6] Paul Güttinger, *Das Verhalten von Atomen im magnetischen Drehfeld*, 1932; Pauli, *Prinzipien der Wellenmechanik*, 1933, p. 162; Hans Gustav Adolf Hellmann (1903 - 1938), *Einführung in die Quantenchemie*, 1937, p. 285; Feynman, *Forces in molecules*, 1939 – see also the criticism of the standard proof in David Carfi, *The Pointwise Hellmann-Feynman Theorem*, 2010 (all cited after http://el.science.wikia.com/wiki/ Θεώρημα_Hellmann-Feynman).

CHAPTER 12

Equation of Motion

Abstract: Having got the quantum-mechanical equation of change of stationary states, it is rather straightforward to derive the quantum-mechanical equation of motion. As anticipated in the foregoing chapter, this is the time-dependent Schrödinger equation. Moreover, Huygens' principle is shown to apply to QM, as obtained by Feynman in another way.

Keywords: Chapman-Kolmogorov equation, Feynman, Huygens' principle, quantum-mechanical equation of motion, Time-dependent wave function, Time-dependent Schrödinger equation.

12.1. TIME-DEPENDENT SCHRÖDINGER EQUATION. II

"Sobald aber V [the potential energy] die Zeit explizite enthält, ist es offenbar unmöglich, der Gleichung (1) bzw. (1') zu genügen durch eine Funktion ψ, welche nur nach (2) von der Zeit abhängt."[1]

While an equation-of-stationary-state-change describes an infinitesimal change of a stationary-state function, an equation of motion calculates finite changes of non-stationary state quantities, here, of the wave functions.

For this, I express the time development of the wave function in the same manner as in the stationary case, see eq. (11.9).

$$\psi(x, t_2) = \widehat{U}(x; t_2, t_1)\psi(x, t_1) \tag{12.1}$$

$\widehat{U}(x; t_2, t_1)$ is called the time-development, or evolution operator. Its definition (12.1) immediately implies the following properties.

$$\widehat{U}(x; t, t) = 1 \tag{12.2a}$$

---

[1] Schrödinger, *4th Commun.*, p. 110. – Engl.: But as soon as V contains explicitly the time, it is obviously *impossible* to obey equation

$$\Delta\psi - \frac{2(E-V)}{E^2}\frac{\partial^2\psi}{\partial t^2} = 0 \tag{1}$$

resp.

$$\Delta\psi + \frac{8\pi^2}{h^2}(E - V)\psi = 0 \tag{1'}$$

through a function ψ, which depends on time solely through

$$\psi \sim P \cdot R \cdot \exp\left\{\pm\frac{2\pi i E t}{h}\right\} \tag{2}$$

$$\hat{U}(x; t_2, t_1)^{-1} = \hat{U}(x; t_1, t_2) \tag{12.2b}$$

$$\hat{U}(x; t_2, t_1)\hat{U}(x; t_1, t_0) = \hat{U}(x; t_2, t_0) \tag{12.2c}$$

Eq. (12.2c) is a Chapman-Kolmogorov equation and thus indicates the applicability of Huygens' principle, see next section.

Then,

$$d\psi(x, t) = d\hat{U}(x; t, t)\psi(x, t); \quad d\hat{U}(x; t, t) \equiv \frac{\partial \hat{U}(x; t', t)}{\partial t'}|_{t'=t}dt \tag{12.3}$$

and the condition (11.23) becomes

$$(< d\hat{U}\psi|\hat{H}|\psi > +< \psi|\hat{H}|d\hat{U}\psi >) < \psi|\psi >$$

$$= (< d\hat{U}\psi|\psi > +< \psi|d\hat{U}\psi >) < \psi|\hat{H}|\psi > \tag{12.4}$$

or,

$$< d\psi|\hat{U}^+\hat{H} + \hat{H}d\hat{U}|\psi >< \psi|\psi >=< \psi|d\hat{U}^+ + d\hat{U}|\psi >< \psi|\hat{H}|\psi > \tag{12.5}$$

$d\hat{U}^+$ is the adjoint operator to $d\hat{U}$; the Hamiltonian being self-adjoint: $\hat{H}^+ = \hat{H}$. The solution to that equation reads

$$d\hat{U} = s(\hat{H})dt \tag{12.6}$$

where $s(\hat{H})$ is an imaginary-valued entire function.

Compatibility with the stationary case, eqs. (11.11) and (11.10), implies $s(\hat{H}) = 1/i\hbar\hat{H}(x, t)$ and the equations

$$d\hat{U}(x; t, t) = \frac{1}{i\hbar}\hat{H}(x, t)dt = \frac{1}{i\hbar}\hat{H}(x, t)\hat{U}(x; t, t)dt \tag{12.7a}$$

$$i\hbar\frac{\partial \hat{U}(x; t_2, t_1)}{\partial t_2} = \hat{H}(x, t_2)\hat{U}(x; t_2, t_1) \tag{12.7b}$$

It is not too surprising, that $\hat{U}$ obeys the same equation of motion as $\hat{U}_0$ does, *viz*, the time-dependent Schrödinger equation. The initial condition is the same, too: $\hat{U}(x; t, t) = 1$.

Notice, that, if $s(\widehat{H})$ were not a linear function of $\widehat{H}$, there would be a universal constant of dimension 'energy'. Then, all energy spectra should be discretized in terms of that constant (like spin and orbital angular momenta being discretized in terms of $\hbar$). This is, obviously, not the case.

For finite time steps, we can use the Chapman-Kolmogorov equation (12.2c) to obtain a product integral after Volterra.[2]

$$\widehat{U}(x;t,0) = \lim_{\delta t \to 0} \widehat{U}(x;t,t-\delta t)\widehat{U}(x;t-\delta t,t-2\delta t)\cdots\widehat{U}(x;\delta t,0) \qquad \text{(12.8a)}$$

$$= \lim_{\delta t \to 0}\left(1-\frac{i}{\hbar}\widehat{H}(x,t-\delta t)\delta t\right)\cdots\left(1-\frac{i}{\hbar}\widehat{H}(x,0)\delta t\right) \qquad \text{(12.8b)}$$

$$= \prod_{\tau=0}^{t}\left(1-\frac{i}{\hbar}\widehat{H}(x,\tau)d\tau\right) \qquad \text{(12.8c)}$$

Feynman has used such a 'time slicing' for the construction of his path integral, see next section.

Eq. (12.7b) for $\widehat{U}$ is equivalent to the (special) linear Volterra integral equation of second kind[3],

$$\widehat{U}(x;t_2,t_1) = 1 - \frac{i}{\hbar}\int_{t_1}^{t_2}\widehat{H}(x,t)\widehat{U}(x;t,t_1)dt \qquad \text{(12.9)}$$

Iteration w.r.t. powers of $\widehat{H}$ yields the Dyson series

$$\widehat{U}(x;t_2,t_1) = 1 + \left(\frac{-i}{\hbar}\right)\int_{t_1}^{t_2}\widehat{H}(x,t)dt + \left(\frac{-i}{\hbar}\right)^2\int_{t_1}^{t_2}\widehat{H}(x,t)\int_{t_1}^{t}\widehat{H}(x,t')dt'dt +$$
$$\cdots \qquad \text{(12.10a)}$$

$$= 1 + \left(\frac{-i}{\hbar}\right)\int_{t_1}^{t_2}\widehat{H}(x,t)dt + \left(\frac{-i}{\hbar}\right)^2\int_{t_1}^{t_2}\int_{t_1}^{t_2}\mathcal{T}\widehat{H}(x,t)\widehat{H}(x,t')dt'dt + \cdots \quad \text{(12.10b)}$$

$$= \mathcal{T}\exp\left(\frac{-i}{\hbar}\int_{t_1}^{t_2}\widehat{H}(x,t)dt\right) \qquad \text{(12.10c)}$$

Here, $\mathcal{T}$ denotes Dyson's time-ordering operator.[4] If $\widehat{H}$ does not depend on time, formula (12.10c) simplifies to formula (11.9).

---

[2] https://de.wikipedia.org/wiki/Zeitentwicklungsoperator (15.09.2016)
[3] Vito Volterra (1860 - 1940), *Sulla inversione degli integrali definiti*, 1896; *Sopra alcune questioni di inversione di integrali definiti*, 1897; Traian Lalescu (1882 - 1929), *Introduction à la théorie des équations intégrales*, 1912; see also Volterra, *Leçons sur les équations intégrales et les équations intégro-différentielles*, 1913.
[4] *Cf* Freeman John Dyson (*1923), *The Radiation Theories of Tomonaga, Schwinger, and Feynman*, 1949, expression (29).

$\widehat{U}(x; t', t)$ is a unitary time-evolution operator, too, since

$$\widehat{U}^{+}(x; t', t) = \widehat{U}^{-1}(x; t', t) = \widehat{U}(x; t, t') \tag{12.11}$$

This solution immediately leads to the famous time-dependent Schrödinger equation for $\psi(x, t)$.

$$i\hbar \frac{\partial \psi(x,t)}{\partial t} = \widehat{H}(x, t)\psi(x, t) \tag{12.12a}$$

For a remarkable appraisal of it, see Fig. (**12.1**)[5].

In the momentum representation, it reads

$$i\hbar \frac{\partial \phi(p,t)}{\partial t} = \widehat{H}(p, t)\phi(p, t) \tag{12.12b}$$

Both representations are equivalent *equations of motion*.

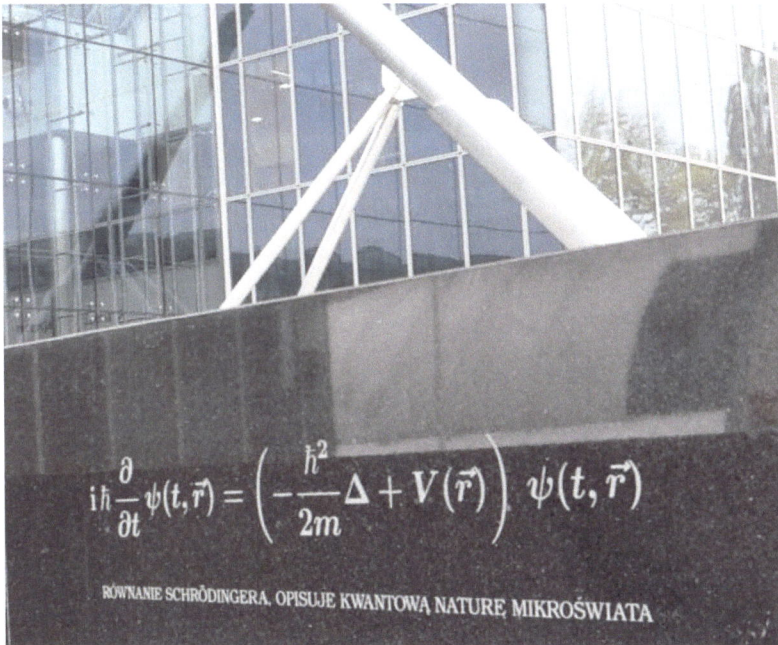

$$i\hbar \frac{\partial}{\partial t}\psi(t, \vec{r}) = \left(-\frac{\hbar^2}{2m}\Delta + V(\vec{r})\right)\psi(t, \vec{r})$$

RÓWNANIE SCHRÖDINGERA, OPISUJE KWANTOWĄ NATURĘ MIKROŚWIATA

Fig. 12.1. "Schrödinger's equation describing the quantum nature of the microworld" – Center of New Technologies, University of Warsaw, Poland

---

[5] Detail of a photo by Halibutt / Wikimedia Commons / CC BY-SA 3.0, https://commons.wikimedia.org/w/index.php?curid=44380786.

As in the stationary case, the homogeneity in ψ corresponds to the fact, that only the relative values of $\psi(x,t)$ have got a physical meaning, see eq. (11.17a).

Problem 12.1. *(i), Show, that* $<\psi|\psi>$ *and* $<\phi|\phi>$ *are time independent! Remark: We will return to this issue in Section 13.1. (ii), Show, that the equations-of-state-change (11.21) are fulfilled!*

Problem 12.2. *[*] Rewrite the time-dependent Schrödinger equation (2.19) in terms of* $\mathfrak{R}\psi$ *and* $\mathfrak{I}\psi$*! What is the physical reason of the occurrence of the matrix* $\hat{\Omega}_2$ *defined in the classical eq. (2.45)?*

Problem 12.3 (*). *Can this treatment be improved by means of Stones theorem[6]?*

12.2. HUYGENS' PRINCIPLE. II*

Applying formula (12.11) twice one can write

$$\psi(x,t_2) = \hat{U}(x;t_2,t_1)\psi(x,t_1)$$

$$= \hat{U}(x;t_2,t_1)\hat{U}(x;t_1,t_0)\psi(x,t_0)$$

$$= \hat{U}(x;t_2,t_0)\psi(x,t_0) \tag{12.13}$$

Hence, for each fixed value of x, the set $\{\hat{U}(x;t_2,t_1)| -\infty < t_2, t_1 < +\infty\}$ forms a group. The unity element is $\hat{U}(x;t,t) = \hat{1}$, the inverse element of $\hat{U}(x;t_2,t_1)$ equals $\hat{U}(x;t_1,t_2)$.[7]

Eq. (12.13) is another version of the Chapman-Kolmogorov equation, *cf* eq. (9.7). This means, that Huygens' principle applies to Schrödinger wave mechanics, too.

The same conclusion has been reached by Feynman within his original path-integral approach[8]. He obtains for the wave function the expression

---

[6] Stone, *Linear Transformations in Hilbert Space. III. Operational Methods and Group Theory*, 1930; *On one-parameter unitary groups in Hilbert space*, 1932

[7] *Cf* also Pauli, *Die allgemeinen Prinzipien der Wellenmechanik*, 1933, § I.8.

[8] Feynman, *Space-Time Approach to Non-Relativistic Quantum Mechanics*, 1948; see also Dirac, *The Lagrangian in Quantum Mechanics*, 1933, eqs. (10), (11).

$$\psi(x_{k+1}, t + \epsilon) = \int \exp\left[\frac{i}{\hbar} S(x_{k+1}, x_k)\right] \psi(x_k, t)\frac{dx_k}{A} \qquad \textbf{(12.14)}$$

(*ibid.*, eq. (18)), ε being infinitesimal, S the classical action (to be provided by CM (!) – *ibid.*, p. 372), x_k and x_{k+1} intermediate position variables, A a normalization factor. The wave function at position x_{k+1} and time $t + \varepsilon$ is calculated from the wave functions at time t on the 'surface' x_k.

In view of the fundamental role of Huygens' principle for transport and propagation processes, let me step aside to remark the following. Kirchhoff's diffraction formula[9] is mathematically correct, but hides the physical essence of the Chapman-Kolmogorov equation.

As a matter of fact, the Green's function[10] of the wave equation does *not* obey the Chapman-Kolmogorov equation. This fact has seduced Feynman to state, that "Huygens' principle is not correct in optics." (*loc. cit.*, p. 377) He correctly observes, "that the wave equation in optics is second order in the time. The wave equation in quantum mechanics is first order in the time". And this is the clue!

Exactly because d'Alembert's wave equation is of second order in time, it is *not* a *fundamental* equation, *cf* Newton's 2nd axiom, eq. (2.6), Euler's principles of state change and the separation of internal and external system parameters in Subsection 9.1.1. Accordingly, that deficiency can be healed through decomposing it into two equations of first order in time. The (matrix) Green's function of that system of equations *does* obey a Chapman-Kolmogorov equation[11].

Problem 12.4 (*). *Feynman's 1948 approach bases on Dirac's 1933 paper* The Lagrangian in Quantum Mechanics *(fn. 1 and p. 378) and Dirac's 1945 paper* On the Analogy Between Classical and Quantum Mechanics *(fn. 2).[12] W.r.t. Huygens' principle, he refers to "the very interesting remarks of Schroedinger" (ibid., fn. 16). There (1926, 2nd Commun., p. 489, fn.), Schrödinger writes, that the wave function, ψ, is "not really related to the action function of a certain motion as in eq. (2) of the 1st Commun.[13] – Whereas the nexus of the wave equation and the*

---

[9] Gustav Robert Kirchhoff (1824 - 1887), *Zur Theorie der Lichtstrahlen*, 1882

[10] George Green (1793 - 1841), *Essay on the application of mathematical analysis on the theories of electricity and magnetism*, 1828; see also Dyson, *George Green and physics*, 1993.

[11] For more details, see Enders, *Huygens' principle as universal model of propagation*, 2009; *Physical, metaphysical and logical thoughts about the wave equation and the symmetry of space-time*, 2011.

[12] *Cf* also Yong Gwan Yi, *Lagrangian Approaches of Dirac and Feynman to Quantum Mechanics*, 2006.

[13] This equation reads

$$S = K \lg \psi \qquad (2)$$

variational problem is, of course, most real: the integrand of the stationary integral is the Lagrangian for the wave process." Does the latter one refer to the *"Hamiltonian integral"*

$$\int \mathrm{d}\tau \left\{ K^2 T \left(q, \frac{\partial \psi}{\partial q} \right) + \psi^2 V \right\} \tag{23}$$

(*1st Commun.*, Addendum during the correction dated 28. II. 1926, p. 376)? *If yes, show it in detail!*

Problem 12.5. *What is the physical meaning of the volume element* $\mathrm{d}\tau$ *in integral (23) of the foregoing problem? Hint: Compare* $\mathrm{d}\tau$ *with the Hertzian line element in eq. (5.3)!*

12.3. LAGRANGIAN FORMULATION

Let us consider the wave function, $\psi(x,t)$, to be the generalized coordinate as in point-mechanics. Which is the generalized momentum, $\pi_\psi(x,t)$? In contrast to CM, there is no momentum from the very beginning.

Although the Schrödinger equation determines at once both the wave function, $\psi(x,t)$, and its complex-conjugated function, $\bar{\psi}(x,t)$, the two functions are linearly independent. This suggests the generalized momentum, $\pi_\psi(x,t)$, to be related to $\bar{\psi}(x,t)$, but how?

At this stage, there are two possibilities.

1. One guesses a Lagrangian, where Lagrange's equation of motion reproduces the time-dependent Schrödinger equation,

2. One guesses a Hamiltonian, where Hamilton's equations of motion reproduce the time-dependent Schrödinger equation.

As a matter of fact, we have got a Hamilton operator, $\hat{H}$, where the space-time dependence of the dynamical variable, $\psi(x,t)$, suggests us to deal with a field theory, see Chapter 20 below. For dimensional reasons I guess the Hamiltonian to equal

---

K is a constant to be introduced for dimensional reasons.

$$H(t) = \int \mathfrak{H}(x,t)\mathrm{d}x; \quad \mathfrak{H}(x,t) = \bar{\psi}(x,t)\hat{H}(x,t)\psi(x,t) \tag{12.15}$$

Accordingly, Hamilton's equations of motion would read

$$\frac{\partial\psi(x,t)}{\partial t} = \frac{1}{i\hbar}\hat{H}(x,t)\psi(x,t) = \frac{\delta H(t)}{\delta\pi_\psi(x,t)} \tag{12.16a}$$

$$\frac{\partial\pi_\psi(x,t)}{\partial t} = -\frac{\delta H(t)}{\delta\psi(x,t)} = -\hat{H}(x,t)\bar{\psi}(x,t) = i\hbar\frac{\partial\bar{\psi}(x,t)}{\partial t} \tag{12.16b}$$

Then, the canonical momentum density equals

$$\pi_\psi(x,t) = i\hbar\bar{\psi}(x,t) \tag{12.17}$$

The Lagrangian density follows from a Legendre transform as

$$\mathfrak{L}(x,t) = \frac{\partial\mathfrak{H}(x,t)}{\partial\pi_\psi(x,t)}\pi_\psi(x,t) - \mathfrak{H}(x,t)$$

$$= \frac{\partial\psi(x,t)}{\partial t}i\hbar\bar{\psi}(x,t) - \bar{\psi}(x,t)\hat{H}(x,t)\psi(x,t) = 0 \tag{12.18}$$

by virtue of the time-dependent Schrödinger equation.

As a consequence, the action vanishes, too.[14]

$$S = \iint \mathfrak{L}(x,t)\,\mathrm{d}x\,\mathrm{d}t = 0 \tag{12.19}$$

Here, the value '0' suggests S to be dimensionless (Ralston, *loc. cit.*, p. 2). However, it actually means $0 \times [action]$.

In contrast, if the wave function, $\psi(x,t)$, is considered to be a *field*, Lagrange's equation of motion reads

$$\frac{\partial\mathfrak{L}}{\partial\psi} - \frac{\partial}{\partial t}\frac{\partial\mathfrak{L}}{\partial\frac{\partial\psi}{\partial t}} - \frac{\partial}{\partial x}\frac{\partial\mathfrak{L}}{\partial\frac{\partial\psi}{\partial x}} = 0 \tag{12.20}$$

The corresponding Lagrangian equals ($\dot{\psi} \equiv \partial\psi/\partial t$)

$$\mathfrak{L}(\psi,\nabla\psi,\dot{\psi}) = \frac{i\hbar}{2}\left(\bar{\psi}\dot{\psi} - \dot{\bar{\psi}}\psi\right) - \frac{\hbar^2}{2m}\nabla\bar{\psi}\nabla\psi - V(\vec{r},t)\bar{\psi}\psi \tag{12.21}$$

---

[14] John P. Ralston, *Quantum Theory without Planck's Constant*, 2012, p. 17

CHAPTER 13

Conservation Laws

Abstract: The conservation law for the weight (probability) is sketched along Schrödinger's (1926) standard way, where $\psi\bar{\psi}$ represents the weight (probability) density. In contrast, there are various expressions for the energy density. I pose the following requirements: (i), The change of total energy should be proportional to $\partial V_{\text{ext}}/\partial t$, as within CM (see principle *CS1* in Subsection 2.5.4); (ii) the current (flux) density should exhibit the hydrodynamic form 'density times velocity'.

Keywords: Conservation of energy, Conservation of probability, Conservation of weight, Eeight flux density, Energy density, Energy flux density, Equation of continuity, Schrödinger, Weight density.

Conservation laws build a crucial element of all branches of physics. They are tightly related to the symmetry of a system (see Part IV) and thus put additional restrictions onto the possible motions of a system, see Subsection 4.1.4. Within our approach, the conservation of weight (probability) and energy are built-in into the theory.

Within point mechanics, a quantity is conserved, if it is time-independent. For instance, this happens for the Hamiltonian, $H(p, x, t)$, if it does not explicitly depend on time.

$$\frac{\mathrm{d}H}{\mathrm{d}t}(p(t), x(t), t)$$

$$= \frac{\partial H}{\partial p}(p(t), x(t), t)\frac{\mathrm{d}p}{\mathrm{d}t}(t) + \frac{\partial H}{\partial x}(p(t), x(t), t)\frac{\mathrm{d}x}{\mathrm{d}t}(t) + \frac{\partial H}{\partial t}(p(t), x(t), t)$$

$$= \frac{\partial H}{\partial t}(p(t), x(t), t) \tag{13.1}$$

by virtue of Hamilton's equation of motion, see Subsection 2.5.6. Here, despite of its functional dependence; $H = H(p, x, t)$, the Hamiltonian is finally considered to depend solely on time, t.

In contrast, QM is a field theory in that the wave functions are fields over the (momentum) configuration space: $\psi(x, t) \neq \psi(x(t), t)$, $\phi(p, t) \neq \phi(p(t), t)$. For this, the balance equations concern not only the time variable – as in eq. (13.1) –,

Peter Enders

but also the space variable. Thus, a distributed quantity, U, is conserved, if its density, $u(\vec{r}, t)$, obeys the equation of continuity.

$$\frac{\partial u}{\partial t}(x, t) + \nabla \cdot \vec{j}_u(x, t) = 0 \qquad (13.2)$$

where $\vec{j}_u(x, t)$ is the current density of u.

In this chapter, I will examine the conservation of total weight (probability) and energy, respectively.

13.1. CONSERVATION OF TOTAL WEIGHT

Multiplying the time-dependent Schrödinger equation (12.12a) with $\bar{\psi}$ and its complex-conjugated form with $-\psi$ and adding both resulting equations yields[1]

$$\frac{\partial}{\partial t}(\psi\bar{\psi}) = \frac{\hbar}{2mi}(\psi\Delta\bar{\psi} - \bar{\psi}\Delta\psi) = \frac{\hbar}{2mi}\nabla \cdot (\psi\nabla\bar{\psi} - \bar{\psi}\nabla\psi) \qquad (13.3)$$

This is "the *equation of continuity* of the weight function" (*ibid.*, p. 137), $\psi\bar{\psi}$. Notice, that the normalization $\iiint \psi\bar{\psi}d^3r = 1$ makes $\mathfrak{F} \equiv \psi\bar{\psi}$ actually to be the weight *density* in configuration space.

The quantity

$$\vec{j}_{\mathfrak{F}} = \frac{\hbar}{2mi}(\bar{\psi}\nabla\psi - \psi\nabla\bar{\psi}) \qquad (13.4a)$$

$$= \frac{\hbar}{m}(\mathfrak{R}\psi\nabla\mathfrak{I}\psi - \mathfrak{I}\psi\nabla\mathfrak{R}\psi) \qquad (13.4b)$$

$$= R^2\frac{\nabla S}{m}; \quad \psi = R\exp\left(i\frac{S}{\hbar}\right) \qquad (13.4c)$$

is the "weight current density" (*ibid.*).

The polar form (13.4c) of the wave function has been introduced in eq. (11.19). It allows for writing the equation of continuity (13.3) in hydrodynamic (Eulerian) form as

---

[1] *Cf* Schrödinger, 4th *Commun.*, eqs. (39) … (41); note, that this is a reversal of Problem 12.1.

$$\frac{\partial \mathfrak{F}}{\partial t} + \nabla \cdot \vec{J}_{\mathfrak{F}} = 0 \qquad (13.5)$$

with the weight current density (13.4c), *i.e*,

$$\vec{J}_{\mathfrak{F}} = \mathfrak{F}\vec{v}_{\mathfrak{F}}, \qquad (13.6)$$

the weight density,

$$\mathfrak{F} = R^2 = |\psi|^2 \qquad (13.7)$$

and the 'weight velocity',

$$\vec{v}_{\mathfrak{F}} = \frac{1}{m}\nabla S \qquad (13.8)$$

Bohm has interpreted $\vec{v}_{\mathfrak{F}}(\vec{r}, t)$ to be the real velocity of a particle at position $\vec{r}$, see Subsection 2.3.3.

The corresponding Lagrangian form of the equation of continuity reads[2]

$$\frac{D_{\mathfrak{F}}}{D_{\mathfrak{F}}t}\ln\mathfrak{F} + \nabla \cdot \vec{v}_{\mathfrak{F}} = 0 \qquad (13.9)$$

where

$$\frac{D_{\mathfrak{F}}}{D_{\mathfrak{F}}t} \equiv \frac{\partial}{\partial t} + \vec{v}_{\mathfrak{F}} \cdot \nabla \qquad (13.10a)$$

$$\frac{D_{\mathfrak{F}}\vec{v}_{\mathfrak{F}}}{D_{\mathfrak{F}}t} = \frac{\partial \vec{v}_{\mathfrak{F}}}{\partial t} + (\vec{v}_{\mathfrak{F}} \cdot \nabla)\vec{v}_{\mathfrak{F}} = \frac{\partial \vec{v}_{\mathfrak{F}}}{\partial t} + (\nabla \times \vec{v}_{\mathfrak{F}}) \times \vec{v}_{\mathfrak{F}} + \nabla\frac{v_{\mathfrak{F}}^2}{2} \qquad (13.10b)$$

is the material (Lagrangian) time derivative.

While formula (13.4a) is reproduced in virtually all textbooks on QM, the following equivalent representation is not.

$$\vec{J}_{\mathfrak{F}} = \frac{\hbar}{2mi}(\bar{\psi}\nabla\psi - \psi\nabla\bar{\psi}) \qquad (13.11a)$$

---

[2] Eyal Heifetz & Eliahu Cohen, *Toward a thermo-hydrodynamic like description of Schrödinger equation via the Madelung formulation and Fisher information*, 2015, eq. (4)

$$= \frac{\hbar}{2mi}\left(\bar{\psi}\psi\frac{\nabla\psi}{\psi} - \psi\bar{\psi}\frac{\nabla\bar{\psi}}{\bar{\psi}}\right) \tag{13.11b}$$

$$= |\psi|^2\frac{\hbar}{m}\Im\frac{\nabla\psi}{\psi} \tag{13.11c}$$

In view of $|\psi|^2 = \mathfrak{F}$, this formula suggests $\frac{\hbar}{m}\Im\frac{\nabla\psi}{\psi}$ to be interpreted as a velocity, again.

The calculation of the weight current density in momentum configuration space is left to the reader, see Problem 13.1.

Mathematically, one can add to the weight densities, $\mathfrak{F} \equiv |\psi(\vec{r},t)|^2$ and $\mathfrak{G} \equiv |\phi(\vec{p},t)|^2$, the divergence of a continuous vector field vanishing at the boundaries of the domain under consideration, without changing the total weights, $\iiint \mathfrak{F}d^3r = \iiint \mathfrak{G}d^3p = 1$. However, this would contradict the physical meaning of the weight functions, $F(\vec{r},t)$ and $G(\vec{p},t)$.

Problem 13.1 (Momentum space). *Calculate the weight current density in momentum configuration space!*

13.2. CONSERVATION OF TOTAL ENERGY

While the weight densities in configuration, $\mathfrak{F} \equiv |\psi(\vec{r},t)|^2$, and momentum configuration spaces, $\mathfrak{G} \equiv |\phi(\vec{p},t)|^2$, are axiomatically fixed by the weight functions, $F(\vec{r},t)$ and $G(\vec{p},t)$, the corresponding energy densities do *not* immediately follow from the primary non-classical representation (7.4) of the total energy, *i.e.*,

$$E = \iiint V(\vec{r})|\psi(\vec{r},t)|^2d^3r + \iiint T(p)|\phi(\vec{p},t)|^2d^3p \tag{13.12}$$

This formula uniquely defines the densities of the potential energy in configuration space and of the kinetic energy in momentum configuration space as

$$\mathfrak{V}(\vec{r},t) = |\psi(\vec{r},t)|^2V(\vec{r},t); \quad \mathfrak{T}(\vec{p},t) = |\phi(\vec{p},t)|^2T(\vec{p},t) \tag{13.13}$$

But how to translate $\mathfrak{T}(\vec{p},t)$ to the configuration space and $\mathfrak{V}(\vec{r},t)$ to the momentum configuration space?

Formulae (8.1) and (8.4), *ie*,

$$E = \iiint \bar{\psi}(\vec{r},t)\hat{H}(\vec{r},t)\psi(\vec{r},t)d^3r = \iiint \bar{\phi}(\vec{p},t)\hat{H}(\vec{p},t)\phi(\vec{p},t)d^3p \quad \textbf{(13.14)}$$

suggest the corresponding total energy densities to equal

$$\mathfrak{E}(\vec{r},t) = \mathfrak{R}\,\bar{\psi}(\vec{r},t)\hat{H}(\vec{r},t)\psi(\vec{r},t) = -\mathfrak{I}\,\bar{\psi}(\vec{r},t)\frac{\partial\psi(\vec{r},t)}{\partial t} \quad \textbf{(13.15a)}$$

$$\mathfrak{E}(\vec{p},t) = \mathfrak{R}\,\bar{\phi}(\vec{p},t)\hat{H}(\vec{p},t)\psi(\vec{p},t) = -\mathfrak{I}\,\bar{\phi}(\vec{p},t)\frac{\partial\phi(\vec{p},t)}{\partial t} \quad \textbf{(13.15b)}$$

In configuration space, the corresponding equation of continuity reads

$$\frac{\partial\tilde{\mathfrak{E}}}{\partial t}(\vec{r},t) + \nabla\cdot\vec{j}_{\tilde{\mathfrak{E}}}(\vec{r},t) = \frac{\partial V}{\partial t}(\vec{r},t)\mathfrak{F}(\vec{r},t) \quad \textbf{(13.16)}$$

where – omitting the argument $(\vec{r},t)$ –

$$\vec{j}_{\tilde{\mathfrak{E}}} = \frac{i\hbar}{4m}\left(\psi\nabla\hat{H}\bar{\psi} - (\nabla\psi)\hat{H}\bar{\psi} - \bar{\psi}\nabla\hat{H}\psi + (\nabla\bar{\psi})\hat{H}\psi\right) \quad \textbf{(13.17)}$$

is the corresponding energy current density.

However, formula (13.15a) for the energy density in configuration space exhibits the following deficiency. If a system assumes the stationary state E, it means, that

$$\tilde{\mathfrak{E}}_E(\vec{r}) = \bar{\psi}_E(\vec{r})\hat{T}(\vec{r})\psi_E(\vec{r}) + |\psi_E(\vec{r})|^2 V(\vec{r}) = |\psi_E(\vec{r})|^2 E \quad \textbf{(13.18)}$$

As a consequence, in domains, where $E < V(\vec{r})$ ('tunneling', see Section 10.1), the kinetic energy density, $\tilde{\mathfrak{T}}_E(\vec{r}) = \bar{\psi}_E(\vec{r})\hat{T}(\vec{r})\psi_E(\vec{r})$ becomes negative.[3]

For this, I propose the energy density in configuration space to equal

$$\mathfrak{E}(\vec{r},t) = \frac{\hbar^2}{2m}|\nabla\psi(\vec{r},t)|^2 + |\psi(\vec{r},t)|^2 V(\vec{r}) \quad \textbf{(13.19a)}$$

$$= R^2\left[\frac{\hbar^2}{2m}\left(\frac{\nabla R}{R}\right)^2 + \frac{(\nabla S)^2}{2m} + V\right] \quad \textbf{(13.19b)}$$

---

[3] This argument does not apply to the Hamiltonian density, $\mathfrak{H} = \bar{\psi}(\vec{r},t)\hat{H}(\vec{r},t)\psi(\vec{r},t)$, for which the Hamiltonian equations of motion yield the time-dependent Schrödinger equation, see Donald H. Kobe (1934 - 2014), *Lagrangian Densities and Principle of Least Action in Nonrelativistic Quantum Mechanics*, 2007, Sect. 4.

It is compatible with the free energy density in the Ginzburg-Landau theory of superconductivity[4], with the energy density in the Gross–Pitaevskii equation[5] being relevant for Bose-Einstein condensates, and close to the standard Lagrange density[6],

$$\mathfrak{L} = \frac{i\hbar}{2}\left(\bar{\psi}\frac{\partial\psi}{\partial t} - \psi\frac{\partial\bar{\psi}}{\partial t}\right) - \frac{\hbar^2}{2m}|\nabla\psi|^2 - |\psi|^2 V_{\text{ext}} \tag{13.20}$$

It obeys the equation of continuity

$$\frac{\partial\mathfrak{E}}{\partial t}(\vec{r},t) + \nabla\cdot\vec{j}_{\mathfrak{E}}(\vec{r},t) = \frac{\partial V}{\partial t}(\vec{r},t)|\psi(\vec{r},t)|^2 \tag{13.21}$$

with the energy current density

$$\vec{j}_{\mathfrak{E}} = \frac{\hbar}{2mi}\left(\nabla\psi\hat{H}\bar{\psi} - \nabla\bar{\psi}\hat{H}\psi\right) \tag{13.22a}$$

$$= -\frac{\hbar^2}{2m}\left(\nabla\psi\frac{\partial\bar{\psi}}{\partial t} + \nabla\bar{\psi}\frac{\partial\psi}{\partial t}\right) \tag{13.22b}$$

$$= -\frac{\hbar^2}{m}\left(\nabla R\frac{\partial R}{\partial t} + \frac{R^2}{\hbar^2}\nabla S\frac{\partial S}{\partial t}\right) \tag{13.22c}$$

Now, the action of the momentum operator upon the wave function, $\psi(\vec{r},t)$, can be written as

$$\hat{\vec{p}}\psi = -i\hbar\nabla\left(Re^{\frac{i}{\hbar}S}\right) = \left(-i\hbar\frac{\nabla R}{R} + \nabla S\right)\psi \equiv \vec{p}\psi \tag{13.23}$$

Here, ∇S is Bohm's (Cartesian) momentum of a quantum particle, see Subsection 2.3.3. This suggests to define a complex-valued velocity as[7]

$$\vec{v} = \vec{v}_r + i\vec{v}_i \tag{13.24a}$$

[4] Vitaly Lazarevich Ginzburg (1916 – 2009, Nobel Award 2003) & Landau, *On the Theory of Superconductivity* 1950; Kleinert, *Gauge Fields in Condensed Matter*, Vol. I, Sect. 3.4
[5] Eugene P. Gross (1926 – 1991), *Structure of a quantized vortex in boson systems*, 1961; Lev Petrovich Pitaevskii (*1933), *Vortex lines in an imperfect Bose gas*, 1961
[6] J. Rogel-Salazar, *The Gross-Pitaevskii Equation and Bose-Einstein condensates*, 2013, eq. (18); Kobe, *Lagrangian Densities and Principle of Least Action in Nonrelativistic Quantum Mechanics*, 2007, eq. (22); Erasmo Recami (*1939) & Giovanni Salesi, *Kinematics and hydrodynamics of spinning particles*, 1998, eq. (1)
[7] See Heifetz & Cohen, *Toward a thermo-hydrodynamic like description of Schrödinger equation via the Madelung formulation and Fisher information*, 2015, eq. (26), and references therein.

$$\vec{v}_r = \frac{\nabla S}{m} \qquad\qquad (13.24b)$$

$$\vec{v}_i = -\frac{\hbar}{m}\frac{\nabla R}{R} = -\frac{\hbar}{m}\nabla \ln R = -\frac{\hbar}{2m}\nabla \ln R^2 \qquad (13.24c)$$

In terms of this velocity, the energy density (13.19b) assumes a form that is quite close to the corresponding point-mechanical expression for the total energy.

$$\mathfrak{E}(\vec{r},t) = R^2\left(\frac{v_i^2}{2m} + \frac{v_r^2}{2m} + V\right) \qquad\qquad (13.25)$$

The energy current density (13.22) becomes

$$\vec{j}_{\mathfrak{E}} = \frac{1}{2}R^2\left(\vec{v}_i\,\hbar\,\frac{\partial R}{R\,\partial t} + \vec{v}_r\,\frac{\partial S}{\partial t}\right) \qquad\qquad (13.26)$$

The reader is encouraged to explore, whether this expression can be simplified and brought into the hydrodynamic form 'density times velocity'.

On the other hand, I would like to point out, that – according to Pitaevskii (e-mail to me dated 11.08.2016) – the energy density has got an intrinsic relevance solely within the General Theory of Relativity, where, perhaps, formula (13.19) would be the correct expression of it.

CHAPTER 14

Applications

Abstract: This chapter considers, (i), the time-dependent 'tunnel effect' (as generalization of the stationary one in Section 10.1) and, (ii), coherent states as providing a *smooth* transition to the classical limit case, $F_{cl}(x,t) = x_{ch}\delta(x - x(t))$, $G_{cl}(p,t) = p_{ch}\delta(p - p(t))$. The reader is encouraged to study several ansatzes for the propagation of a time-dependent wave function into the classically forbidden region. Schrödinger's 1926 calculations on the classical limit of wave mechanics by means of the invented by him coherent wave functions are completed for the harmonic oscillator. Here, the physical meaning of the weight function $F_E(x)$ (see Chapter 3) is most helpful, again.

Keywords: Classical limit, Coherent states, Harmonic oscillator, Limiting function, Schrödinger, Time-dependent tunneling, Tunnel effect.

14.1. ABOUT THE 'TUNNEL EFFECT'. II. TIME-DEPENDENT CASE*

*Dedicated to Günther Nimtz (*1936)*

The common treatment of the 'tunnel effect' (see Section 10.1) pays all attention to the penetration into the *classical* barrier: $V(x) > E$. Recall, however, that the actual *quantum* potential energy function *aka* quantum work storage is never larger than the total energy: $V_E(x) \equiv F_E(x)V(x) < E$, see Section 7.1.

14.1.1. General Considerations

In the time-dependent 'tunnel effect', the most interesting feature is the traversal time trough the barrier. Since the position of a moving quantum particle is ill defined, its very definition has long been discussed.[1] The most spectacular result is the one by Hartman, that, for increasing barrier thickness, the traversal time becomes *independent* of the thickness of the barrier[2] (*Hartman effect*). Surprisingly enough, it is undisputed in the literature, although it implies *superluminal* propagation. On the contrary, the corresponding experimental results were objected for exactly that reason. Nimtz compellingly argues, that the transmission of

---

[1] See, *e.g.*, S. Collins, David Lowe & J. R. Barker, *The quantum mechanical tunnelling time problem–revisited*, 1987; Eivind Hiis Hauge & J. A. Støvneng, *Tunneling times: a critical review*, 1989; Mark Yakovlev Azbel', *Time, tunneling and turbulence*, 1998; Horst Aichmann & Nimtz, *On the Traversal Time of Barriers*, 2014. – Within Bohmian mechanics, the particle velocity within the barrier is smaller than that outside the barrier, see Oliver Passon (*1969), *Bohmsche Mechanik*, 2004, p. 91, Fig. (**7.8**).
[2] Hartman, *Tunneling of a Wave Packet*, 1962

Mozart's music makes any discussion on waveforms and signal characteristics to be superfluous. Moreover, superluminal propagation is not necessarily a violation of special relativity and causality.[3]

Hartman considers the wave packet (11.16) moving against the rectangular barrier (*loc. cit.*, eq. (1))

$$V(x) = \begin{cases} 0, & x \leq 0, & \text{(region I)} \\ V_0, & 0 < x < a, & \text{(region II)} \\ 0, & a \leq x, & \text{(region III)} \end{cases} \tag{14.1}$$

His calculations are straightforward, but quite laborious. For this, I encourage the reader to search for other wave functions and/or barriers.

The fundamental solution to the time-dependent Schrödinger equation for a free particle,

$$i\hbar \frac{\partial}{\partial t} \psi_{\text{free}}(x,t) = -\frac{\hbar^2}{2m} \frac{\partial^2}{\partial x^2} \psi_{\text{free}}(x,t) \tag{14.2}$$

is a Gaussian with imaginary width, see Problem 14.1. It is not localized and hence not suitable to describe the penetration into a barrier. For this, I propose to use localized solutions.

Example 14.1. *The plane wave*

$$\psi_k(x,t) = \begin{cases} \sqrt{\frac{k}{\pi}}\, e^{i(kx-\omega t)}; & 0 \leq kx - \omega t \leq \pi \\ 0; & \text{elsewhere} \end{cases}; \quad \hbar\omega = \frac{\hbar^2 k^2}{2m} = E \tag{14.3}$$

is also a solution to eq. (14.2). The reader is encouraged to investigate the 'tunneling' of this wave through the rectangular potential barrier

---

[3] See also Daniela Mugnai, Anedio Ranfagni & L. Ronchi, *The question of tunneling time duration: A new experimental test at microwave scale*, 1998; A. A. Stahlhofen & Nimtz, *Evanescent modes are virtual photons*, 2006; Recami, *Superluminal Motions? A Bird's-Eye View of the Experimental Situation*, 2001; Mugnai & Ranfagni, *Microwave Experiments on Tunneling Time*, 2008 (I thank Daniela Mugnai for providing me with a copy of this and of the foregoing reference *via* ResearchGate); Eugene V. Stefanovich (*1960), *Causality of the Coulomb field of relativistic electron bunches*, 2016.

$$V(x) = \begin{cases} V; & 0 \le x \le a \\ 0; & \text{elsewhere} \end{cases}; \quad V > E \tag{14.4}$$

Example 14.2. *Alternatively, one may consider a potential wall within a finite interval,*

$$V(x) = \begin{cases} \infty; & x \le -a \\ 0; & -a \le x \le 0 \\ V; & 0 < x < a \\ \infty; & x \ge a \end{cases}; \quad \frac{\hbar^2 \pi^2}{2ma^2} < V \tag{14.5}$$

and an initial wave function, which is localized on the left-hand side of the wall, e.g.,

$$\psi_a(x,0) = \begin{cases} \sqrt{\frac{2}{a}} \sin\left(\pi \frac{x}{a}\right); & -a \le x \le 0 \\ 0; & 0 \le x \le a \end{cases} \tag{14.6}$$

Moreover, I would like to encourage the reader to tackle the following problems.[4]

Problem 14.1. *Solve eq. (14.2) analogously to the fundamental solution to the diffusion equation!*

Problem 14.2. *Obtain $\psi_k(x,t)$ (14.3) from the ansatz*

$$\psi_{\text{free}}(x,t) = A\sin(kx - \omega t) + B\cos(kx - \omega t); \quad 0 \le kx - \omega t \le \pi$$

Problem 14.3. *Many numerical methods for solving wave equations contain intrinsic speeds of propagation given by the ratio of spatial and temporal grid widths. Can they – nevertheless – be exploited for calculating the speed in the 'tunneling' region?[5]*

Problem 14.4. *For the problems above, the standard ansatz,*

$$\psi(x,t) = \sum_{n=1}^{\infty} c_n(t)\psi_{E_n}(x) \tag{14.7}$$

---

[4] When appropriate, consider first the case, that the wave front has not yet reached the end of the tunnel barrier, *cf* Partha Ghose (*1939) & M. K. Samal, *Lorentz-invariant superluminal tunneling*, 2001.
[5] For a recent review, see Andrei V. Lavrinenko, Jesper Lægsgaard, Niels Gregersen, Frank Schmidt & Thomas Søndergaard, *Numerical Methods in Photonics*, 2015. For simulations by means of electric circuits and related tools, see Ralf-Michael Vetter (*1969), *Simulation von Tunnelstrukturen. Experimentelle und theoretische Untersuchungen an Systemen mit anomaler Dispersion*, 2002.

faces the inconvenience, that the energies, E_n, are the solutions to a transcendental equation. For the potential wall (14.5), try to use trigonometric functions as basis set! Hints*:*

- *Think about simplifications through shifting the x-interval from $[-a, a]$ to $[0, 2a]$;*

- *try to simplify the calculations by means of other potential functions[6];*

- *see also the next problem.*

Problem 14.5. *The foregoing problem resembles the inverse problem, i.e, to calculate $V(x)$ from $\psi_{E_n}(x)$. Set*

$$V(x) = E_n - \frac{\hbar^2}{2m}\frac{g_{n,0}(x)}{g_{n,2}(x)} \tag{14.8}$$

so that the stationary Schrödinger equation becomes

$$g_{n,2}(x)\frac{\partial^2 \psi_{E_n}(x)}{\partial x^2} + g_{n,0}(x)\psi_{E_n}(x) = 0 \tag{14.9}$$

Search for orthogonal polynomials or functions that obey such an equation with expressions for E_n and $\psi_{E_n}(x)$ as simple as possible, but such, that there are 'tunneling intervals' with $g_{n,0}(x)/g_{n,2}(x) < 0$![7] Compute

$$\psi(x, t) = \sum_{n=1}^{\infty} c_n(0)e^{-iE_n t/\hbar}\psi_{E_n}(x) \tag{14.10}$$

with an initial wave function, $\psi(x, 0)$, that is localized *outside the 'tunneling intervals'!* Hint*: Since the functions $\psi_{E_n}(x)$ spread over the 'tunneling intervals', you need perhaps many terms of that sum, in order to obtain a realistic picture of the penetration into the potential barriers ('tunneling intervals').*

14.1.2. Hydrodynamic Treatment

In his *First Communication*, eq. (2), Schrödinger made for the action, S, the ansatz

$$S(\vec{r}, t) = K\ln\psi(\vec{r}, t) \tag{14.11}$$

---

[6] The most comprehensive table of integrals I am aware of is Prudnikov, Brychkov & Marichev, *Integrals and Series. Elementary Functions*, 1981.

[7] See, *e.g.*, Danos & Rafelski, *Pocketbook of mathematical functions*, 1984, 22.6.

where K is a constant (of dimension 'action') being necessary for dimensional reasons. In his *Second Communication*, fn. 1, Schrödinger stresses, that there is actually *no* relationship between wave function and action in the form of eq. (14.11). Instead, he refers to Hamilton's explorations on the relationships between CM and optics, in particular, to the Hamilton-Jacobi equation (called by him "Hamilton's partial differential equation (H. P.)").

In the spirit of the latter one, Madelung[8] has written the wave function in the form

$$\psi(\vec{r}, t) = R(\vec{r}, t)\exp\left\{\frac{i}{\hbar}S(\vec{r}, t)\right\} \tag{14.12}$$

The reader is encouraged to explore the 'tunnel effect' in terms of that form of the wave function, where she should consider not only the weight (probability) velocity (13.8), but also the energy current densities discussed in Section 13.2.

14.2. CLASSICAL LIMIT. II. COHERENT STATES

With the non-stationary wave functions, $\psi(x, t)$ and $\phi(p, t)$, and the corresponding weight amplitudes,

$$f(x, t) = \sqrt{x_{ch}}\psi(x, t); \ \sqrt{p_{ch}}\phi(p, t) \tag{14.13}$$

where x_{ch} (p_{ch}) is a *ch*aracteristic length in (momentum) configuration space, we are in the position to consider the classical limit (14.14) of the weight functions in more detail.

$$F_{cl}(x, t) = |f_{cl}(x, t)|^2 = x_{ch}\delta(x - x(t)) \tag{14.14a}$$

$$G_{cl}(p, t) = |g_{cl}(p, t)|^2 = p_{ch}\delta(p - p(t)) \tag{14.14b}$$

14.2.1. Schrödinger's Treatment

Immediately after having invented his equation, Schrödinger has sought for "the smooth transition from micro- to macro-mechanics."[9] In view of Bohr's

---

[8] Madelung, *Eine anschauliche Deutung der Gleichung von Schrödinger*, 1926; *Quantentheorie in hydrodynamischer Form*, 1927 – see also Bohm, *A Suggested Interpretation of the Quantum Theory in Terms of 'Hidden' Variables*, 1952; Heifetz & Eliahu Cohen, *Toward a thermo-hydrodynamic like description of Schrödinger equation via the Madelung formulation and Fisher information*, 2015.
[9] This is the English translation of the title of Schrödinger's paper *Der stetige Übergang von der Mikro- zur Makromechanik*, 1926.

correspondence principle, a wave group of high quantum numbers should be appropriate.

Now, for very large values of z, the function $f(n) = z^n/n!$ exhibits a sharp maximum at $n = z$. This suggests to exploit the generating function of the Hermite polynomials, H_n.[10]

$$\sum_{n=0}^{\infty} \frac{s^n}{n!} H_n(z) = e^{-s^2 + 2sz} \tag{14.15}$$

For the Hermite polynomials enter the stationary wave functions of the linear harmonic oscillator, *cf* formula (9.34c).

$$\psi_n(x,t) = \sqrt[4]{\frac{m\omega}{\pi\hbar}}\, \frac{1}{\sqrt{2^n n!}}\, e^{-\frac{m\omega}{2\hbar}x^2} H_n(\sqrt{m\omega\hbar} \cdot x) e^{i(n+1/2)\omega t} \tag{14.16}$$

Basing on that, Schrödinger (*ibid.*) "builds the following aggregate", were I add the normalization factor, $\exp(-A^2/4)$.

$$\psi_{\text{coh}}(x,t) = e^{-\frac{1}{4}A^2} \sum_{n=0}^{\infty} \frac{A^n}{\sqrt{2^n n!}}\, \psi_n(x,t) \tag{14.17a}$$

$$= \sqrt[4]{\frac{m\omega}{\pi\hbar}}\, e^{-\frac{1}{4}A^2 - \frac{m\omega}{2\hbar}x^2 + \frac{i}{2}\omega t} \sum_{n=0}^{\infty} \left(\frac{A}{2}e^{i\omega t}\right)^n \frac{1}{n!} H_n(\sqrt{m\omega\hbar} \cdot x) \tag{14.17b}$$

$$= \sqrt[4]{\frac{m\omega}{\pi\hbar}}\, e^{-\frac{1}{4}A^2 - \frac{m\omega}{2\hbar}x^2 + \frac{i}{2}\omega t}\, e^{-\frac{1}{4}A^2 e^{2i\omega t} + A e^{i\omega t}\sqrt{\frac{m\omega}{\hbar}}x} \tag{14.17c}$$

$$= \sqrt[4]{\frac{m\omega}{\pi\hbar}}\, e^{-\frac{1}{4}A^2 - \frac{m\omega}{2\hbar}x^2}\, e^{-\frac{1}{4}A^2 e^{2i\omega t} + A e^{i\omega t}\sqrt{\frac{m\omega}{\hbar}}x + \frac{i}{2}\omega t} \tag{14.17d}$$

He shows, that, for $A \ggg 1$, the real part, $\Re\psi_{\text{coh}}(x,t)$, exhibits a sharp peak at $x_{\text{peak}}(t)$, that oscillates harmonically.

$$x_{\text{peak}}(t) = \sqrt{\frac{\hbar}{m\omega}}\, A\cos(\omega t) \tag{14.18}$$

---

[10] Schrödinger, *ibid.*, eq. (7), referring to Richard Courant (1888 - 1972) & Hilbert, *Methoden der mathematischen Physik I*, 1924, ch. II, § 10,4, eq. (58). *Cf* Danos & Rafelski, *Pocketbook of mathematical functions*, 1984, 22.9.17 and 22.11.7. See also the (somewhat more general) formula 5.12.1.1 in Prudnikov, Brychkov & Marichev, *Integrals and Series. Special Functions*, 1983.

The associated classical energy equals[11]

$$E_{cl} = \frac{m}{2}\omega^2 \left(\sqrt{\frac{\hbar}{m\omega}}A\right)^2 = \frac{\hbar\omega}{2}A^2 = n\hbar\omega \tag{14.19}$$

Schrödinger concludes, "The correspondence is thus a perfect one in this respect, too." (*Ibid.*)

14.2.2. Classical Limit of the Weight Function $F(x,t)$

The weight function corresponding to the wave function (14.17) equals

$$F_{coh}(x,t) = x_{ch}|\psi_{coh}(x,t)|^2 \tag{14.20a}$$

$$= x_{ch}\sqrt{\frac{m\omega}{\pi\hbar}}e^{-\frac{1}{2}A^2-\frac{m\omega}{\hbar}x^2}\exp\left\{-\frac{1}{2}A^2\cos(2\omega t) + 2A\sqrt{\frac{m\omega}{\hbar}}x\cos(\omega t)\right\} \tag{14.20b}$$

$$= x_{ch}\sqrt{\frac{m\omega}{\pi\hbar}}\exp\left\{-A^2\cos^2(\omega t) + 2A\sqrt{\frac{m\omega}{\hbar}}x\cos(\omega t) - \frac{m\omega}{\hbar}x^2\right\} \tag{14.20c}$$

$$= x_{ch}\sqrt{\frac{m\omega}{\pi\hbar}}\exp\left\{-\left[A\cos(\omega t) - \sqrt{\frac{m\omega}{\hbar}}x\right]^2\right\} \tag{14.20d}$$

Like above, I set

$$A = \sqrt{\frac{m\omega}{\hbar}}x_{max} = \sqrt{\frac{E_{cl}}{E_0}}; \quad x(t) = x_{max}\cos(\omega t) \tag{14.21}$$

where x_{max} is the classical oscillation amplitude, $E_{cl} = \frac{m}{2}\omega^2 x_{max}^2$ the classical total energy and $E_0 = \frac{1}{2}\hbar\omega$ the quantum zero-point energy of the oscillator. Then, with $x_{ch} = x_{max}$, I obtain

$$F_{coh}(x,t) = \sqrt{\frac{E_{cl}}{\pi E_0}}\exp\left\{-\frac{E_{cl}}{E_0}\left[\frac{x(t)-x}{x_{max}}\right]^2\right\} \tag{14.22}$$

For macroscopic bodies, we have $E_{cl} \gg E_0$, and $F_{coh}(x,t)$ becomes an extremely sharp Gaussian around $x(t)$.

---

[11] For very large values of A, the factor $A^n/\sqrt{2^n n!}$ exhibits a sharp maximum at $n = A^2/2$.

Using the heat kernel (Dirac sequence[12])

$$\delta_\varepsilon(z) = \sqrt{\frac{1}{2\pi\varepsilon}} \exp\left\{-\frac{z^2}{2\varepsilon}\right\} \tag{14.23}$$

the classical limit becomes

$$F_{cl}(x,t) = \lim_{\hbar\to 0} F_{coh}(x,t) = \lim_{E_0\to 0} F_{coh}(x,t) = x_{max}\delta(x - x(t)) \tag{14.24}$$

This is exactly the formula that reverts the non-classical expression of the potential energy into its classical one, see eq. (7.4). The analogous calculations for the kinetic energy are left to the reader (Problem 14.6).

Problem 14.6. *Repeat the calculations above for the wave function in momentum space to obtain* $G_{coh}(p,t)$ *and* $G_{cl}(p,t)!$ *Hint: Exploit the symmetry of the Hamiltonian in x and p!*

Problem 14.7. *Why* $x(t)$ *in* $F_{cl}(x,t)$ *contains no initial conditions,* $x(0)$ *and* $v(0)?$

Problem 14.8. *Apply this treatment to the Pöschl-Teller potential[13]! Hint: Calculate states like* (14.17a), *e.g, by means of Laguerre polynomials[14]!*

14.2.3. Coherent States*

After the rediscovery of the states (14.17a) at the beginning of the 1960's for describing coherent (laser) radiation and the classical-quantum correspondence, coherent states have got wide applications not only within general quantum mechanics and quantum field theory, but also in quantum optics, thermodynamics, solid-state physics, nuclear physics, and other fields.[15] For this, I strongly

---

[12] en.wikipedia.org/wiki/Dirac_delta_function; de.wikipedia.org/wiki/Delta-Distribution; the mathematical subtleties I am discarding are sketched there, too.
[13] G. [Herta] Pöschl & Edward Teller (1908 - 2003), *Bemerkungen zur Quantenmechanik des anharmonischen Oszillators*, 1933
[14] P. Kayupe Kikodio & Z. Mouayn, *New coherent states with Laguerre polynomials coefficients for the symmetric Pöschl-Teller oscillator*, 2015
[15] See, *e.g.*, John R. Klauder (*1932) & Bo-Sture Skagerstam, *Coherent States – applications in physics and mathematical physics*, 1985 (a primer and collection of 69 pioneering papers); Askold M. Perelomov, *Generalized Coherent States and Their Applications*, 1986; Jean-Pierre Gazeau, *Coherent States in Quantum Physics*, 2009.

recommend the reader to tackle the following problems.[16]

Problem 14.9. *Translate the calculations of the foregoing sections into the occupation number representation!*

Problem 14.10. *What is the meaning of Schrödinger's coherent wave function, $\psi_{coh}(x, t)$ (14.17a), in the static case ($\omega = 0$)?*

Problem 14.11. *Show, that Schrödinger's coherent wave function(14.17a) is an eigenfunction of the annihilation operator, $\hat{b}$ (10.11), with the boundary condition $\psi_{coh}(\pm\infty, t) = 0$!*

Problem 14.12. *Use the coherent weight functions $F_{coh}(x, t)$ (14.20d), (14.22) and $G_{coh}(p, t)$ (see Problem 14.6) to calculate*

$$\langle x^k \rangle = \int_{-\infty}^{+\infty} x^k F_{coh}(x, t) dx; \quad \langle p^k \rangle = \int_{-\infty}^{+\infty} p^k G_{coh}(p, t) dp; \quad k = 1,2 \quad \textbf{(14.25)}$$

Then, show that

$$(\langle x^2 \rangle - \langle x \rangle^2)(\langle p^2 \rangle - \langle p \rangle^2) = \hbar/2 \quad \textbf{(14.26)}$$

This is often interpreted as 'minimum uncertainty'.

Problem 14.13. *The experimental proof of the Bose-Einstein condensation[17] of atomic gases in the 1990-ies has risen considerable sensation.[18] Discuss the relationship between coherent states and Bose-Einstein condensation!* Hint: *See also Problem 14.15.*

Problem 14.14. *Which coherent states describe harmonic oscillators of displaced*

---

[16] *Cf* Haken, *Quantenfeldtheorie des Festkörpers*, 1973, § 6; Enders, *Von der klassischen Physik zur Quantenphysik*, 2006, § 6.2.

[17] Satyendranath Bose (1894 - 1974), *Plancks Gesetz und Lichtquantenhypothese*, 1924; Einstein, *Anmerkung zu S. N. Boses Abhandlung Plancks Gesetz und Lichtquantenhypothese*, 1924; see also Einstein, *Quantentheorie des einatomigen idealen Gases. Zweite Abhandlung*, 1925; Fritz London (1900 - 1954), *On the Bose-Einstein condensation*, 1938; Alan Griffin (1939 - 2011), David W. Snoke & Sandro Stringari (*1949) (Eds.), *Bose-Einstein condensation*, 1995; Jürgen Schnakenberg (*1937), *Thermodynamik und Statistische Physik*, 2002, Sect. 17.1.

[18] Carl Edwin Wieman (*1951, Nobel Award 2001), *The creation and study of Bose-Einstein condensation in a dilute atomic vapour*, 1997; Wolfgang Ketterle (*1957, Nobel Award 2001), *Experimental studies of Bose-Einstein condensation*, 1999; Eric Allin Cornell (*1961, Nobel Award 2001) & Wieman, *Die Bose-Einstein-Kondensation*, 2003; Oliver Morsch, *Licht und Materie*, 2003

midpoints, $< x > \neq 0$?

Problem 14.15. *How the 'displaced harmonic oscillators' of Problem 14.14 are related to crystalline phase transitions?*[19]

Problem 14.16 (*). *The standard coherent states belong to systems with equidistant energy levels, notably, to the harmonic oscillator. In order to overcome this restriction, Popov has developed what he calls the 'diagonal operator ordering technique'.*[20] *How this approach is related to the recursion formulae used in Section 10.2 to define rising/creation and lowering/annihilation operators, respectively?* Hint: *Examine first the infinite potential well, the Pöschl-Teller*[21] *and Morse potentials*[22]*, for which Popov provides explicit results!*

---

[19] *Cf* Walter Kohn (1923-2016, Nobel Award 1998) & David Sherrington, *Two kinds of bosons and Bose condensation*, 1970, and also Enders, *loc. cit.*, § 6.3.3.

[20] Dušan Popov, *Coherent States of Systems with Non-Equidistant Energy Levels*, 2017

[21] Pöschl & Teller (1908 - 2003), *Bemerkungen zur Quantenmechanik des anharmonischen Oszillators*, 1933

[22] Philip McCord Morse (1903 - 1985), *Diatomic Molecules According to the Wave Mechanics. II. Vibrational Levels*, 1929.

Part IV

Symmetry

CHAPTER 15

General Considerations

Abstract: This chapter introduces the issue of symmetry through rather general notions and theorems. I propose a symmetry theorem for the stationary-state functions that underpins their importance.

Keywords: Conservation laws, Gauge symmetry, Group, von Neumann's theorem, Nöther's theorem, Stationary-state functions, Symmetry

> "La symétrie caractéristique d'un phénomène est la symétrie maxima compatible avec l'existence du phénomène.
>
> Un phénomène peut exister dans un milieu qui possède sa symétrie caractéristique ou celle d'un des intergroupes de sa symétrie caractéristique.
>
> …C'est la dissymétrie qui crée le phénomène."[1]

Something is called symmetric, if it is unchanged by a certain action upon it. The corresponding equations are called invariant against that action. Elementary examples are the geometric symmetries of regular figures like the square, where the actions are rotations by 90° or various reflections.

Obviously, the square is a simpler figure than the rectangular as it needs less quantities to describe it (one side length against two ones). Indeed, symmetries simplify. Due to that, the general theory of crystals is much simpler than that of plasmas, although the ingredients: electrons and ions, are one and the same.[2]

Mathematically, the set of all actions leaving something unchanged – also called *symmetry operations* – form a mathematical group.[3] This way, group theory represents a most general and hence most powerful mathematical tool. Here, however, it is not necessary to go into the details of group theory, because I will concentrate on the benefits from exploring the stationary-state functions.

---

[1] Pierre Curie (1859 - 1906, Nobel Award 1903), *Sur la symétrie dans les phénomènes physiques, symétrie d'un champ électrique et d'un champ magnétique*, 1894, p. 400 – for a collection of English translations of that and other related papers, see Joseph Rosen (*1937), *Symmetry in Physics*, 1982.

[2] For a broad review, see, *e.g,* John Horton Conway (*1937), Heidi Burgiel & Chaim Goodman-Strauss, *The Symmetries of Things*, 2008.

[3] Hans Wußing (1927 - 2011), *Die Genesis des abstrakten Gruppenbegriffs*, 1969

Fig. 15.1. Sculpture *One of N* by Chaim Goodman-Strauss (design) and Eugene Sargent (construction); photo: http://www.eugenesargent.com/jobs/thecube.htm

Within physics, symmetry is related to conservation laws.[4] This is emphasized in

Theorem 15.1 (Noether). *Any differentiable symmetry of the action of a physical system has a corresponding conservation law.*[5]

and in

Theorem 15.2 (von Neumann). *The observables exhibit the symmetry of the system.*[6]

---

[4] For the methodological role of the latter ones, see, *e.g*, Ernst Schmutzer (*1930), *Symmetrien und Erhaltungssätze der Physik*, 1972; Rosen (Ed.), *Symmetry in Physics: Selected Reprints*, 1982; *Symmetry Rules: How Science and Nature Are Founded on Symmetry*, 2008; Enders, *Physical, metaphysical and logical thoughts about the wave equation and the symmetry of space-time*, 2011; Peter Schmelcher (*1959), *Symmetrien diktieren nicht alles*, 2012.

[5] Emmy Noether (1882 - 1935), *Invariante Variationsprobleme*, 1918 – See Naas & Schmid, *Mathematisches Wörterbuch*, 1974, for the full original formulation. For a more general and formal treatise of that topic, see Walter Greiner (*1935) & Joachim Reinhardt (*1952), *Feldquantisierung*, 1993, pp. 46ff. For detailed calculations and examples, see also Ulrich Schröder, *Noether's Theorem and the Conservation Laws in Classical Field Theories*, 1968. For interesting extensions to open systems and to the symmetrization of the canonical energy-momentum tensor, see Yvette Kosmann-Schwarzbach (*1941), *The Noether Theorems. Invariance and Conservation Laws in the Twentieth Century*, 2011.

[6] v. Neumann, *Mathematische Grundlagen der Quantenmechanik*, 1932; see also Arthur P. Cracknell, *Group Theory*, 1971, Sect. 2.10.

Therefore, symmetry also represents a most powerful methodical principle, and symmetry considerations can largely simplify reasonings.[7]

In this part, I wish to show that this approach largely facilitates the treatment of symmetry within (non-relativistic) QM. This is due to the following

Theorem 15.3.

- For a quantum system in stationary state E, the stationary-state functions, $Z_\psi =_{def} |\psi_E(x)|^2$ and $Z_{<\hat{H}>} =_{def} < \psi_E(x)|\hat{H}(x)|\psi_E(x) >$, have got the same symmetry as the classical Hamiltonian, $H(p,x)$.

- For a non-stationary quantum system, the non-stationary-state functions, $Z_\psi(t) =_{def} |\psi(x,t)|^2$ and $Z_{<\hat{H}>}(t) =_{def} < \psi(x,t)|\hat{H}(x,t)|\psi(x,t) >$, have got the same symmetry as the classical Hamiltonian, $H(p,x,t)$.

For the proof, see Problems 16.1f.

Within QM, most symmetries refer to the spatial symmetry of a time-independent potential energy, $V(x)$, of conservative systems, but the exchange symmetry (permutation of equal bodies/particles).[8] For this, I will spend most attention to that two cases.

Surprisingly enough, gauge symmetry being central to modern quantum field theory will emerge almost as a byproduct.

---

[7] See, *e.g*, Jakob Schwichtenberg, *Physics from Symmetry*, 2015.
[8] For broad literature reviews, see David Allen Park (1919 - 2012), *Resource Letter SP-1 on Symmetry in Physics*, 1968; Rosen, *Resource letter SP-2: Symmetry and group theory in physics*, 1981; classic papers are reprinted, *eg*, in Cracknell, *Applied group theory*, 1968; Rosen, *Symmetry in Physics*, 1982.

CHAPTER 16

Space-time Symmetries

Abstract: This chapter is devoted to the perhaps simplest kind of symmetry. The manner of deriving QM from CM presented above allows to equal the space-time symmetries of the quantum-mechanical state functions to that of the classical ones. On this basis, spatial and temporal symmetries of the wave functions are dealt with, (i), in general (Wigner's theorem), (ii), w.r.t. space inversion (parity), (iii) external time periodicity (Floquet theory), (iv), spatial periodicity (Bloch's theorem), (v), internal time periodicity (time crystals), (vi), combined space-time periodicity (choreographic crystals). The symmetry breaking of homogeneity of space during freezing and the corresponding Nambu-Goldstone modes[1] are only mentioned here.

Keywords: Bloch's theorem, Floquet theory, Hamiltonian, Limiting function, Localization, Parity symmetry, Periodically driven systems, Spatially periodic potential, Stationary-state function, Symmetry, Time crystal, Wave function, Weight function, Wigner's theorem.

16.1. SYMMETRY OF THE STATIONARY-STATE FUNCTIONS

Let me first recall the stationary case, where the system is in state E, E being its total energy. By definition, $F_E(\vec{r})$ is the weight of the configuration, $\vec{r}$, such that $V_E(\vec{r}) \equiv F_E(\vec{r})V(\vec{r})$ equals its contribution to the *partial* non-classical work storage; $\int V_E(\vec{r})\mathrm{d}^3\vec{r} / \int F_E(\vec{r})\mathrm{d}^3\vec{r}$ being the total work storage in the stationary state E.

Now, there is no reason for the spatial distribution of the *non*-classical work storage, $V_E(\vec{r})$, to be *less* symmetric than that of the classical work storage, $V(\vec{r})$. This suggests $F_E(\vec{r})$ to exhibit the same spatial symmetry as $V(\vec{r})$. For instance, if $V(\vec{r})$ is mirror-symmetric: $V(-\vec{r}) = V(\vec{r})$, $V_E(\vec{r})$ is expected to be mirror-symmetric, too: $V_E(-\vec{r}) = V_E(\vec{r})$. Therefore, $F_E(\vec{r})$ should also be mirror-symmetric: $F_E(-\vec{r}) = F_E(\vec{r})$.

---

[1] Yoichiro Nambu (1921 - 2015, Nobel Prize 2008), *Quasiparticles and Gauge Invariance in the Theory of Superconductivity*, 1960; *Broken Symmetry*, 1995; Jeffrey Goldstone (*1933), *Field Theories with Superconductor Solutions*, 1961; Goldstone, Abdus Salam (1926 - 1996, Nobel Award 1996), & Weinberg, *Broken Symmetries*, 1962; see also http://www.scholarpedia.org/article/Englert-Brout-Higgs-Guralnik-Hagen-Kibble_mechanism#Proof_of_the_theorem.

The kinetic energy operator, $\hat{T} = -\dfrac{\hbar^2}{2m}\Delta$, is homogeneous and isotropic. Hence, the symmetry of $G_E(\vec{p})$ follows – *via* the Fourier transform between $\psi(\vec{r})$ and $\phi(\vec{p})$ – from that of $F_E(\vec{r})$.

As a consequence, the quantum-mechanical stationary-state functions,

$$Z_{\psi_E(x)} = F_E(x) \tag{16.1a}$$

$$Z_{\langle\hat{H}(\vec{r})\rangle} = \langle\psi_E(\vec{r})|\hat{H}(\vec{r})|\psi_E(\vec{r})\rangle \tag{16.1b}$$

$$= Z_{\langle\hat{H}(\vec{p})\rangle} = \langle\phi_E(\vec{p})|\hat{H}(\vec{p})|\psi_E(\vec{p})\rangle \tag{16.1c}$$

should exhibit the same spatial symmetry as the classical Hamiltonian, $H(p, x) = E$, being a classical stationary-state function (see Subsection 2.5.4).

The same applies *cum grano salis* to time symmetries in the non-stationary case. If the classical Hamiltonian, $H(\vec{p}, \vec{r}, t)$, exhibits a time symmetry, the quantum-mechanical state functions,

$$Z_{\psi(\vec{r},t)} = F(\vec{r}, t); \quad Z_{\phi(\vec{p},t)} = G(\vec{p}, t) \tag{16.2a}$$

$$Z_{\langle\hat{H}(\vec{r},t)\rangle} = \langle\psi(\vec{r}, t)|\hat{H}(\vec{r}, t)|\psi(\vec{r}, t)\rangle \tag{16.2b}$$

$$= Z_{\langle\hat{H}(\vec{p},t)\rangle} = \langle\phi(\vec{p}, t)|\hat{H}(\vec{p}, t)|\psi(\vec{p}, t)\rangle \tag{16.2c}$$

should exhibit the same time symmetry.

Notice, however, that the actual symmetries of a state and of the motion of the system under consideration depend not only on the symmetries of the causes of that state and motion (forces/potentials), but also on the symmetry of the initial conditions. The latter one will turn out to be essential for understanding localization in spatially periodic potentials (crystals) as well as for the fact, that the symmetry of Kepler orbits is less than that of Newtonian spherically symmetric gravitational potentials to be dealt with in the final chapter of this Part.

Problem 16.1. *Prove Theorem 15.3 for the case of spatial symmetry!* Hint: *Exploit the relationship between conservation of total (angualar) momentum and spatial symmetry!*

Problem 16.2. *Generalize your arguing in the foregoing problem by means of Noether's theorem 15.1!*

16.2. SYMMETRY OF THE WAVE FUNCTION. WIGNER'S THEOREM

The stationary state E exhibits the symmetry R, if, $F_E(x)$ is invariant against R.

$$\hat{R}\bar{\psi}_E \hat{R}\psi_E = \bar{\psi}_E \psi_E \qquad (16.3)$$

where $\hat{R}$ is the operator expressing the symmetry R.

If the energetic spectrum is discrete, wave functions for different energies are orthogonal each to another.

$$|\langle \psi_{E'}|\psi_E\rangle| = 0 \text{ if } E' \neq E \qquad (16.4)$$

It is natural to assume that the same holds true for the transformed wave functions.

$$|\langle \hat{R}\psi_{E'}|\hat{R}\psi_E\rangle| = 0 \text{ if } E' \neq E \qquad (16.5)$$

Then,

$$|\langle \hat{R}\psi_{E'}|\hat{R}\psi_E\rangle| = |\langle \psi_{E'}|\psi_E\rangle| \; \forall \; \psi_E, \psi_{E'} \in \mathcal{H} \qquad (16.6)$$

where $\mathcal{H}$ is the Hilbert space[2] ψ_E and $\psi_{E'}$ are belonging to. This allows for the application of the following

Theorem 16.1 (Wigner). *A surjective (not necessarily linear) map: $R: \mathcal{H} \to \mathcal{H}$ on a complex Hilbert space $\mathcal{H}$ that satisfies*

$$|\langle \hat{R}\psi_1|\hat{R}\psi_2\rangle| = |\langle \psi_1|\psi_2\rangle| \;\; \forall \psi_{1,2} \in \mathcal{H} \qquad (16.7)$$

has the form

$$\hat{R}\psi = e^{i\varphi}\hat{U}\psi \qquad (16.8)$$

---

[2] Hilbert, Nordheim & von Neumann, *Über die Grundlagen der Quantenmechanik*, 1927

where φ is a real-valued phase and $U: \mathcal{H} \to \mathcal{H}$ is either unitary: $\langle \widehat{U}\psi_1 | \widehat{U}\psi_2 \rangle = \langle \psi_1 | \psi_2 \rangle$, or anti unitary: $\langle \widehat{U}\psi_1 | \widehat{U}\psi_2 \rangle = \overline{\langle \psi_1 | \psi_2 \rangle}$.[3]

In the subsequent section, we will see, that the most general form (16.8) of the operator $\widehat{R}$ is restricted by the invariance of the total energy against the symmetry R.

16.3. TRANSFORMATION OF THE HAMILTONIAN, $\widehat{H}(x)$

The stationary state E exhibits the symmetry R, if the Hamiltonian, $\widehat{H}$, is invariant against $\widehat{R}$.[4]

$$\widehat{H}\psi_E(x) = E\psi_E(x) \Rightarrow \widehat{H}\widehat{R}\psi_E(x) = E\widehat{R}\psi_E(x) \qquad (16.9)$$

Generally speaking, the symmetry R may be an accidental symmetry of one or several particular stationary states, which are well known for the Kepler respectively Coulomb potential, see Subsection 19.4.4. In such cases, this equation is redundant.

If, however, R is a symmetry of the system such, that this equation holds true for *all* stationary states, it implies the formula

$$\widehat{R}^{-1}\widehat{H}\widehat{R} = \widehat{H} \qquad (16.10)$$

Obviously, this describes the invariance of the Hamiltonian, $\widehat{H}$, against the symmetry operation, $\widehat{R}$, *i.e*, the very symmetry of $\widehat{H}$.

The expression $\widehat{R}^{-1}\widehat{H}\widehat{R}$ may look strange for a *single* symmetry operation. It becomes best comprehensible within matrix mechanics, where $\widehat{H}$ and $\widehat{R}$ are represented by matrices (and ψ_E by a vector).

Corollary 16.1. *Eq. (16.10) means, that $\widehat{H}$ and $\widehat{R}$ commute.*

$$\widehat{H}\widehat{R} = \widehat{R}\widehat{H} \qquad (16.11)$$

---

[3] Wigner, *Group theory and its application to the quantum mechanics of atomic spectra*, 1959, pp. 233-236; the proof has been completed by Valentine (Valya) Bargmann (1908 - 1989), *Note on Wigner's Theorem on Symmetry Operations*, 1964. For proofs fitting the needs of introductory QM courses, see Messiah, *Quantum Mechanics*, 1999, § XV.6, or Hendrik van Hees, *Diskrete Symmetrietransformationen*, 1998. I follow largely the formulation in L. Molnár, *An Algebraic Approach to Wigner's Unitary-Antiunitary Theorem*, 1998.
[4] This is also called Wigner's theorem, see Cracknell, *Applied Group Theory*, 1967, § 2.10.

Corollary 16.2. *If*

$$\hat{H}\hat{R} = r\hat{H}; \quad \hat{R}^{-1}\hat{H} = \frac{1}{r}\hat{H} \tag{16.12}$$

then, $\psi_E(x)$ depends also on r and, additionally, obeys the equation

$$\hat{R}\psi_{E;r}(x) = r\psi_{E;r}(x) \tag{16.13}$$

Mathematically, this represents an eigenvalue equation. If the symmetry is related to another quantization – such as that of the angular momentum – another non-classical solution method should apply.[5]

Problem 16.3. *Suppose, that the operators $\hat{A}$ and $\hat{B}$ commute: $\hat{A}\hat{B} = \hat{B}\hat{A}$. Show, that, if $\psi(x)$ obeys the equation $\hat{A}\psi(x) = a\psi(x)$, then, $\hat{B}\psi(x)$ does so as well! In which cases this implies that, moreover, $\psi(x)$ obeys the equation $\hat{B}\psi(x) = b\psi(x)$?*

Problem 16.4 (*). *Is there a relationship between the possibility of the wave function to obey several eigenvalue equations and the fact, that the excitation of quantum systems proceeds parametrically?*

16.4. PARITY SYMMETRY

Parity symmetry is one of the simplest and thus rather common symmetries; it is the invariance against mirroring the configurations: $\vec{r} \to -\vec{r}$. The harmonic oscillator: $V(\vec{r}) = \frac{\kappa}{2}r^2 = V(-\vec{r})$, and the Coulomb or Kepler potential: $V(\vec{r}) = V(-\vec{r})\sim 1/r$ are mirror symmetric. Accordingly,

$$|\psi_E(-\vec{r})| = |\psi_E(\vec{r})| \tag{16.14a}$$

$$\langle\psi_E(-\vec{r})|\hat{H}(\vec{r})|\psi_E(-\vec{r})\rangle = \langle\psi_E(\vec{r})|\hat{H}(-\vec{r})|\psi_E(\vec{r})\rangle \tag{16.14b}$$

$$= \langle\psi_E(\vec{r})|\hat{H}(\vec{r})|\psi_E(\vec{r})\rangle \tag{16.14c}$$

In view of Wigner's theorem above, the parity operator, $\hat{P}$, acts as

---

[5] For the Bohr orbitals, see Schrödinger, *A Method of Determining Quantum-Mechanical Eigenvalues and Eigenfunctions*, 1940, § 2. He presents a SUSY-like partial factorization of the Hamiltonian, which interrelates solutions to different energies. In contrast to our arguing, he seeks the agreement with the known energy spectrum and does not exploit the absence of *perpetua mobilia*. In *Further Studies on Solving Eigenvalue Problems by Factorization*, 1940, he generalizes this approach to general linear ordinary differential equations of second order. The reader is encouraged to combine both methods!

$$\hat{P}\psi = e^{i\varphi}\hat{U}\psi \qquad\qquad (16.15)$$

Now, by virtue of

$$\langle\psi_1(-x)|\psi_2(-x)|\psi_1(-x)|\psi_2(-x)\rangle$$

$$= \langle\hat{U}\psi_1(x)|\hat{U}\psi_2(x)|\hat{U}\psi_1(x)|\hat{U}\psi_2(x)\rangle \, \langle\psi_1(x)|\psi_2(x)|\psi_1(x)|\psi_2(x)\rangle \quad (16.16)$$

(see Problem 16.5), the operator $\hat{U}$ is unitary.

Problem 16.5. *Show that* $\langle\psi_1(-\vec{r})|\psi_2(-\vec{r})\rangle = \langle\psi_1(\vec{r})|\psi_2(\vec{r})\rangle$!

16.5. PERIODICALLY DRIVEN SYSTEMS. FLOQUET THEOREMS

This section will highlight further benefits of our Newton-Eulerian state approach.

Periodically driven systems are systems, the Hamiltonian of which is cyclic (temporally periodic): $\hat{H}(t + T) = \hat{H}(t)$, T being the cycle duration. This symmetry of the Hamiltonian implies the following

Theorem 16.2 (Floquet 1). *If the function* $\psi(\vec{r}, t)$ *solves the time-dependent Schrödinger equation,*

$$i\hbar\frac{\partial}{\partial t}\psi(\vec{r}, t) = \hat{H}(t)\psi(\vec{r}, t) \qquad\qquad (16.17)$$

in which $\hat{H}(t + T) = \hat{H}(t)$, *then, the function* $\psi(\vec{r}, t + T)$ *does so as well.*

Consequently, the functions $\psi(\vec{r}, t)$ and $\psi(\vec{r}, t + T)$ differ at most by a time-independent factor. Since both functions are normalized to unity, this factor is a phase factor, say, $e^{-i\omega T}$ (see Problem 16.6).

As a consequence, the state functions are periodic in time, too.

$$Z_{\langle\hat{H}\rangle}(t) = \langle\psi(x, t)|\hat{H}(x, t)|\psi(x, t)\rangle = Z_{\langle\hat{H}\rangle}(t + T) \qquad (16.18a)$$

$$Z_\psi(t) = |\psi(x, t)|^2 = Z_\psi(t + T) \qquad\qquad (16.18b)$$

This is exactly that, what one would expect.

In turn, the relation $\psi(\vec{r}, t + T) = e^{-i\omega T}\psi(\vec{r}, t)$ is a direct consequence of the periodicity of stationary-state functions. This underpins their fundamental role.

Furthermore, from the relation $\psi(\vec{r}, t + T) = e^{-i\omega T}\psi(\vec{r}, t)$, the simplest variant of Floquet's theorem follows.

Theorem 16.3 (Floquet 2). *If the Hamiltonian is time-periodic:* $\hat{H}(t + T) = \hat{H}(t)$, *then, the time-dependent Schrödinger equation,*

$$i\hbar \frac{\partial}{\partial t} \psi(\vec{r}, t) = \hat{H}(t)\psi(\vec{r}, t) \qquad (16.19)$$

has got solutions of the form 'non-periodic phase factor times periodic amplitude'.

$$\psi_\omega(\vec{r}, t) = e^{-i\omega t}u_\omega(\vec{r}, t); \quad u_\omega(\vec{r}, t + T) = u_\omega(\vec{r}, t) \qquad (16.20)$$

The solutions (16.20) have been termed *Floquet states*[6], or *Floquet solutions*; the angular frequency, ω, being called *characteristic exponent*.

Inserting the Floquet solution (16.20) into the time-dependent Schrödinger equation (16.19) yields the so-called *quasi-eigenvalue equation*.

$$\left[\hat{H}(t) - i\hbar \frac{\partial}{\partial t}\right] u_\omega(\vec{r}, t) = \hbar\omega u_\omega(\vec{r}, t) \qquad (16.21)$$

$\hbar\omega$ has been termed 'quasi-energy', in analogy to the term 'quasi-momentum' in spatially periodic systems, such as crystals (see next section).[7] The name 'quasi-eigenvalue equation' originates from the analogy to the stationary problem.

The phase factor, $e^{-i\omega t}$, does not change, when ω is replaced with $\omega + 2n\pi$ (n integer). It is convenient to restrict ω to the interval $[0, 2\pi/T)$ or $[-\pi/T, \pi/T)$ and to refer to all other ω-intervals through the index n: $u_\omega(\vec{r}, t) \rightarrow u_{n\omega}(\vec{r}, t)$. This periodicity in ω is analogous to the reciprocal lattice in crystals, see Subsection 16.6.4.

To be historically fair, the mathematics of differential equations with periodic coefficients have been developed (and partially independently rediscovered) by Mathieu, Hill, Floquet and Lyapunov[8]. This variety demonstrates the fascination of

---

[6] Jon H. Shirley, *Interaction of a Quantum System with a Strong Oscillating Field*, 1963; *Solution of the Schrödinger Equation with a Hamiltonian Periodic in Time*, 1965

[7] Yakov Borisovich Zel'dovich (1914 - 1987), *The quasienergy of a quantum-mechanical system subjected to a periodic action*, 1967

[8] Émile Léonard Mathieu (1835 - 1890), *Mémoire sur Le Mouvement Vibratoire d'une Membrane de forme Elliptique*, 1868; George William Hill (1838 - 1914), *On the Part of the Motion of Lunar Perigee Which is a Function of the Mean Motions of the Sun and Moon*, 1886; Achille Marie Gaston Floquet (1847 - 1920), *Sur*

this subject. An important application is the bounded motion of ions in radio frequency quadrupole fields, so-called Paul traps[9].

Problem 16.6. *Show, that the phase factor is indeed linear in T!* Hint: *Cf Subsection 16.6.3 below!*

16.6. SPATIALLY PERIODIC POTENTIALS

Analogously to the foregoing section, one of the fundamental theorems of the electronic theory of crystals is quite a natural consequence of our Newton-Eulerian approach basing on *stationary*-state functions. For this, it is sufficient to consider Bravais lattices[10]. Physically, this are crystals with atoms or molecules solely on the lattice places.

Problem 16.7 (Quasi-crystals and quasicrystals*). *Can the concepts of quasi-crystal and quasicrystal benefit from this approach?*

16.6.1. Few Mathematical Stuff

Geometrically, the lattice consists of the discrete point set

$$\{\vec{G}\} = \{n_1\vec{a}_1 + n_2\vec{a}_2 + n_3\vec{a}_3\}; \quad n_1, n_2, n_3 \in \mathbb{Z} \tag{16.22}$$

Here, $\vec{a}_1, \vec{a}_2, \vec{a}_3$ are three linearly independent vectors from any lattice point to its next neighbors and $\mathbb{Z}$ the set of all integer numbers. The three vectors span a hexahedron. This hexahedron is a *unit cell* of the lattice (crystal), because it fills the whole space without gaps and exhibits the smallest possible volume, $V_{\text{uc}} = \vec{a}_1 \cdot (\vec{a}_2 \times \vec{a}_3)$.

Below, it will prove useful to change from the Cartesian coordinate system x, y, z with the basis $(1,0,0), (0,1,0), (0,0,1)$ to the (possibly, non-Cartesian) coordinate system x_1, x_2, x_3 with the basis $\vec{a}_1, \vec{a}_2, \vec{a}_3$. In the latter one, the radius vectors are given as

$$\vec{r} = x_1\frac{\vec{a}_1}{a_1} + x_2\frac{\vec{a}_2}{a_2} + x_3\frac{\vec{a}_3}{a_3}; \quad a_{1(2,3)} \equiv |\vec{a}_{1(2,3)}| \tag{16.23}$$

---

les équations différentielles linéaires à coefficients périodiques, 1883; Alexander Mihailovich Lyapunov (1857 - 1918), *The General Problem of the Stability of Motion*, 1892

[9] Wolfgang Paul (1913 - 1993, Nobel Award 1989) & Helmut Steinwedel, *Ein neues Massenspektrometer ohne Magnetfeld*, 1953

[10] Auguste Bravais (1811 - 1863), *Mémoire sur les systèmes formés par les points gulièrement sur un plan ou dans l'espace*, 1850

For any lattice periodic function, $f(\vec{r})$: $f(\vec{r} + \vec{G}) = f(\vec{r})$, can be written as

$$f(\vec{r}) = f\left(x_1 \frac{\vec{a}_1}{a_1} + x_2 \frac{\vec{a}_2}{a_2} + x_3 \frac{\vec{a}_3}{a_3}\right) = g(x_1, x_2, x_3) \tag{16.24}$$

The function $g(x_1, x_2, x_3)$ is *periodic in all three arguments*. This largely simplifies its Fourier transform, as we will see soon.

To each lattice (16.22) a (virtual) reciprocal lattice is associated, the lattice vectors of which equal[11]

$$\{\vec{K}\} = \{n_1 \vec{g}_1 + n_2 \vec{g}_2 + n_3 \vec{g}_3\}; \ n_1, n_2, n_3 \in \mathbb{Z} \tag{16.25}$$

Its constituting vectors, $\vec{g}_{1(2,3)}$, are defined as $\vec{g}_i \cdot \vec{a}_j = 2\pi\delta_{j,i}$. Thus,

$$\vec{K} \cdot \vec{r} = (n_1 \vec{g}_1 + n_2 \vec{g}_2 + n_3 \vec{g}_3) \cdot \left(x_1 \frac{\vec{a}_1}{a_1} + x_2 \frac{\vec{a}_2}{a_2} + x_3 \frac{\vec{a}_3}{a_3}\right)$$

$$= 2\pi\left(x_1 \frac{n_1}{a_1} + x_2 \frac{n_2}{a_2} + x_3 \frac{n_3}{a_3}\right) \tag{16.26}$$

Using that, the Fourier representation of any lattice periodic function, $f(\vec{r})$, equals the Fourier series

$$f(\vec{r}) = \sum_{\vec{K}} h(\vec{K}) \, e^{i\vec{K}\cdot\vec{r}} \tag{16.27a}$$

$$h(\vec{K}) = \int_0^{a_1} \frac{dx_1}{a_1} \int_0^{a_2} \frac{dx_2}{a_2} \int_0^{a_3} \frac{dx_3}{a_3} \, g(x_1, x_2, x_3) \, e^{-i\vec{K}\cdot\vec{r}}$$

$$= \frac{1}{\vec{a}_1 \cdot (\vec{a}_2 \times \vec{a}_3)} \iiint_{\text{unit cell}} f(\vec{r}) \, e^{-i\vec{K}\cdot\vec{r}} \, d^3\vec{r} \tag{16.27b}$$

Within solid-state physics, the Wigner-Seitz cell[12] is preferred. It is the locus of all spatial points being closer to a given lattice point than to any other lattice point.[13]

---

[11] Peter Paul Ewald (1888 - 1985), *Zur Theorie der Interferenzen der Röntgenstrahlen in Kristallen*, 1913; *Das reziproke Gitter in der Strukturtheorie*, 1921
[12] Wigner & Frederick Seitz (1911 - 2008), *On the constitution of metallic sodium*, 1933/1934
[13] In 2D, it represents a special case of the Voronoi diagrams, see Georgy Feodosevich Voronoy (1868 - 1908), *Nouvelles applications des paramètres continus à la théorie de formes quadratiques*, 1908, and also Thomas M. Liebling & Lionel Pournin, *Voronoi Diagrams and Delaunay Triangulations: Ubiquitous Siamese Twins*, 2012.

Hence, it is a unit cell, too, contains exactly one lattice point and is thus a primitive cell. The Wigner-Seitz cell of the reciprocal lattice is the (first) Brillouin zone[14].

Problem 16.8. *Verify the calculations and statements above! Hint: The Jacobian determinant between the two coordinate systems under consideration equals* $a_1 a_2 a_3 / [\vec{a}_1 \cdot (\vec{a}_2 \times \vec{a}_3)]$ *(https://en.wikipedia.org/wiki/Fourier_series).*

16.6.2. Symmetry of the Stationary-State Functions

By virtue of their very definition, (ideal) crystals have got a spatially periodic potential energy: $V(\vec{r}) = V(\vec{r} + \vec{G})$, $\vec{G}$ being a lattice vector. Be $\hat{R}_{\vec{G}}$ the operator that translates the position vector, $\vec{r}$, to $\vec{r} + \vec{G}$, such, that

$$\hat{R}_{\vec{G}} \psi_E(\vec{r}) = \psi_E(\vec{r} + \vec{G}) \qquad (16.28)$$

Then, 'spatially periodic potential' means (see Problem 16.9)

$$\hat{R}_{\vec{G}}^{-1} V(\vec{r}) \hat{R}_{\vec{G}} = V(\vec{r} + \vec{G}) = V(\vec{r}) \qquad (16.29)$$

Since the standard kinetic energy (operator) is invariant against any spatial translation of the coordinate system, we have also

$$\hat{R}_{\vec{G}}^{-1} \hat{H}(\vec{r}) \hat{R}_{\vec{G}} = \hat{H}(\vec{r} + \vec{G}) = \hat{H}(\vec{r}) \qquad (16.30)$$

As argued above, the lattice periodicity (2.38) of the Hamiltonian implies that of the stationary-state functions.

$$\left| \hat{R}_{\vec{G}} \psi_E(\vec{r}) \right| = \left| \psi_E(\vec{r} + \vec{G}) \right| = \left| \psi_E(\vec{r}) \right| \qquad (16.31)$$

$$\langle \psi_E(\vec{r} + \vec{G}) | \hat{H}(\vec{r} + \vec{G}) | \psi_E(\vec{r} + \vec{G}) \rangle = \langle \psi_E(\vec{r}) | \hat{H}(\vec{r}) | \psi_E(\vec{r}) \rangle \qquad (16.32)$$

Problem 16.9 *Reason formula (16.29)! Hint: Use the stationary Schrödinger equation!*

---

[14] Léon Nicolas Brillouin (1889 - 1969), *Les électrons dans les métaux et le classement des ondes de de Broglie correspondantes*, 1930

16.6.3. Symmetry of the Wave Functions

As a consequence and analogously to the time-periodic case in the foregoing section, the action of the operator $\hat{R}_{\vec{G}}$ upon the wave function consists at most in the change of its phase.

$$\psi_E(\vec{r} + \vec{G}) = e^{iw(\vec{G})}\psi_E(\vec{r}); \quad w(\vec{0}) = 1 \tag{16.33}$$

At once, we have

$$\psi_E(\vec{r}) = e^{i\widetilde{w}(\vec{r})}u_E(\vec{r}); \quad u_E(\vec{r} + \vec{G}) = u_E(\vec{r}) \tag{16.34}$$

Hence,

$$\widetilde{w}(\vec{r} + \vec{G}) = w(\vec{G}) + \widetilde{w}(\vec{r}) \tag{16.35}$$

Since $\vec{G}$ represents a variety of values, this equation implies both functions to be linear.

$$\widetilde{w}(\vec{r}) = w(\vec{r}) = \vec{k} \cdot \vec{r} \tag{16.36}$$

Here, $\vec{k}$ is a new quantity to be specified next.

The wave function thus obeys the relation

$$\psi_{\vec{k},E(\vec{k})}(\vec{r} + \vec{G}) = e^{i\vec{k}\cdot\vec{G}}\psi_{\vec{k},E(\vec{k})}(\vec{r}) \tag{16.37}$$

The dependence of ψ on $\vec{k}$ follows from the necessity of both sides of an equation to depend on the same parameters. The dependence of E on $\vec{k}$ preserves the influence of all elementary cells on E. From the stationary Schrödinger equation,

$$\hat{H}(\vec{r})\psi_{\vec{k},E(\vec{k})}(\vec{r}) = E(\vec{k})\psi_{\vec{k},E(\vec{k})}(\vec{r}) \tag{16.38}$$

follows, that the wave functions does not explicitly depend on $\vec{k}$, since the Hamiltonian, $\hat{H}(\vec{r})$, is independent of $\vec{k}$: $\psi_{\vec{k},E(\vec{k})}(\vec{r}) = \psi_{E(\vec{k})}(\vec{r})$.

The *novel* parameter, $\vec{k}$, emerges as being characteristic for spatially periodic systems. In the limit case of vanishing potential, the Bloch function (16.42) below is expected to become the wave function of a free particle, where $u_{n\vec{k}}(\vec{r})$ becomes a constant and $\hbar\vec{k}$ its momentum. Consequently, for finite lattice potentials, $\hbar\vec{k}$ is a somehow modified momentum; it has been termed *quasi-momentum*.

16.6.4. Reciprocal Lattice

If $\vec{k} = \vec{K}$ with $\vec{K}$ being a reciprocal lattice vector: $\vec{K} \cdot \vec{G} = 2\pi l$ (l entire, see Subsection 16.6.1), we have

$$\psi_{E(\vec{K})}(\vec{r} + \vec{G}) = e^{2\pi i l}\psi_{E(\vec{K})}(\vec{r}) = \psi_{E(\vec{K})}(\vec{r}) \tag{16.39}$$

This means, that, for $\vec{k} = \vec{K}$, the wave function itself is invariant against the translation by any lattice vector, $\vec{G}$. For this, the set $\{\vec{K}\}$ forms the reciprocal lattice.[15] What about the intermediate values?

Be $\vec{k} = \vec{k}_1 + \vec{K}$. Then,

$$\psi_{E(\vec{k}_1+\vec{K})}(\vec{r} + \vec{G}) = e^{i\vec{k}_1 \cdot \vec{G}}\psi_{E(\vec{k}_1+\vec{K})}(\vec{r}) \tag{16.40}$$

$\psi_{E(\vec{k}_1+\vec{K})}(\vec{r})$ transforms as $\psi_{E(\vec{k}_1)}(\vec{r})$. Hence, the energy is threefold periodic in $\vec{k}$, the periods being given by the three $\vec{K}$-vectors that span an elementary cell of the reciprocal lattice.

$$E(\vec{k} + \vec{K}) = E(\vec{k}) \tag{16.41}$$

Usually, the first Brillouin zone (see p. 177) is chosen as elementary cell.

16.6.5. Bloch's Theorem

All that is condensed in

Theorem 16.4 (Bloch). *If the Hamiltonian is spatially periodic:* $\hat{H}(\vec{r} + \vec{G}) = \hat{H}(\vec{r})$, *then, there are solutions to the stationary Schrödinger equation,* $\hat{H}(\vec{r})\psi(\vec{r}) = E\psi(\vec{r})$, *in form of a Bloch wave.*

---

[15] It has been invented by Ewald, *Zur Theorie der Interferenzen der Röntgenstrahlen in Kristallen*, 1913; *Das reziproke Gitter in der Strukturtheorie*, 1921, in order to explain the x-ray interference patterns of crystals.

$$\psi_{n\vec{k}}(\vec{r}) = e^{i\vec{k}\cdot\vec{r}} u_{n\vec{k}}(\vec{r}); \quad u_{n\vec{k}}(\vec{r} + \vec{G}) = u_{n\vec{k}}(\vec{r}) \tag{16.42}$$

The corresponding energy is a periodic function of $\vec{k}$.

$$E = E_n(\vec{k}) = E_n(\vec{k} + \vec{K}); \quad \vec{K}\cdot\vec{G} = 2\pi l; \; l \text{ entire} \tag{16.43}$$

n labels the periodicity branches (energy bands).[16]

The modulation of the envelope (plane wave), $e^{i\vec{k}\cdot\vec{r}}$, by the lattice-periodic *Bloch factor*, $u_{n\vec{k}}(\vec{r})$, is depicted in Fig. (**16.1**)[17].

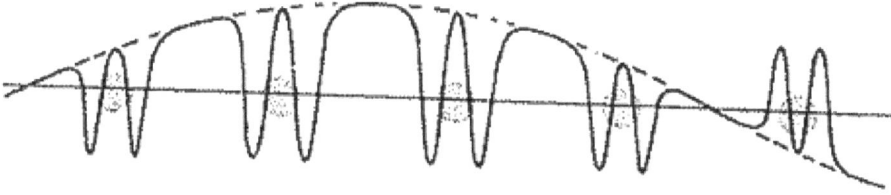

Fig. 16.1. Schematic of the real-valued part of a Bloch wave in 1D; dotted line: envelope e^{ikx}; light circles: atoms

Strictly speaking, the corresponding particle, the Bloch electron, is not an ordinary particle like an electron or proton, but a *quasi-particle*, *i.e.*, an *elementary excitation* that behaves like a single particle.[18]

16.6.6 Hellmann-Feynman Theorem with Bloch Factors. Group Velocity in Photonic Crystals

Let us return to the Hellmann-Feynman Theorem in Section 11.5. If the Hamiltonian $\hat{H}(\lambda)$ is of the form

$$\hat{H}(\vec{p},\vec{r};\lambda) = T(\hat{\vec{p}}) + V(\vec{r};\lambda) \tag{16.44}$$

the Hellmann-Feynman theorem (11.28) simplifies as follows.

---

[16] Felix Bloch (1905 - 1983, Nobel award 1952), *Über die Quantenmechanik der Elektronen in Kristallgittern*, 1928

[17] Source: "Bloch function" by Praveen Thappily / PravySoft - Own work. Licensed under CC BY-SA 3.0 *via* Commons - commons.wikimedia.org/wiki/File:Bloch_function.jpg#/media/File:Bloch_function.jpg (27.10. 2015)

[18] This fundamental concept is due to Landau, *Diamagnetismus der Metalle*, 1930.

$$\frac{dE(\lambda)}{d\lambda} = \left\langle u_{n\vec{k}}(\vec{r};\lambda) \left| \frac{d}{d\lambda} V(\vec{r};\lambda) \right| u_{n\vec{k}}(\vec{r};\lambda) \right\rangle \tag{16.45}$$

Since both the Bloch factor, $u_{n\vec{k}}(\vec{r};\lambda)$, and the potential energy, $V(\vec{r};\lambda)$, are lattice-periodic, the integral over the whole crystal on the r.h.s. reduces to an integral over one of the N unit cells.

$$\left\langle u_{n\vec{k}} \left| \frac{dV}{d\lambda} \right| u_{n\vec{k}} \right\rangle = \iiint_{\text{crystal}} \left| u_{n\vec{k}} \right|^2 \frac{dV}{d\lambda} d^3r = N \iiint_{\text{unitcell}} \left| u_{n\vec{k}} \right|^2 \frac{dV}{d\lambda} d^3r \tag{16.46}$$

In this form, the Hellmann-Feynman theorem has been applied to a classical eigenvalue problem in photonic crystals as will shortly be sketched in the rest of this subsection.

Photonic crystals are materials that form a lattice for electromagnetic waves.[19] Let us consider a non-magnetic material, in which the photonic lattice is formed by the spatially periodic dielectric function, $\varepsilon(\vec{r})$, and a time-harmonic electromagnetic field propagating through it. The amplitude, $\vec{H}(\vec{r})$, of the magnetic field strength, $\vec{H}(\vec{r})e^{-i\omega t}$, obeys the equation

$$\nabla \times \left(\frac{1}{\varepsilon(\vec{r})} \nabla \times \vec{H}(\vec{r}) \right) = \left(\frac{\omega}{c} \right)^2 \vec{H}(\vec{r}) \tag{16.47}$$

(*ibid.*, eq. (7)). Since the dielectric function is lattice-periodic: $\varepsilon(\vec{r}) = \varepsilon(\vec{r} + \vec{G})$, the magnetic field strength can be written in form of a Bloch function (16.42) (*ibid.*, eq. (10)).

$$\vec{H}(\vec{r}) = e^{i\vec{k}\cdot\vec{r}} \vec{u}_{n\vec{k}}(\vec{r}); \quad \vec{u}_{n\vec{k}}(\vec{r} + \vec{G}) = \vec{u}_{n\vec{k}}(\vec{r}) \tag{16.48}$$

Accordingly, the angular frequency of the mode $n\vec{k}$ depends on n and $\vec{k}$: $\omega = \omega_n(\vec{k})$. The Bloch factor, $\vec{u}_{n\vec{k}}(\vec{r})$, obeys the equation

$$\widehat{\vec{\Theta}}_{\vec{k}} \vec{u}_{n\vec{k}}(\vec{r}) = \frac{\omega_n^2(\vec{k})}{c^2} \vec{u}_{n\vec{k}}(\vec{r}) \tag{16.49a}$$

---

[19] I largely follow John D. Joannopoulos (*1947), Steven G. Johnson, Joshua N. Winn & Robert D. Meade, *Photonic Crystals. Molding the Flow of Light*, 2008. For additional quantum effects, see the recent review by Alexandre M. Zagoskin, Didier Felbacq & Emmanuel Rousseau, *Quantum metamaterials in the microwave and optical ranges*, 2016.

$$\widehat{\widehat{\Theta}}_{\vec{k}} \equiv (i\vec{k} + \nabla) \times \frac{1}{\varepsilon(\vec{r})} (i\vec{k} + \nabla) \times \qquad \textbf{(16.49b)}$$

(*ibid.*, eq. (11)). From this equation, the group velocity of that wave,

$$\vec{v}_n(\vec{k}) = \frac{\partial}{\partial \vec{k}} \omega_n(\vec{k}) \qquad \textbf{(16.50)}$$

can be calculated. However, the Hellmann-Feynman theorem (16.45) with $\lambda \to \vec{k}$ and $E = \omega_n^2(\vec{k})/c^2$ immediately yields the formula (*ibid.*, eq. (26))

$$\vec{v}_n(\vec{k}) = \frac{c^2}{2\omega_n(\vec{k})} \left\langle u_{n\vec{k}}(\vec{r}) \left| \frac{\partial}{\partial \vec{k}} \widehat{\widehat{\Theta}}_{\vec{k}} \right| u_{n\vec{k}}(\vec{r}) \right\rangle \qquad \textbf{(16.51)}$$

The reader is encouraged to explore those thoughts along his lines of interest!

16.6.7. Localization by Destructive Interference

The periodicity of the modulus of the Bloch wave functions, $|\psi_{n\vec{k}}(\vec{r})|$, suggests the Bloch electrons to occupy *all* elementary cells in the *same* manner. In special lattices, however, destructive interference can localize them and other elementary excitations within a volume of the order of the elementary cell.[20] This applies even to photons in appropriate photonic lattices.[21]

Problem 16.10 (*). *Does this phenomenon disprove our statement, that the stationary-state functions are lattice-periodic?*

Problem 16.11 (*). *Show, that the Bloch factor,* $u_{n\vec{k}}(\vec{r})$*, is complex-valued, if* $\vec{k} \cdot \nabla u_{n\vec{k}}(\vec{r}) \neq 0$*!*

---

[20] Seminal papers are Chen Zeng & Veit Elser, *Numerical studies of antiferromagnetism on a Kagomé net*, 1990, and Andreas Mielke & Hal Tasaki, *Ferromagnetism in the Hubbard model*, 1993; for a recent review, see Oleg V. Derzhko (*1960), Johannes Richter & Mykola Maksymenko, *Strongly correlated flat-band systems: The route from Heisenberg spins to Hubbard electrons*, 2015.

[21] Rodrigo A. Vicencio, Camilo Cantillano, Luis Morales-Inostroza, Bastián Real, Cristian Mejía-Cortés, Steffen Weimann, Alexander Szameit & Mario I. Molina, *Observation of Localized States in Lieb Photonic Lattices*, 2015; Sebabrata Mukherjee, Alexander Spracklen, Debaditya Choudhury, Nathan Goldman, Patrik Öhberg, Erika Andersson & Robert R. Thomson, *Observation of a Localized Flat-Band State in a Photonic Lieb Lattice*, 2015. – The Lieb lattice is named after Elliott Hershel Lieb (*1932).

16.7. TIME CRYSTALS

> "One of the most powerful ideas in modern physics is that the Universe
> is governed by symmetry."[22]

As a matter of fact, in our universe, we observe a dialectics of constancy and change. Constancy is intimately connected with symmetry. But even change can be connected with symmetry, notably in the case of spontaneous symmetry breaking[23].

> "He was a magician, he would pull one rabbit out of the hat, and
> another, and then suddenly the rabbits would arrange themselves in a
> pattern and start dancing in a way you'd never seen before. Where he
> got the idea, you could never imagine."[24]

A long-known example is crystallization, where the homogeneity and isotropy of liquid or gaseous matter breaks down to the crystal lattice symmetry.

The particularities of spatial periodicity have led to an own branch of physics: crystallography. Therefore, it is tempting to explore temporal periodicity as well.[25] Thus, in contrast to Section 16.5, we deal with potential energies and thus Hamiltonians that are explicitly time *in*dependent. This is an essential difference to ordinary crystals, the spatial periodicity of which is explicitly given in the potential energy.

The following argument seems to exclude the existence of time crystals from the very beginning.[26] The ground state expectation value, $\langle \hat{A} \rangle_0$, of any observable, $\hat{A}$, of a *finite* system is time *in*dependent. For $\frac{d}{dt} \langle \hat{A} \rangle_0 = \frac{i}{\hbar} \langle \psi_0 [\hat{H}, \hat{A}] \psi_0 \rangle = 0$, because $\hat{H} \psi_0 \rangle = E_0 \psi_0 \rangle$. The same arguing applies to any stationary state E.[27]

However, this argument merely shows, that time crystals are formed not by systems in equilibrium, but at most by systems in a non-equilibrium steady state .

---

[22] *Physicists Predict The Existence of Time Crystals*, https://www.technologyreview.com/s/426917/physicists-predict-the-existence-of-time-crystals/

[23] Nambu, *Quasiparticles and Gauge Invariance in the Theory of Superconductivity*, 1960; *Broken symmetry. Selected papers*, 1995; see also Vladimir A. Miransky, *Dynamical Symmetry Breaking in Quantum Field Theories*, 1994.

[24] Peter George Oliver Freund (*1936) about Yoichiro Nambu, quoted on https://www.nytimes.com/2015/07/18/us/yoichiro-nambu-nobel-winning-physicist-dies-at-94.html?_r=0

[25] Wilczek, *Quantum Time Crystals*, 2012; Shapere & Wilczek, *Classical Time Crystals*, 2012

[26] Patrick Bruno (*1964), *Comment on "Space-Time Crystals of Trapped Ions": And Yet it Moves* Not!, 2012

[27] Wilczek, *Quantum Time Crystals*, 2012, eq. (1)

16.7.1. Classical Time Crystals

Periodicity in time without external excitation means some internal motion. This motion be described by the dynamical variable $\phi(t)$. Then, there should be a stable state with $\dot{\phi} \neq 0$.

One of the simplest Lagrangians to obtain such a state reads[28]

$$L(\phi, \dot{\phi}) = -\frac{\kappa}{2}\dot{\phi}^2 + \frac{\lambda}{4}\dot{\phi}^4 \tag{16.52}$$

The corresponding energy function equals

$$h(\phi, \dot{\phi}) = \dot{\phi}\frac{\partial L}{\partial \dot{\phi}} - L = -\frac{\kappa}{2}\dot{\phi}^2 + \frac{3\lambda}{4}\dot{\phi}^4 \tag{16.53}$$

It exhibits minima at $\dot{\phi} = \pm\sqrt{\kappa/3\lambda}$. This ambiguity could indicate a bifurcation[29], but actually it points to the following problem.

The canonical momentum equals

$$p_\phi = \frac{\partial L}{\partial \dot{\phi}} = -\kappa\dot{\phi} + \lambda\dot{\phi}^3 \tag{16.54}$$

Obviously, the function $\dot{\phi}(p_\phi)$ is not unique. Moreover, the total energy as function of momentum is *multivalued* and exhibits *cusps* (Shapere & Wilczek, *loc. cit.*, Fig. **1**).

Anyway, $\dot{\phi}(t)$ should not be constant. For this, Shapere and Wilczek ingeniously discuss various physical solutions for special models. Instead of sketching them here, I encourage the reader to explore non-quadratic Lagrangians being free of those difficulties, like that in Problem 16.12.

Problem 16.12 *Start with the Lagrangian (in arbitrary units)*

$$L(\phi, \dot{\phi}) = \sin\dot{\phi}, 0 \leq \dot{\phi} \leq \pi \tag{16.55}$$

In contrast to the model above,

---

[28] Shapere & Wilczek, *Classical Time Crystals*, 2012, eqs. (3) f.
[29] Henri Poincaré (1854 - 1912), *L'équilibre d'une masse fluide animée d'un mouvement de rotation*, 1885

- *the energy function,*

$$h(\phi, \dot{\phi}) = \dot{\phi}\cos\dot{\phi} - \sin\dot{\phi}, 0 \leq \dot{\phi} \leq \pi \qquad (16.56)$$

  - *is unique,*

  - *exhibits not only the (non-wanted) minimum at $\dot{\phi} = 0$ (no internal motion), but also one unique minimum at $\dot{\phi} = \pi$.*

- *The canonical momentum,*

$$p_\phi = \cos\dot{\phi}, 0 \leq \dot{\phi} \leq \pi, -1 \leq p \leq 1 \qquad (16.57)$$

  is a *one-to-one* function of $\dot{\phi}(p_\phi)$.

Find a potential energy function, $V(\phi)$, for which a cyclic $\phi(t)$ yields the minimum of the total energy!

16.7.2. Quantum Time Crystals

In his pioneering paper 'Quantum Time Crystals' (2012), Wilczek examines first a model borrowed from superconductivity, because a superconductor can support a stable current.[30] *Viz,* "a particle with charge q and unit mass, confined to a ring of unit radius that is threaded by [magnetic] flux $2\pi\alpha/q$." ($\alpha = e^2/4\pi\varepsilon_0\hbar c$ being the famous fine-structure, or Sommerfeld constant[31])

16.8. CHOREOGRAPHIC CRYSTALS*

"a nice meshing of group theory and periodic dynamics"[32]

Crystals are usually defined as orderly arrays of static components, such as atoms, ions and/or molecules. But recently, Boyle, Khoo & Smith have proposed a new kind of crystals, in which the order comes instead from a combination of temporal

---

[30] Heike Kamerlingh Onnes (1853 – 1926, Nobel Prize 1913), *Further experiments with liquid helium. D. On the change of electric resistance of pure metals at very low temperatures, etc. V. The disappearance of the resistance of mercury,* 1911

[31] Sommerfeld, *Die Feinstruktur der Wasserstoff- und der Wasserstoff-ähnlichen Linien,* 1915/1916; *Zur Quantentheorie der Spektrallinien,* 1916

[32] James Patrick Crutchfield (*1955), quoted by Philip Ball, *Focus: New Crystal Type is Always in Motion,* 2016.

and spatial symmetries, "the orchestrated movements of the components, such as orbiting satellites."[33]

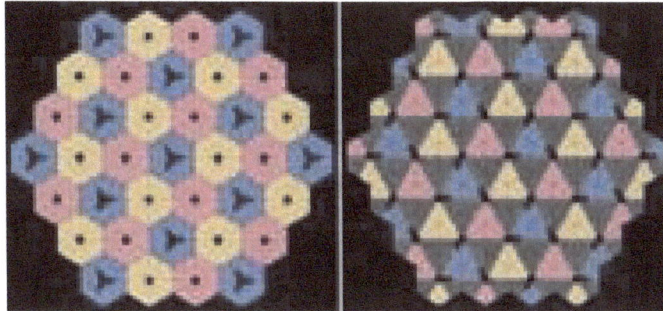

Fig. 16.2. "Dance diagram for math geeks." (Ball, *loc. cit.*) Two planar choreographic crystals; black dots: initial positions of the particles; arrows: directions of their motion; colored tiling: the "dance" proceeds from blue to yellow to pink, in repeating pattern. Source: Boyle *et al.*, *loc. cit.*

An example are four satellites ($\alpha = 0,1,2,3$) on circular orbits of equal radius. The four corners of a regular tetrahedron build the spatially most symmetric three-dimensional ensemble the satellites can form. They be given through the twelve coordinates

$$q_\alpha^j = (-1)^{1+\delta_{0,\alpha}+\delta_{j,\alpha}}; \quad \alpha = 0,1,2,3; \quad j = 1,2,3 \tag{16.58}$$

(Boyle *et al.*, *loc. cit.*, eq. (1)), *i.e*, through the four coordinate vectors

$$\vec{q}_0 = \begin{pmatrix} 1 \\ 1 \\ 1 \end{pmatrix}; \; \vec{q}_1 = \begin{pmatrix} 1 \\ -1 \\ -1 \end{pmatrix}; \; \vec{q}_2 = \begin{pmatrix} -1 \\ 1 \\ -1 \end{pmatrix}; \; \vec{q}_3 = \begin{pmatrix} -1 \\ -1 \\ 1 \end{pmatrix} \tag{16.59}$$

From them the orbits are constructed as

$$r_\alpha^j(t) = q_\alpha^j \cos(t - 2\pi j/3) \tag{16.60}$$

(*ibid.*, eq. (2), where I have replaced their $p_\alpha^j(t)$ with $r_\alpha^j(t)$), *i.e*,

---

[33] Latham Boyle, Jun Yong Khoo & Kendrick Smith, *Symmetric Satellite Swarms and Choreographic Crystals*, 2016

$$\vec{r}_0(t) = \begin{pmatrix} \cos(t - \frac{2\pi}{3}) \\ \cos(t + \frac{2\pi}{3}) \\ \cos(t) \end{pmatrix}; \; \vec{r}_1(t) = \begin{pmatrix} \cos(t - \frac{2\pi}{3}) \\ \cos(t - \frac{\pi}{3}) \\ \cos(t - \pi) \end{pmatrix};$$

$$\vec{r}_2(t) = \begin{pmatrix} \cos(t + \frac{\pi}{3}) \\ \cos(t + \frac{2\pi}{3}) \\ \cos(t + \pi) \end{pmatrix}; \; \vec{r}_3(t) = \begin{pmatrix} \cos(t + \frac{\pi}{3}) \\ \cos(t - \frac{\pi}{3}) \\ \cos(t) \end{pmatrix} \tag{16.61}$$

Here, additionally, the time is shifted, so that the symmetry operations are combinations of spatial and temporal operations.[34]

Problem 16.13 (*). *Construct a corresponding quantum-mechanical model! Hint: Use the Hamiltonian of the rigid rotor (I – moment of inertia).*[35]

$$\hat{H} = -\frac{\hbar^2}{2I}\left[\frac{1}{\sin(\theta)}\frac{\partial}{\partial\theta}\left(\sin(\theta)\frac{\partial}{\partial\theta}\right) + \frac{1}{\sin^2(\theta)}\frac{\partial^2}{\partial\phi^2}\right] \tag{16.62}$$

- *Construct coherent states with weight functions, $F_E(\vec{r})$, being maximum along the classical orbits of the satellites (cf Subsection 14.2.2);*

- *assume the satellites to act upon a quantum particle of mass m. The Hamiltonian be*

$$\hat{H}(\vec{r}, t) = -\frac{\hbar^2}{2m}\Delta + V(\vec{r} - \vec{r}_0(t)) + V(\vec{r} - \vec{r}_1(t)) + V(\vec{r} - \vec{r}_2(t)) + V(\vec{r} - \vec{r}_3(t)) \tag{16.63}$$

What can be said about the symmetry of the weight function, $F_E(\vec{r})$?

---

[34] For an animation of such a motion, see Ball, *loc. cit.*, https://physics.aps.org/articles/v9/4.
[35] Lucy (Lucie) Mensing (1901 - unknown), *Die Rotations-Schwingungsbanden nach der Quantenmechanik*, 1926; Schrödinger, 2nd Commun., 1926, § 3, Section 3

Classical Mechanics and Quantum Mechanics, 2019, 199-212

CHAPTER 17

Gauge Symmetry

Abstract: Gauge symmetry respectively gauge invariance plays a crucial role in modern quantum field theory. Historically, the first locally gauge-invariant equations of motion are Maxwell's original equations for the electrical charges and currents and the electromagnetic fields created by them. In order to keep the relation to classical mechanics as close as possible, this chapter starts with a critical examination of global gauge invariance, basing on Helmholtz's interpretation of the potential energy as "disposable work storage". Then, Helmholtz's explorations of the relationships between forces and energies are sketched and extended. Their generalization to quantum mechanics leads to the gauge invariance of the stationary and time-dependent Schrödinger equations. An important application is the Ehrenberg-Siday-Aharonov-Bohm effect and its gravito-electromagnetic analog. For the sake of the unity of physics, it is shown, how the (gravito-)electromagnetic equations can be deduced. Here, Newton's imaginations about the field of gravity are also exploited.

Keywords: Ehrenberg-Siday-Aharonov-Bohm effect, Energy, Field, Force, Gauge invariance, Gauge symmetry, Gravito-electromagnetism, Helmholtz, Maxwell's equations, Newton, Potential Energy, Schrödinger equation, Unity of physics, Work storage.

Gauge symmetry respectively gauge invariance, *i.e.*, the invariance against a gauge transformation, plays a crucial role in modern quantum field theory.[1] The first locally gauge-invariant equations of motion are Maxwell's original set of equations[2]. The electrical field strength, $\vec{E}$, and the magnetic induction, $\vec{B}$, are given in terms of the scalar, Φ, and vector potentials, $\vec{A}$, as

$$\vec{E} = -\nabla\Phi - \frac{\partial\vec{A}}{\partial t} \qquad (17.1a)$$

$$\vec{B} = \nabla \times \vec{A} \qquad (17.1b)$$

There is no direct relation between Φ and $\vec{A}$.[3] Hence, there is a principal ambiguity known as gauge freedom. The values of $\vec{E}$ and $\vec{B}$ – and hence all phenomena described by Maxwell's theory – are unchanged, when the potentials are 'regauged' as

---

[1] See, *eg*, Weinberg, *The Quantum Theory of Fields*, 1995…2000; Taichiro Kugo, *Eichtheorie*, 1997.
[2] Maxwell, *On physical lines of force*, 1861f.
[3] See, *eg*, Enders, *Underdeterminacy and Redundance in Maxwell's Equations. Origin of Gauge Freedom – Transversality of Free Electromagnetic Waves – Gauge*free *Canonical Treatment* without *Constraints*, 2009.

Peter Enders

$$\Phi = \Phi' + \frac{\partial \chi}{\partial t}; \ \vec{A} = \vec{A}' - \nabla \chi \qquad\qquad (17.2)$$

where $\chi(r, t)$ is an arbitrary sufficiently smooth function.

$$\vec{E} = -\nabla \Phi - \frac{\partial \vec{A}}{\partial t} = -\nabla \Phi' - \frac{\partial \vec{A}'}{\partial t} \qquad\qquad (17.3a)$$

$$\vec{B} = \nabla \times \vec{A} = \nabla \times \vec{A}' \qquad\qquad (17.3b)$$

Weyl has elevated this ambiguity to a principle, in order to derive novel results from it.[4] In this chapter, the gauge invariance of the Schrödinger equation will emerge from a generalization of Helmholtz's explorations of the relationship between forces and energies.

17.1. GLOBAL GAUGE. ABSOLUTE VALUE OF ENERGY

"energy is essentially positive"[5]

The gauge function, $\chi(r, t)$, in eqs. (17.2) describes a punctual gauge of the potentials Φ and $\vec{A}$ in space-time. In eqs. (2.37) and (2.43) the vector $\vec{b}(\vec{v}, \vec{r})$ represents a local gauge of the momentum. Invariance against the latter one is sometimes postulated, in order to get the Hamiltonian of a charged particle in an electromagnetic field. For me, Helmholtz's explorations of the relationships between forces and energies (see next section) is an axiomatically more satisfying approach.

Global gauge means one and the same value of change for the whole space-time.

Often is it claimed, that the momentum, $\vec{p}$, of a body, the total energy, E, of a (conservative) system, or, within electrostatics, the electric (scalar) potential, Φ, can be changed ('regauged') by a constant value, without affecting the physics, *e.g*, the equations of motion. However, this point of view neglects the fact, that the total energy, E, and the potential energy, $q\Phi$ (q – electrical charge), represent the ability to deliver work. As argued above, the values $E = 0$ and $q\Phi = 0$ belong to the ground state, in which no work can be delivered to the environment.

---

[4] Weyl, *Gravitation und Elektrizität*, 1918; *Raum – Zeit – Materie*, 1918. In *Gruppentheorie und Quantenmechanik*, 1928, § 12, he has limited the "principle of gauge invariance" to the interaction of electromagnetic and Schrödinger fields. For the relation of gauge symmetry to charge conservation and Noether's theorem, see also Kathcrine A. Brading, *Which symmetry? Noether, Weyl, and conservation of electric charge*, 2002.

[5] Maxwell, *A Dynamical Theory of the Electromagnetic Field*, 1865, § 82

For pedagogical reasons, it is customary to assign the value $E = 0$ to the boundary between bound and unbound stationary states, notably, when dealing with Kepler and Coulomb systems, respectively. This setting is facilitated by the fact, that the underlying model potential energy, $V(r) = -const/r$, lacks a minimum.

Problem 17.1. *Real Kepler systems are built by bodies of* finite *diameter. Show, that the minimum of the potential energy is located within a body and equals zero!*

Problem 17.2. (*) *How the finiteness of the ground state energy can be realized for Coulomb systems?*

17.2. THE RELATIONSHIP BETWEEN FORCES AND ENERGIES AFTER HELMHOLTZ

Helmholtz[6] asked,

1. which forces constitute – together with the bodies they act upon – a conservative system?

2. which forces leave the kinetic energy of a body unchanged?

17.2.1. Gradient Forces – Constant Total Energy

The answer to the first question reads (Helmholtz, *loc. cit.*),

- central forces between the bodies: $\vec{F}(\vec{r}_1, \vec{r}_2) = \vec{F}_{central}(\vec{r}_1, \vec{r}_2) = \vec{F}(|\vec{r}_1 - \vec{r}_2|)$, or, more generally,

- static gradient fields: $\vec{F}(\vec{r}) = \vec{F}_{grad}(\vec{r}) = -\nabla V(\vec{r})$.

In both cases, the Hamiltonian equals the total energy, E.

$$H(\vec{p}, \vec{r}) = \frac{1}{2m} p^2 + V(\vec{r}) = E = const \qquad (17.4)$$

17.2.2. Lipschitz Forces – Constant Kinetic Energy

The answer to Helmholtz's second question has been given by Lipschitz[7] through the expression

---

[6] Helmholtz, *Über die Erhaltung der Kraft*, 1847, II; 1881, Addendum 3; *Vorlesungen über die Dynamik discreter Massenpunkte*, 1911, § § 48f.

[7] Rudolf Otto Sigismund Lipschitz (1832 - 1903), *priv. commun. to Helmholtz*; in: Helmholtz, *Über die Erhaltung der Kraft*, Addendum 3 to the 1881 ed., v. Helmholtz, *Wissenschaftliche Abhandlungen I*, p. 70; Planck, *Das Prinzip von der Erhaltung der Energie*, 1908, p. 182

$$\vec{F}_{\text{Lip}}(t,\vec{r},\vec{v},\vec{a},\dots) = \vec{v} \times \vec{K}(t,\vec{r},\vec{v},\vec{a},\dots) \tag{17.5}$$

where $\vec{K}(t,\vec{r},\vec{v},\vec{a},\dots)$ is a rather arbitrary function of time, t, position, $\vec{r}$, velocity, $\vec{v}$, acceleration, $\vec{a}$, and higher time-derivatives of $\vec{r}$. Due to $\vec{F}_{\text{Lip}} \cdot \vec{v} \equiv 0$, this 'Lipschitz force' deflects a body without changing its kinetic energy, and this independently of its current position and velocity.

Without the presence of other forces, the Lipschitz force gets a place within Hamiltonian mechanics only, if $\vec{K}$ is a solenoidal field depending solely on $\vec{r}$.

$$\vec{K}(t,\vec{r},\vec{v},\vec{a},\dots) = \nabla \times \vec{J}(\vec{r}) \tag{17.6}$$

This looks rather surprising, but there is a well-known example: a magnetic field without electrical field is a static one.

In this case, the Hamiltonian equals the kinetic energy.

$$H(\vec{p},\vec{r}) = \frac{1}{2m}\left(\vec{p} - \vec{J}(\vec{r})\right)^2 = \frac{m}{2}v^2 = const \tag{17.7}$$

$\vec{J}$, the vector potential (17.6) of the 'Lipschitz field', $\vec{K}$, represents the Lipschitz force, $\vec{F}_{\text{Lip}}$, in the Hamiltonian, H, just in such a manner, that the kinetic energy is still determined by the modulus of the velocity vector, v. In other words, the deviation of the canonical momentum, $\vec{p}$, from the Cartesian momentum, $m\vec{v}$, is just compensated. This complies with the very definition of the Lipschitz force and will thus persist in the time-dependent case considered below.

Problem 17.3. *Prove the form (17.6) of a single Lipschitz force within Hamiltonian mechanics!*[8]

17.2.3 Combination of Lipschitz and Static Gradient Forces: Helmholtzian Conservative Systems

In order to free us from the restriction of static Lipschitz force fields, $\vec{K} = \vec{K}(\vec{r})$, let us consider the combination of time-independent gradient and Lipschitz forces.

$$\vec{F} = \vec{F}_{grad}(\vec{r}) + \vec{F}_{\text{Lip}}(t,\vec{r},\vec{v},\vec{a},\dots) \tag{17.8}$$

---

[8] Enders, *Towards the Unity of Classical Physics*, 2009

However, $\vec{K}$ turns out to be time-independent and solenoidal: $\vec{K} = \nabla \times \vec{J}(\vec{r})$, again; $\vec{F}$ being still static.

$$\vec{F} = -\nabla V(\vec{r}) + \vec{v} \times \nabla \times \vec{J}(\vec{r}) \tag{17.9}$$

The best known analog of such a force is the static (Maxwell-)Lorentz force created by static electrical and magnetic fields. The Hamiltonian is the sum of the kinetic and potential energies and does not explicitly depend on time, again.

$$H(\vec{p}, \vec{r}) = \frac{1}{2m} \left(\vec{p} - \vec{J}(\vec{r}) \right)^2 + V(\vec{r}) = T(\vec{v}(\vec{p})) + V(\vec{r}) = E = const \tag{17.10}$$

Conservative systems described by such a Hamiltonian are proposed to be termed *Helmholtzian conservative systems*.

17.2.4. Extension to Time-Dependent Forces

The action of external influences upon simple conservative systems with Hamiltonians $H_0 = T(\vec{p}) + V(\vec{r}) = E = const$ is described by potential terms, $V_{ext}(\vec{r}, t)$, see Subsections 2.5.4ff. and 11.2.1. Accordingly, the action of external influences upon Helmholtzian conservative systems is expected to be described by both such a term, $V_{ext}(\vec{r}, t)$, and by a time-dependent Lipschitzian vector potential, $\vec{J}_{ext}(\vec{r}, t)$.

Thus, let us consider the Hamiltonian

$$H(\vec{p}, \vec{r}, t) = \frac{1}{2m} \left(\vec{p} - \vec{J}(\vec{r}, t) \right)^2 + V(\vec{r}, t) \tag{17.11}$$

The corresponding Hamiltonian equations of motion read

$$\frac{d\vec{r}}{dt} \equiv \vec{v} = \frac{\partial H}{\partial \vec{p}} = \frac{\vec{p} - \vec{J}(\vec{r}, t)}{m} \tag{17.12a}$$

$$\frac{d\vec{p}}{dt} = -\frac{\partial H}{\partial \vec{r}} = (\vec{v} \cdot \nabla)\vec{J} + \vec{v} \times \nabla \times \vec{J} - \nabla V = m\frac{d\vec{v}}{dt} + (\vec{v} \cdot \nabla)\vec{J} + \frac{\partial \vec{J}}{\partial t} \tag{17.12b}$$

Newton's equation of motion follows immediately as

$$m\frac{d^2\vec{r}}{dt^2} = \vec{F}_{ext} = \vec{v} \times \nabla \times \vec{J} - \frac{\partial \vec{J}}{\partial t} - \nabla V \tag{17.13}$$

As a matter of fact (see Problem 17.4), the r.h.s. represents the most general external force of the form, that fits into Hamiltonian mechanics. It is the sum of a Lipschitz force of the form

$$\vec{F}_{\text{Lip}}(\vec{v},\vec{r},t) = \vec{v} \times \vec{K}(\vec{r},t) = \vec{v} \times \nabla \times \vec{J}(\vec{r},t) \tag{17.14}$$

and of a 'generalized gradient force' of the form

$$\vec{F}(\vec{r},t) = \vec{F}_{\text{e}}(\vec{r},t) \equiv -\partial \vec{J}(\vec{r},t)/\partial t - \nabla V(\vec{r},t) \tag{17.15}$$

For later use, I notice, that the Lipschitz field,

$$\vec{K}(\vec{r},t) = \nabla \times \vec{J}(\vec{r},t) \tag{17.16}$$

is sourceless.

$$\nabla \cdot \vec{K}(\vec{r},t) \equiv 0 \tag{17.17}$$

And the two force terms, $\vec{F}_{\text{Lip}}$ and $\vec{F}_{\text{e}}$, are interrelated as

$$\nabla \times \vec{F}_{\text{e}} = -\nabla \times \frac{\partial \vec{J}}{\partial t} = -\frac{\partial \vec{K}}{\partial t} \tag{17.18}$$

This formula is analogous to the induction law between the electrical field strength, $\vec{E}$, and the magnetic induction, $\vec{B}$: $\nabla \times \vec{E} = -\vec{B}$, see next subsection.

Moreover, $\vec{F}_{\text{e}}(\vec{r},t)$ and $\vec{F}_{\text{Lip}}(\vec{v},\vec{r},t)$ are unaffected by the gauge transformation

$$V(\vec{r},t) = V'(\vec{r},t) + \frac{\partial}{\partial t} x(\vec{r},t) \tag{17.19a}$$

$$\vec{J}(\vec{r},t) = \vec{J}'(\vec{r},t) - \nabla x(\vec{r},t) \tag{17.19b}$$

where the gauge function, $x(\vec{r},t)$, is rather arbitrary. The analogy to the electromagnetic gauge transformation (17.2) is obvious. This will be deepened in the following subsection.

Problem 17.4. *Show, that a force of the form*

$$\vec{F} = \vec{F}(\vec{r},t) + \vec{F}_{\text{Lip}}(t,\vec{r},\vec{v},\vec{a},\ldots) \tag{17.20}$$

fits into Hamiltonian mechanics, if and only if[9]

$$\vec{F}_{\text{Lip}}(t, \vec{r}, \vec{v}, \vec{a}, \dots) = \vec{v} \times \nabla \times J(\vec{r}, t) \tag{17.21}$$

17.2.5. Sidestep to the Unity of Classical Physics: The (Gravito-) Electromagnetic Equations *In Vacuo**

In order to illustrate the power of this approach, let me sidestep to the unity of classical physics and show, that eqs. (17.17) and (17.18) provide a *unified* and *axiomatic* derivation of the homogeneous electromagnetic and gravito-electromagnetic equations.

According to Newton (*Principia*, Definitions), the force of gravity, $\vec{F}_g$, equals a body-dependent factor: the gravitating mass, m_g, times a geometric factor, $\vec{E}_g$: $\vec{F}_g = m_g \vec{E}_g$.[10] This suggests settings as in Table **17.1** below.

Table 17.1. Exemplifications of the general classical force quantities, $\vec{F}_e,\dots,\vec{J}$, in electromagnetism and in gravito-electromagnetism; q – electrical charge, m_g – gravitating mass

Force quantity	Body quantity	Field quantity
$\vec{F}_e(\vec{r}, t) = m_g \vec{E}_g(\vec{r}, t)$	m_g	$\vec{E}_g(\vec{r}, t)$ – gravito-electrical field strength
$\vec{F}_e(\vec{r}, t) = q \vec{E}(\vec{r}, t)$	q	$\vec{E}(\vec{r}, t)$ – electrical field strength
$V(\vec{r}, t) = m_g \Phi_g(\vec{r}, t)$	m_g	$\Phi_g(\vec{r}, t)$ – gravito-electrical potential
$V(\vec{r}, t) = q \Phi(\vec{r}, t)$	q	$\Phi(\vec{r}, t)$ – electrical potential
$\vec{K}(\vec{r}, t) = m_g \vec{B}_g(\vec{r}, t)$	m_g	$\vec{B}_g(\vec{r}, t)$ – gravito-magnetic induction
$\vec{K}(\vec{r}, t) = q \vec{B}(\vec{r}, t)$	q	$\vec{B}(\vec{r}, t)$ – magnetic induction
$\vec{J}(\vec{r}, t) = m_g \vec{A}_g(\vec{r}, t)$	m_g	$\vec{A}_g(\vec{r}, t)$ – gravito-magnetic vector potential
$\vec{J}(\vec{r}, t) = q \vec{A}(\vec{r}, t)$	q	$\vec{A}(\vec{r}, t)$ – magnetic vector potential

---

[9] Enders, *Towards the Unity of Classical Physics*, 2009
[10] For more details, see Enders, *Precursors of force fields in Newton's Principia*, 2010.

Then, eqs. (17.17) and (17.18) read, (i),

$$\nabla \cdot \vec{B}_g = 0 \tag{17.22a}$$

$$\nabla \times \vec{E}_g = -\frac{\partial \vec{B}_g}{\partial t} \tag{17.22b}$$

This is Heaviside's homogeneous gravito-electromagnetic equations[11]. (ii),

$$\nabla \cdot \vec{B} = 0 \tag{17.23a}$$

$$\nabla \times \vec{E} = -\frac{\partial \vec{B}}{\partial t} \tag{17.23b}$$

This is the homogeneous Maxwell-Lorentz equations. Eq. (17.16) becomes

$$\vec{B} = \nabla \times \vec{A} \tag{17.24}$$

This is eqs. (A) in Maxwell's *Treatise*, No. 591.

The inhomogeneous equations can be obtained through additional, but also rather general reasonings.[12]

Problem 17.5. (*) *A magnetic field, $\vec{B}$, does not transfer energy to a point-like charge. Is this an explanation for the formula $\nabla \cdot \vec{B} = 0$?*

Problem 17.6. (*) *Does the velocity, $\vec{v}$, in the Lipschitz force (17.21) represent Euler's state variable, or merely an abbreviation for $d\vec{r}/dt$?*

Problem 17.7. (*) *Within Maxwell's electromagnetism[13], there is no primary connection between the 'field strengths' of the electric, $\vec{E}$, and magnetic fields, $\vec{H}$, despite of their similar names. In contrast, the absence of a primary connection between the 'dielectric displacement', $\vec{D}$, and the 'magnetic induction', $\vec{B}$, complies with their disjunct naming. Discuss this observation in terms of 'accuracy of notions'![14]*

---

[11] Oliver Heaviside (1850 - 1925), *A gravitational and electromagnetic analogy*, 1893
[12] Enders, *Towards the Unity of Classical Physics*, 2009
[13] Maxwell, *A Dynamical Theory of the Electromagnetic Field*, 1965; *A Treatise on Electricity & Magnetism*, 1891 – I do *not* second Hertz's view about Maxwell's theory (*Die Kräfte elektrischer Schwingungen, behandelt nach der Maxwell'schen Theorie*, 1888, *Gesammelte Werke 2*, 1894, p. 148).
[14] See also Enders, *Electromagnetic Momentum Balance in Maxwell's and Hertz's Works*, 2012.

Problem 17.8 (*). *In the full set of Heaviside's gravito-electromagnetic equations, the field source is the rest mass. For this, it is Lorentz covariant in the same manner as the Maxwell-Lorentz equations are. However, this contradicts the equality of inertial and gravitational masses. And it deviates from a corresponding linearization of the equations of general relativity. Can these two deficiencies be healed – while retaining Lorentz invariance –, e.g, by means of extensions and rescalings?*

17.3. GENERALIZATION TO WAVE MECHANICS

When asking for the kind of interactions, which leaves the stationary quantum states unchanged, we will be led back to classical (gravito-)electromagnetism.

This treatment is semi-classical in that those interactions are not quantized. The quantization of the interactions leads to quantum field theory (QFT). In case of the electromagnetic fields, this is quantum electrodynamics (QED). QED represents the most successful theory of these days as it makes the most accurate predictions the current technique can verify. It is beyond the scope of this book, however.

Nevertheless, in view of the lacking unification of QFT and general theory of relativity, it may be allowed to develop this section along lines that are open to both, electromagnetic and gravitational interaction.

17.3.1. Interactions Leaving the Stationary States Unchanged

An obvious generalization of Helmholtz's questions on p. 189 is the

Helmholtzian question:

Which interactions leave the stationary-state quantities $|\psi_E(\vec{r})|^2$ and $E = \langle \psi_E | \hat{H} | \psi_E \rangle / \langle \psi_E | \psi_E \rangle$ unchanged?

Generalizing the cases of spatial-temporal symmetries in the foregoing chapter, the invariance of $|\psi_E(\vec{r})|^2$ implies such an interaction to enter the wave function at most through a phase factor, $\varphi(\vec{r})$.

$$\psi_E(\vec{r}; \varphi) = e^{i\varphi(\vec{r})}\psi_E(\vec{r}; 0) \tag{17.25}$$

Then, the Hamiltonian, $\hat{H}(\vec{r}; \varphi)$, for which

$$\hat{H}(\vec{r}; \varphi)\psi_E(\vec{r}; \varphi) = \hat{H}(\vec{r}; 0)\psi_E(\vec{r}; 0) = E\psi_E(\vec{r}; 0) \tag{17.26}$$

holds true, equals

$$\hat{H}(\vec{r};\varphi) = T\left(\hat{\vec{p}} - \hbar\nabla\varphi(\vec{r})\right) + V(\vec{r}); \quad T(p) = \frac{p^2}{2m} \tag{17.27}$$

In other words, we have the

Theorem 17.1. (Gauge invariance, time-independent) *The solution,* $\psi_E(\vec{r})$, *to the stationary Schrödinger equation*

$$\left[T(\hat{\vec{p}} - \hbar\nabla\varphi(\vec{r})) + V(\vec{r})\right]e^{i\varphi(\vec{r})}\psi_E(\vec{r}) = E\,e^{i\varphi(\vec{r})}\psi_E(\vec{r}) \tag{17.28}$$

is independent of the function $\varphi(\vec{r})$.

This is the gauge invariance of the stationary Schrödinger equation.[15]

Problem 17.9. (Quantization of $p^m x^n$ **II*)** *The quantization of products like* $p^m x^n$ *is ambiguous, see Problem 4.4. Can theorem (17.28), or*

$$e^{-i\varphi(\vec{r})}\left(\hat{\vec{p}} - \hbar\nabla\varphi(\vec{r})\right)^2 e^{i\varphi(\vec{r})} = \hat{p}^2 \tag{17.29}$$

be exploited for resolving this ambiguity, at least for (gravito-)magnetic interactions?

Problem 17.10. *Using theorem (17.28), calculate the energy spectrum of an electron in a static magnetic field to obtain the Landau levels*[16]*!*

17.3.2. The Ehrenberg-Siday-Aharonov-Bohm Effect and its Gravito-Electromagnetic Analog

The most spectacular appearance of the gauge invariance of the stationary Schrödinger equation (17.28) is, perhaps, the Ehrenberg-Siday-Aharonov-Bohm effect[17]. Our approach shows, that there is an analogous effect within *any type* of

---

[15] *Cf* Fock, *Zur Schrödingerschen Wellenmechanik*, 1926.

[16] Landau, *Diamagnetismus der Metalle*, 1930

[17] Ehrenberg & Siday, *The Refractive Index in Electron Optics and the Principles of Dynamics*, 1949; Aharonov & Bohm, *Significance of Electromagnetic Potentials in the Quantum Theory*, 1959; *Further considerations of electromagnetic potential in the quantum theory*, 1961. For a recent demonstration in connection with 'tunneling', see Atsushi Noguchi, Yutaka Shikano, Kenji Toyoda & Shinji Urabe, *Aharonov-Bohm effect in the tunnelling of a quantum rotor in a linear Paul trap*, 2014.

that interaction, in particular, within gravito-electromagnetism (where it is of very much smaller magnitude, though).

The term $\hbar \nabla \varphi(\vec{r})$ in the quantum Hamiltonian (17.28) corresponds to the vector potential $\vec{J}(\vec{r})$ in the classical Hamiltonian (17.10). Suppose, that $\vec{J}(\vec{r}) = \hbar \nabla \varphi(\vec{r})$. Such a vector potential makes no contribution to the classical force (17.13); hence, we deal with a pure quantum effect.

Let me rewrite the solution (17.25) in terms of $\vec{J}(\vec{r})$.

$$\psi_E(\vec{r}; \vec{J}) = \exp\left\{ i \left[\frac{1}{\hbar} \int_{\vec{r}_0}^{\vec{r}} \vec{J}(\vec{r}') \cdot d\vec{r}' + \varphi(\vec{r}_0) \right] \right\} \psi_E(\vec{r}; 0) \qquad \textbf{(17.30)}$$

The crucial difference to formula (17.25) consist in the following. The line integral over a gradient field along a *closed* path does *not* necessarily vanish, if the path runs through a *multiply* connected domain (the coil in Fig. **17.1** below creates such a domain). The curl theorem[18], however, still applies. For this, when the integral path in $\psi_E(\vec{r}; \vec{J})$ (17.30) is closed, the wave function becomes

$$\psi_E(\vec{r}; \vec{K}) = \exp\left\{ \frac{i}{\hbar} \iint \vec{K}(\vec{r}') \cdot d^2\vec{r}' \right\} \psi_E(\vec{r}; 0) \qquad \textbf{(17.31)}$$

The integral is the flux of the field $\vec{K}(\vec{r})$ through the surface built by the closed path $(\vec{r}_0 \to \vec{r}_0)$.

Now, $\vec{K}(\vec{r})$ may vanish along the closed path, but not in all parts of the area enclosed by it. In this case, the flux of $\vec{K}(\vec{r})$ shifts the phase of the wave function.

Such a situation exists in the Ehrenberg-Siday-Aharonov-Bohm setup, see Fig. **(17.1)**[19].

If only slit 1 is open, the wave function equals (*e* denoting the electron charge)

$$\psi_1(\vec{r}, t; \vec{A}) = \psi_1(\vec{r}, t; 0) e^{i\frac{e}{\hbar} \int_{\text{path1}} \vec{A}(\vec{r}') \cdot d\vec{r}'} \qquad \textbf{(17.32a)}$$

If only slit 2 is open, the wave function equals

---

[18] Victor J. Katz (* 1943), *The History of Stokes' Theorem*, 1979
[19] Taken from commons.wikimedia.org/wiki/File:Aharonov-Bohm_effect.svg.

$$\psi_2(\vec{r}, t; \vec{A}) = \psi_2(\vec{r}, t; 0)e^{i\frac{e}{\hbar}\int_{\text{path2}} \vec{A}(\vec{r}')\cdot d\vec{r}'} \tag{17.32b}$$

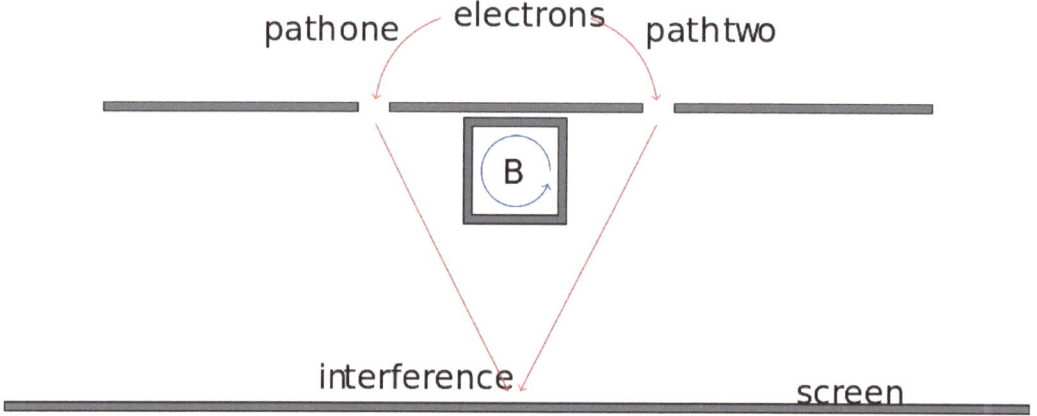

Fig. 17.1. Ehrenberg-Siday-Aharonov-Bohm setup: Outside the coil, where the electrons move from the double slit to the interference screen (red arrows), the magnetic field vanishes: $\vec{B} \equiv \vec{0}$, but the vector potential does not: $\vec{A} \neq \vec{0}$.

If both slits are open, both wave functions superpose to the wave function

$$\psi_{1+2}(\vec{r}, t; \vec{A}) = \psi_1(\vec{r}, t; 0)e^{i\frac{e}{\hbar}\int_{\text{path1}} \vec{A}(\vec{r}')\cdot d\vec{r}'} + \psi_2(\vec{r}, t; 0)e^{i\frac{e}{\hbar}\int_{\text{path2}} \vec{A}(\vec{r}')\cdot d\vec{r}'}$$

$$= \left[\psi_1(\vec{r}, t; 0)e^{i\frac{e}{\hbar}\Phi_B} + \psi_2(\vec{r}, t; 0)\right] e^{i\frac{e}{\hbar}\int_{\text{path2}} \vec{A}(\vec{r}')\cdot d\vec{r}'} \tag{17.32c}$$

For

$$\int_{\text{path1}} \vec{A}(\vec{r}') \cdot d\vec{r}' - \int_{\text{path2}} \vec{A}(\vec{r}') \cdot d\vec{r}' = \oint \vec{A}(\vec{r}') \cdot d\vec{r}'$$

$$= \iint_{\text{coil}} (\nabla \times \vec{A})(\vec{r}') \cdot d^2\vec{r}' = \iint_{\text{coil}} \vec{B}(\vec{r}') \cdot d^2\vec{r}' = \Phi_B \tag{17.33}$$

Φ_B being the magnetic flux. The interference pattern is proportional to

$$\left|\psi_{1+2}(\vec{r}, t; \vec{A})\right|^2 = \left|\psi_1(\vec{r}, t; 0)e^{i\frac{e}{\hbar}\Phi_B} + \psi_2(\vec{r}, t; 0)\right|^2$$

$$= |\psi_{1+2}(\vec{r}, t; 0)|^2 + 2\left[\cos\left(\frac{e}{\hbar}\Phi_B\right) - 1\right]\psi_1(\vec{r}, t; 0)\psi_2(\vec{r}, t; 0) \tag{17.34}$$

assuming for simplicity $\psi_1(\vec{r}, t; 0)$ and $\psi_2(\vec{r}, t; 0)$ to be real-valued. It depends harmonically on the magnetic flux in the coil.[20] The period of this dependence is a natural constant as it equals

$$2\pi\frac{\hbar}{e} = \frac{h}{e} = 2\Phi_0 \tag{17.35}$$

Φ_0 denotes the magnetic flux quantum that occurs in superconductivity.[21]

As a consequence of the closed loop, the expression $\frac{e}{\hbar}\oint \vec{A}(\vec{r}') \cdot d\vec{r}'$ can be understood as a *geometric phase*.[22]

Problem 17.11. *Can this effect be described classically? Hint: Explore the correspondig Darwin-Breit Hamiltonian*[23]!

17.3.3. The Time-Dependent Case

Quite analogous calculations lead to the gauge invariance of the time-dependent Schrödinger equation.

Theorem 17.2. (Gauge invariance, time-dependent) *The solution,* $\psi(\vec{r}, t)$, *to the time-dependent Schrödinger equation,*

$$\left[T\left(\hat{\vec{p}} - \hbar\nabla\varphi(\vec{r}, t)\right) + V(\vec{x}, t) - \hbar\frac{\partial}{\partial t}\varphi(\vec{r}, t) \right] e^{i\varphi(\vec{r},t)}\psi(\vec{r}, t)$$
$$= i\hbar\frac{\partial}{\partial t}e^{i\varphi(\vec{r},t)}\psi(\vec{r}, t) \tag{17.36}$$

is independent of the function $\varphi(\vec{r}, t)$.

Here, the gauge concerns not only the phase of the wave function: $\psi \to \psi e^{i\varphi}$, and the momentum: $\vec{p} \to \vec{p} - \hbar\nabla\varphi$, but also the potential energy function: $V \to V - \hbar\,\partial\varphi/\partial t$.

---

[20] For a review of experimental papers, see en.wikipedia.org/wiki/Aharonov%E2%80%93Bohm_effect.
[21] See en.wikipedia.org/wiki/Magnetic_flux_quantum.
[22] *Cf* Di Xiao, Ming-Che Chang & Qian Niu, *Berry Phase Effects on Electronic Properties*, 2009; Arno Bohm (*1936), Ali Mostafazadeh (*1965), Hiroyasu Koizumi, Qian Niu & Josef Zwanziger, *The Geometric Phase in Quantum Systems*, 2003; Shapere & Wilczek, *Geometric Phases in Physics*, 1989.
[23] Stefanovich, *Classical Electrodynamics without Fields and the Aharonov-Bohm effect*, 2008

Correspondingly, there is an *electrical* Ehrenberg-Siday-Aharonov-Bohm effect.[24]

In turn, there is a limitation upon the modeling of certain interactions.

Theorem 17.3. (Gauge invariance, Hamiltonian) *An external field, $w(\vec{r}, t)$, which does not act upon, (i), the modulus of the wave functions, (ii), the characteristic extension(s) and, (iii), in the stationary case, the total energy of a system, enters the system's Hamiltonian as $T(\hat{\vec{p}}) \to T(\hat{\vec{p}} - \hbar \nabla \varphi)$ and $V(\vec{r}, t) \to V(\vec{r}, t) - \hbar \partial \varphi / \partial t$, where $\varphi(\vec{r}, t)$ is a dimensionless function(al) of $w(\vec{r}, t)$.*

Weyl has introduced the "gauge factor", $e^{i\varphi(\vec{r}, t)}$, as a consequence of general relativity and hence considered the electromagnetic field as a "necessity."[25] Here, Weyl's ingenious idea has been shown to be rooted in Helmholtz's explorations on the relations between interactions and conserved quantities, which are independent of Lagrangian treatments and Noether's theorem (see the foregoing section). Notice also, that the gauge factor, $e^{i\varphi(\vec{r}, t)}$, yields only a *gradient* term in the kinetic energy, *i.e, not* the *transverse* (solenoidal) component, $\vec{A}_T$, of the vector potential. This complies with the fact, that the gauge freedom within classical electromagnetism concerns *solely* the scalar potential, Φ, and the *longitudinal* (irrotational) component of the vector potential, $\vec{A}_L$.

---

[24] Alexander van Oudenaarden, Michel H. Devoret, Yu. V. Nazarov, J. E. Mooij, *Magneto-electric Aharonov-Bohm effect in metal rings*, 1998
[25] Weyl, *Elektron und Gravitation. I*, 1929, Section 'Electrical Field', quoted after Lochlainn O'Raifeartaigh (1933 - 2000) & Straumann, *Early History of Gauge Theories and Kaluza-Klein Theories, with a Glance at Recent Developments*, 1999, p. 21

Permutation Symmetry

Abstract: The permutation symmetry of Newtonian stationary-state functions – such as the total momentum and the Hamiltonian – is crucial for correct state counting in classical statistical mechanics, but plays virtually no role for classical many-body systems. In contrast, it yields novel effects for quantum many-body systems of equal/identical particles. In order to avoid the misunderstandings Jaynes (1992) has rightly stressed, basic notions, such as equal and identical particles, (in)distinguishability and exchangeability (the really relevant notion) are discussed first within classical mechanics. Then, they are applied to the weight (limiting) functions. The permutation symmetry of the wave functions follow largely from that of the weight functions. The different effects for bosons and fermions are mentioned and illustrated. For the generalization towards anyons and the fractional quantum Hall effect, some ideas for further exploration are proposed. Entanglement is treated as a consequence of the conservation laws, in particular, for the angular momentum.

Keywords: Angular momentum, Bose-Einstein condensation, Boson, Conservation law, Entanglement, Equal particles, Exchange hole, Exchangeability, Fermion, Fractional quantum Hall effect, Hamiltonian, Identical particles, Indistinguishability, Jaynes, Limiting functions, Many-body system, Newtonian stationary-state function, Pauli ban, Permutation symmetry, Total momentum, Weight functions.

"Some important facts about thermodynamics have not been understood by others to this day, nearly as well as Gibbs understood them over 100 years ago[1]… For 80 years it has seemed natural that, to find what Gibbs had to say about this, one should turn to his Statistical Mechanics. For 60 years, textbooks and teachers (including, regrettably, the present writer) have impressed upon students how remarkable it was that Gibbs, already in 1902, had been able to hit upon this paradox which foretold – and had its resolution only in – quantum theory with its lore about indistinguishable particles, Bose and Fermi statistics, *etc*.

It was therefore a shock to discover that…Gibbs [in *Heterogeneous Equilibrium*] displays a full understanding of this problem, and disposes of it without a trace of that confusion over the "meaning of

---

[1] The author refers to Josiah Willard Gibbs (1839 - 1903), *On the Equilibrium of Heterogeneous Substances*, 1875-78.

entropy" or "operational distinguishability of particles" on which later writers have stumbled. He goes straight to the heart of the matter as a simple technical detail, easily understood as soon as one has grasped the full meanings of the words "state" and "reversible" as they are used in thermodynamics. In short, quantum theory did not resolve any paradox, because there was no paradox.

Today, the universally taught conventional wisdom holds that "Classical mechanics failed to yield an entropy function that was extensive, and so statistical mechanics based on classical theory gives qualitatively wrong predictions of vapor pressures and equilibrium constants, which was cleared up only by quantum theory in which the interchange of identical particles is not a real event". We argue that, on the contrary, phenomenological thermodynamics, classical statistics, and quantum statistics are all in just the same logical position with regard to extensivity of entropy; they are silent on the issue, neither requiring it nor forbidding it."[2]

In this chapter I will sketch, how the Euler-Helmholtzian manner of (stationary) state description adopted in this book, in particular, the non-classical representation of energy (7.15), can be exploited for the description of quantum many-body systems. *No* additional assumptions will be made, but the rather natural one, that – as in single-particle systems – the symmetries of the classical system persist in the corresponding non-classical system (recall, how the use of the classical Hamiltonian has been justified within our approach).

18.1. EQUAL VERSUS IDENTICAL PARTICLES. (IN)DISTINGUISHABLITY AND EXCHANGEABILITY

Two classical bodies as well as two quantum particles are *equal*, if they have got the same mass and charge distributions.[3] It has become customary to call two equal quantum particles 'identical', assuming that they are not distinguishable.[4] This, however, violates Leibniz's famous principle of the identity of indiscernibles[5]. Only recently it has been rediscovered that (in)distinguishablity is not a property of

---

[2] Edwin Thompson Jaynes (1922 - 1998), *The Gibbs Paradox*, 1992
[3] *Cf* v. Helmholtz, *Einleitung zu den Vorlesungen über Theoretische Physik*, 1903.
[4] Władysław Natanson (1864 - 1937), *On the statistical theory of radiation*, 1911; Paul Ehrenfest (1880 - 1933), *Welche Züge der Lichtquantenhypothese spielen in der Theorie der Wärmestrahlung eine wesentliche Rolle?*, 1911
[5] Leibniz, *Discours de métaphysique*, 1686, Sect. 9; after Peter Forrest, *The Identity of Indiscernibles*, 2012

particles, but of states.[6] For instance, the two electrons in the ground state of the *He* atom are indistinguishable in that no individual property can be uniquely assigned to one of them. If one assumes that their spins are anti-parallel (to obtain vanishing total spin), it is impossible to determine which electron exhibits which spin direction. In contrast, two electrons in a solid become distinguishable by their positions, if they are pinned at distant defects. Generally speaking, for particles, the term 'interchangeability' is more appropriate than the term 'indistinguishability', see the example from the snooker game below.

The state concept à la Newton and Euler allows for a classical discussion of (in)distinguishablity that is quite close to the needs of quantum theory.[7]

18.2. PERMUTATION SYMMETRY AND (IN) DISTINGUISHABILITY WITHIN CM

Two classical bodies occupy disjunct places in space, see Newton-Euler's exclusion principle in Subsection 4.1.1. They differ at least by their places and hence are distinguishable by them. As a consequence, the nowadays used Lagrange-Laplacian states of two bodies are not permutation invariant. This means, that the corresponding state functions, (*eg*, the points in phase space), Z, change, if equal bodies are interchanged: $Z(x_2, x_1; v_2, v_1) \neq Z(x_1, x_2; v_1, v_2)$.

Actually, the situation under consideration may be quite different. For instance, in a snooker game, all red balls are (ideally) equivalent, so that an interchange of any two of them does not alter the outcome of the game.[8]

In contrast, Newton's and Euler's notions of (stationary) state do *not* contain the position of a body. The (stationary) states of a single body are described by means of the conserved quantities momentum (Newton) and velocity (Euler), respectively.

---

[6] See, *e.g*, Jaynes, *The Gibbs Paradox*, 1992, quoted above; Alexander Bach, *Indistinguishable Classical Particles*, 1997. For a recent review, see Robert H. Swendsen, *The ambiguity of " distinguishability" in statistical mechanics*, 2015.

[7] Enders, *Equality and Identity and (In)distinguishability in Classical and Quantum Mechanics from the Point of View of Newton's Notion of State*, 2004; *Von der klassischen Physik zur Quantenphysik*, 2006, Sect. 5.1; *Is Classical Statistical Mechanics Self-Consistent? (A paper of honour of C. F. von Weizsäcker, 1912-2007)*, 2007; *Equality and Identity and (In)distinguishability in Classical and Quantum Mechanics from the Point of View of Newton's Notion of State*, 2008; *Gibbs' Paradox in the Light of Newton's Notion of State*, 2009; *State, Statistics and Quantization in Einstein's 1907 Paper, 'Planck's Theory of Radiation and the Theory of Specific Heat of Solids'*, 2009

[8] Far-reaching consequences for classical statistical mechanics have been discussed, *e.g*, in Enders, *loc. cit.*; *Are there physical systems obeying the Maxwell-Boltzmann statistics?*, 2009; *Bose-Einstein versus Maxwell-Boltzmann distributions*, 2011.

The sum of the momenta of two bodies interacting solely among one another is constant (Newton). Knowing solely the total momentum of two equal bodies ($m_1 = m_2 = m$),

$$\vec{p}_{\text{tot}} = m_1\vec{v}_1 + m_2\vec{v}_2 = m(\vec{v}_1 + \vec{v}_2) = m(\vec{v}_2 + \vec{v}_1) \qquad (18.1)$$

it remains unknown, which body has got velocity $\vec{v}_1$ and which one $\vec{v}_2$. The interchange of the bodies, *i.e*, of the velocities, $\vec{v}_1$ and $\vec{v}_2$, does not change the Newtonian (stationary-)state variable, $\vec{p}_{\text{tot}}$.

Similarly, the Hamiltonian of a system of two equal bodies, say,

$$H(\vec{p}_1, \vec{p}_2, \vec{r}_1, \vec{r}_2, t) = \frac{p_1^2}{2m} + \frac{p_2^2}{2m} + V(\vec{r}_1, \vec{r}_2, t) \qquad (18.2)$$

is permutation-symmetric (see Problem 18.1).

$$H(\vec{p}_2, \vec{p}_1, \vec{r}_2, \vec{r}_1, t) = H(\vec{p}_1, \vec{p}_2, \vec{r}_1, \vec{r}_2, t) \qquad (18.3)$$

The generalization to an arbitrary number of equal bodies is obvious.

Within statistical mechanics, the use of Lagrange's and Laplace's rather than Newton's or Euler's notions of state has led to Gibbs' paradox, *cf* the quotation at the beginning of this chapter.

In what follows I will transfer these ideas to the quantum realm.

Problem 18.1. *Show, that, for equal bodies, the potential energy is permutation-symmetric: $V(\vec{r}_2, \vec{r}_1, t) = V(\vec{r}_1, \vec{r}_2, t)$, using qualitative arguments!*

Problem 18.2. *Show, that equal classical particles in physically equivalent states are interchangeable! Conclude, that (in)distinguishablity is not a properties of particles, but of states![9]*

18.3. PERMUTATION SYMMETRY OF THE WEIGHT FUNCTIONS, $F(\vec{r}_1, \vec{r}_2, t)$ And $G(\vec{p}_1, \vec{p}_2, t)$

For the same reasons, for which the classical potential and kinetic energies of systems of two equal bodies are permutation-symmetric, the non-classical ones are.

---

[9] *Cf* also Gibbs, *Elementary Principles in Statistical Mechanics*, 1902, Ch. XV.

$$V_{\text{ncl}}(\vec{r}_2, \vec{r}_1, t) = V_{\text{ncl}}(\vec{r}_1, \vec{r}_2, t) \qquad \textbf{(18.4a)}$$

$$T_{\text{ncl}}(\vec{p}_2, \vec{p}_1, t) = T_{\text{ncl}}(\vec{p}_1, \vec{p}_2, t) \qquad \textbf{(18.4b)}$$

Hence, the weight functions are permutation-symmetric, too.

$$F(\vec{r}_2, \vec{r}_1, t) = F(\vec{r}_1, \vec{r}_2, t) \qquad \textbf{(18.5a)}$$

$$G(\vec{p}_2, \vec{p}_1, t) = G(\vec{p}_1, \vec{p}_2, t) \qquad \textbf{(18.5b)}$$

The generalization to an arbitrary number of equal particles is straightforward.

18.4. PERMUTATION SYMMETRY OF THE WAVE FUNCTIONS, $\psi(x_1, x_2, t)$ And $\phi(p_1, p_2, t)$

"Yeah, it is reminiscent of what distinguishes the good theorists from the bad ones. The good ones always make an even number of sign errors, and the bad ones always make an odd number."[10]

The permutation symmetry (18.5) of the weight functions, $F(x_1, x_2)$ and $G(p_1, p_2, t)$, is equivalent to that of the modulus of the wave functions.

$$|\psi(\vec{r}_2, \vec{r}_1, t)| = |\psi(\vec{r}_1, \vec{r}_2, t)| \qquad \textbf{(18.6a)}$$

$$|\phi(\vec{p}_2, \vec{p}_1, t)| = |\phi(\vec{p}_1, \vec{p}_2, t)| \qquad \textbf{(18.6b)}$$

The two simplest and best known solutions to eqs. (18.6) are

$$\psi^{\pm}(\vec{r}_2, \vec{r}_1, t) = \pm\psi^{\pm}(\vec{r}_1, \vec{r}_2, t) \qquad \textbf{(18.7a)}$$

$$\phi^{\pm}(\vec{p}_2, \vec{p}_1, t) = \pm\phi^{\pm}(\vec{p}_1, \vec{p}_2, t) \qquad \textbf{(18.7b)}$$

The particles described by these wave functions are called bosons (+) and fermions (−), respectively. They obey the Bose-Einstein[11] respectively Fermi-Dirac

---

[10] Anthony Zee (*1945), *Quantum Field Theory in a Nutshell*, 2005, p. 150
[11] Satyendra Nath (Satyendranath) Bose (1894 - 1974), *Plancks Gesetz und Lichtquantenhypothese*, 1924; Einstein, *Anmerkung zu S. N. Boses Abhandlung Plancks Gesetz und Lichtquantenhypothese*, 1924

statistics[12]. Their properties are well known and need not be outlined here. I just wish to present an illustrative example[13].

Two non-interacting particles be confined to the same one-dimensional infinite square well: $V(x) = 0$ for $-L/2 \leq x \leq +L/2$, while $V(x) = \infty$ otherwise. The normalized solutions to the stationary single-particle Schrödinger equation read

$$\psi_n(x) = \sqrt{\frac{2}{L}} \begin{Bmatrix} \sin\left(n\pi\frac{x}{L}\right) \\ \cos\left(n\pi\frac{x}{L}\right) \end{Bmatrix}; \quad n = \begin{Bmatrix} 2,4,\dots \\ 1,3,\dots \end{Bmatrix}; \quad E_n = n^2 \frac{\hbar^2\pi^2}{2mL^2} \qquad (18.8)$$

The corresponding non-symmetric two-particle solutions are built by the product of the wave functions and the sum of the energies.[14]

$$\psi_{nm}(x_1, x_2) = \frac{2}{L} \begin{Bmatrix} \sin\left(\frac{\pi}{L}nx_1\right)\sin\left(\frac{\pi}{L}mx_2\right) \\ \cos\left(\frac{\pi}{L}nx_1\right)\sin\left(\frac{\pi}{L}mx_2\right) \end{Bmatrix}; \quad n,m = \begin{Bmatrix} 2,4,\dots \\ 1,3,\dots \end{Bmatrix} \qquad (18.9a)$$

$$E_{nm} = (n^2 + m^2)\frac{\pi^2\hbar^2}{2mL^2} \qquad (18.9b)$$

Therefore, the (anti-)symmetrized wave functions are[15]

$$\psi_{nm}^{\pm}(x_1, x_2, t) = \frac{1}{\sqrt{2}}[\psi_{nm}(x_1, x_2) \pm \psi_{nm}(x_2, x_1)]; \quad n \neq m \qquad (18.10a)$$

$$\psi_{nn}^{+}(x_1, x_2, t) = \psi_{nn}(x_1, x_2) \qquad (18.10b)$$

$$\psi_{nn}^{-}(x_1, x_2, t) \equiv 0 \qquad (18.10c)$$

The energy still equals E_{nm} (18.9b), where it becomes meaningless for the case $nn\,-$, see eq. (18.10c).

---

[12] Enrico Fermi (1901 - 1954, Nobel Award 1938), *Zur Quantelung des idealen einatomigen Gases*, 1926; Dirac, *On the Theory of Quantum Mechanics*, 1926

[13] The basic idea is taken from Michael Fowler, *Multiparticle Wavefunctions and Symmetry*.

[14] Douglas Rayner Hartree (1897 - 1958), *The wave mechanics of an atom with a non-Coulomb central field. Part I. Theory and methods*, 1927

[15] Slater, *The Theory of Complex Spectra*, 1929; Fock, *Näherungsmethoden zur Lösung des quantenmechanischen Mehrkörperproblems*, 1930. Antisymmetry in the coordinates – to account for Pauli's exclusion principle – had been required already by Heisenberg and Dirac, see Mehra & Rechenberg, *The Historical Development of Quantum Theory*, Vol. 6.1, 2000, p. 156.

Formula (18.10c) is a special case of *Pauli's exclusion principle*[16], also called *Pauli ban*: Two equal fermions never occupy the same one-particle state. Pauli has obtained it through an ingenious analysis of spectroscopic data, shortly before the invention of matrix (1925) and wave mechanics (1926). Within our approach, the permutation symmetry (18.10) of the wave functions is a natural consequence of the permutation symmetry (18.5) of the weight functions.

Problem 18.3. *The transition from the weight to the wave functions contains the characteristic extensions. Show, that these are permutation-invariant, too!* Hint: *Use qualitative arguments as in Problem 18.1.*

18.4.1. Bosons

In order to illustrate the differences between bosonic and fermionic wave functions, I will draw first the bosonic two-particle weight function

$$F_{23}^{+}(x_1, x_2) = 2\left[\sin\left(2\pi\frac{x_1}{L}\right)\cos\left(3\pi\frac{x_2}{L}\right) + \sin\left(2\pi\frac{x_2}{L}\right)\cos\left(3\pi\frac{x_1}{L}\right)\right]^2 \quad \textbf{(18.11)}$$

Its two main maxima lie on the line $x_1 = x_2$, see Fig. **18.1**.

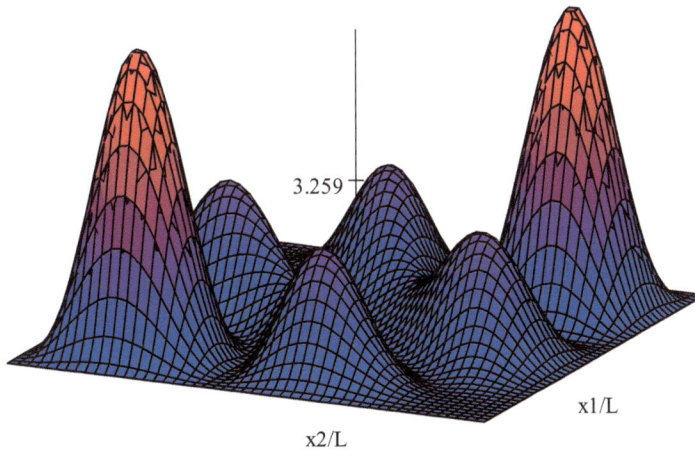

Fig. 18.1. Bosonic weight function $F_{23}^{+}(x_1, x_2)$, formula (18.11)

---

[16] Pauli, *Über den Zusammenhang des Abschlusses der Elektronengruppen im Atom mit der Komplexstruktur der Spektren*, 1925

"…the [bosonic] molecules do not appear as being localized independently of one another, but rather they have a preference to sit together with another molecule in the same cell."[17]

Pictorially speaking,

"…bosons like to be together and imitate one another, which quickly leads to the macroscopic perception of continuous medium, that is, the electromagnetic field for photons."[18]

The number of bosons occupying the same state is unlimited, *cf* formula (18.10b); eventually, a Bose-Einstein condensation happens.[19]

18.4.2. Fermions

"Сего значкочешуйна змия
Сгубила антизимметрию.
Матринодышающий урод
Не будет уже пугать народ."[20]

Consider second the *fermionic* two-particle weight function

$$F_{23}^-(x_1, x_2) = 2\left[\sin\left(2\pi\frac{x_1}{L}\right)\cos\left(3\pi\frac{x_2}{L}\right) - \sin\left(2\pi\frac{x_2}{L}\right)\cos\left(3\pi\frac{x_1}{L}\right)\right]^2 \quad (18.12)$$

Its two main maxima lie on the line $x_1 = -x_2$, see Fig. **18.2**.

Pictorially speaking,

"…fermions prefer to remain aloof, away and distinct from each other. It is this "separatedness" of fermions that makes them manifest in the macroscopic world like little separate chunks of matter." (Meglicki, *ibid.*)

[17] Einstein to Schrödinger, 28 February 1925 (EA 22-002), quoted after Don Howard, *The Early History of Quantum Entanglement, 1905-1935*, 2007
[18] Gustav Meglicki, linkedin.com/groupItem?view= item=5881300260642381824 type=member gid=1892648 trk=eml-b2_anet_digest_weekly-null-13-null fromEmail=fromEmail ut=0InnsI9m0zN6g1 (15.06.2014)
[19] Bose, *Plancks Gesetz und Lichtquantenhypothese*, 1924; Einstein, *Quantentheorie des einatomigen idealen Gases*, 1924f.; Kohn & Sherrington, *Two kinds of bosons and Bose condensation*, 1970
[20] Fock, 1930; quoted after http://hep.phys.spbu.ru/fok/fok_detail_4_r.htm (23.08.2015) – Engl.: This 'sign-flaky' snake / was destroyed by antisymmetry. / The matrix breathing beast / however, will not scare the folks.

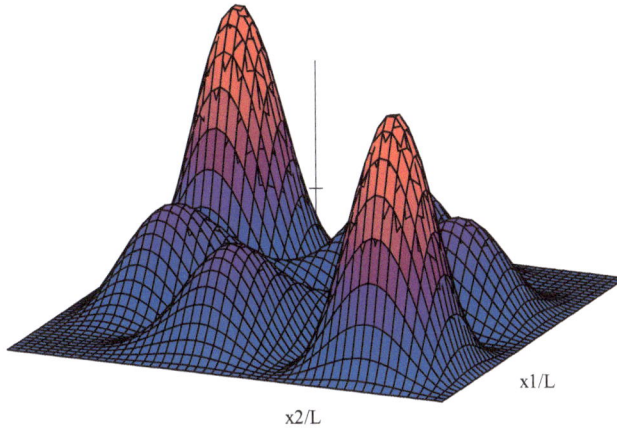

Fig. 18.2. Fermionic weight function $F_{23}^-(x_1, x_2)$, formula (18.12)

In other words, fermions tend to 'repel' each another, so that there is an 'exchange hole'.

18.4.3. GENERALIZATION TOWARDS ANYONS. FRACTIONAL QUANTUM HALL EFFECT

The bosonic and fermionic cases (18.7a) can be written as

$$\psi^\pm(\vec{r}_2, \vec{r}_1, t) = e^{\pi/2 \mp \pi/2}\psi^\pm(\vec{r}_1, \vec{r}_2, t) \tag{18.13}$$

More generally, the phase angle accompanying the particle exchange may depend on position and time.

$$\psi(\vec{r}_2, \vec{r}_1, t) = e^{i\varphi(\vec{r}_2, \vec{r}_1, t)}\psi(\vec{r}_1, \vec{r}_2, t) \tag{18.14}$$

Applying this relation twice yields

$$1 = e^{i\varphi(\vec{r}_2, \vec{r}_1, t)}e^{i\varphi(\vec{r}_1, \vec{r}_2, t)} \tag{18.15}$$

or

$$\varphi(\vec{r}_2, \vec{r}_1, t) + \varphi(\vec{r}_1, \vec{r}_2, t) = 2n\pi; \quad n \text{ integer} \tag{18.16}$$

Therefore, theorem (17.36) generalizes to

Theorem 18.1. (Anyons) *The solution,* $\psi(\vec{r}_1,\vec{r}_2,t)$, *to the time-dependent Schrödinger equation,*

$$\left[T\left(\hat{\vec{p}}_1 - \hbar\nabla_1\varphi(\vec{r}_1,\vec{r}_2,t),\hat{\vec{p}}_2 - \hbar\nabla_2\varphi(\vec{r}_1,\vec{r}_2,t),\right) + V(\vec{r}_1,\vec{r}_2,t) - \hbar\frac{\partial}{\partial t}\varphi(\vec{r}_1,\vec{r}_2,t)\right]$$

$$\times\, e^{i\varphi(\vec{r}_1,\vec{r}_2,t)}\psi(\vec{r}_1,\vec{r}_2,t) = i\hbar\frac{\partial}{\partial t}e^{i\varphi(\vec{r}_1,\vec{r}_2,t)}\psi(\vec{r}_1,\vec{r}_2,t)$$

is *independent* of the function $\varphi(\vec{r}_1,\vec{r}_2,t)$.

Problem 18.4. (Anyons*) *Apply this theorem to anyons[21], which may occur, e.g, in two-dimensional systems![22]*

Problem 18.5. (*) *Utilize this theorem for tackling the fractional quantum Hall effect[23]! Hint: Analyze the Laughlin ansatz[24] for the multi-particle wave function!*

18.5. ENTANGLEMENT

In order to show, that the probabilistic features of QM result from some kind of incompleteness, Einstein, Podolsky & Rosen invented a *gedankenexperiment* being now known as 'Einstein-Podolsky-Rosen (EPR) paradox'. Their famous paper[25] incited Schrödinger to write his celebrated "Referat oder Generalbeichte" (dissertation or general confession – p. 845, fn. 1) on the measurement problem, in which he invented 'Schrödinger's cat' and coined the term 'Verschränkung' (entanglement).[26] His subsequent paper clarifies,

> "When two systems, of which we know the states by their respective representatives, enter into temporary physical interaction due to known forces between them, and when after a time of mutual influence the systems separate again, then they can no longer be described in the same way as before, *viz.* by endowing each of them with a representative of its own. I would not call that *one* but rather *the*

[21] Wilczek, *Quantum Mechanics of Fractional-Spin Particles*, 1982
[22] *Cf* also Jon Magne Leinaas (*1946) & Jan Myrheim (*1948), *On the theory of identical particles*, 1977.
[23] Bertram Marcus Schwarzschild, *Physics Nobel Prize Goes to Tsui, Stormer and Laughlin for the Fractional QuantumHall Effect*, 1998
[24] Robert Betts Laughlin (*1950, Nobel Award 1998), *Anomalous Quantum Hall Effect: An Incompressible Quantum Fluid with Fractionally Charged Excitations*, 1983; *Fractional Quantization*, 1998
[25] Einstein, Boris Podolsky (1896 - 1966) & Nathan Rosen (1909 - 1995), *Can Quantum-Mechanical Description of Physical Reality Be Considered Complete?*, 1935. About the pre-history, see Howard, *The Early History of Quantum Entanglement, 1905-1935*, 2007.
[26] Schrödinger, *Die gegenwärtige Situation in der Quantenmechanik*, 1935

characteristic trait of quantum mechanics, the one that enforces its entire departure from classical lines of thought. By the interaction the two representatives (or ψ-functions) have become entangled."[27]

In other words, entanglement occurs, if the state of a particle can*not* be described *in*dependently of the system it belongs to. This leads to quite strange effects, such as "spooky action-at-distance" (Einstein).

One of the most significant contributions to this issue is still Bell's inequality[28]. Bell's original inequality and its subsequent refinements[29] are about the correlation of measurement results of entangled properties. So far, all theoretical and experimental work has confirmed quantum and rejected classical correlation, *i.e,* Einstein's arguing.

It would lead us too far away to go into the more philosophical aspects of this topic. This would need, first of all, a comprehensive discussion of many basic notions, such as causality, reality, wave-particle dualism. Let me just mention few points.

- Heisenberg 1925 and Schrödinger 1926 have abandoned the classical trajectories from the very beginning of their work. Thus, one cannot measure properties a quantum particle is not exhibiting.

- Bohmian mechanics is not a hidden-variable theory.

- Propagation of signals (information) with speed larger than transverse electromagnetic waves in vacuo ('speed of light', c_0) is not a violation of causality and/or locality, but has actually been measured, though over small distances in special arrangements (evanescent, exponentially decreasing fields)[30].

Usually, the correlation over arbitrary large distances concern quantities, for which a conservation law holds true. For instance, if two electrons are ejected such, that the sum of the z-component of their spins vanishes at the moment of ejection, it

---

[27] Schrödinger, *Discussion of Probability Relations between Separated Systems*, 1935, p. 555; see also *Probability Relations Between Separated Systems*, 1936.

[28] John Stewart Bell (1928 - 1990), *On the Einstein-Podolsky-Rosen Paradox,* 1964; see also *Bertlmann's socks and the nature of reality*, 1981, and other essays in *Speakable and unspeakable in quantum mechanics*, 1987.

[29] See, in particular, John Francis Clauser (*1942), Michael Horne, Abner Shimony (1928 - 2015) & Richard A. Holt, *Proposed Experiment to Test Local Hidden-Variable Theories*, 1969.

[30] A. Enders & Nimtz, *On superluminal barrier traversal*, 1992; *Evanescent-mode propagation and quantum tunneling*, 1993; Mugnai, Ranfagni & L. Ronchi, *The question of tunneling time duration: A new experimental test at microwave scale*, 1998

perseveres so as long as the electrons fly away without external perturbation.[31] This suggests to apply the theorem, that the stationary wave functions depend solely on those parameters, on which the conserved quantities depend.[32] Indeed, the two most simple examples of entangled states are that of two equal bosons and fermions depicted in Figs **18.1** and **18.2** , respectively.

Entanglement were not a mystery, if it would not still be subject to broad debates, including such fundamental questions as the entropy of entangled states[33]. The reader is thus encouraged to follow this most exciting topic, last but not least, because it is at the heart of quantum computing and quantum cryptography.

Problem 18.6. (*) *Tackle the issue of entangled states by means of the theorem just mentioned, that the stationary wave functions depend solely on conserved quantities! Can Bell's and similar inequalities benefit from it?*

---

[31] *Cf* Sergei Vladimirovich Gantsevich & V. L. Gurevich, *Correlation, Entanglement and Locality of Quantum Theory*, 2017.
[32] *Cf* Born, *Quantenmechanik der Stoß vorgänge*, 1926, Introduction.
[33] Margarita A. Man'ko & Vladimir I. Man'ko, *Properties of Nonnegative Hermitian Matrices and New Entropic Inequalities for Noncomposite Quantum Systems*, 2015

'Aequat Causa Effectum' Spherically Symmetric Potentials *versus* Non-spherically Symmetric Orbits and Orbitals

Abstract: 'Aequat causa effectum' (the cause is equal to the effect) is an old metaphysical rule, which seems, however, to be violated for the single trajectories in spherically symmetric force fields, as the Kepler orbits are *not* spherically symmetric. The reason for the latter is the lower symmetry of the initial conditions. The rule is reestablished, if appropriate *sets* of trajectories (*i.e*, sets of initial conditions) are considered. The corresponding hodographs are also explored. The fact, that the Kepler orbits are plane, is derived by means of logical and symmetry arguments, *without* using angular momentum conservation. Spherically symmetric sets of Bohr orbitals are constructed by means of a sum rule for the spherically harmonic functions (Unsöld's theorem). This results in a less known similarity between classical and quantum mechanical systems. Metaphysical rules do not solve concrete physical problems, but encourage to explore them more profoundly.

Keywords: Aequat causa effectum, Hodograph, Initial conditions, Kepler orbits, Metaphysics, Spherically symmetric force fields, Unsöld's theorem.

"From given Equal, Equal follows."[1]

"If an ensemble of causes is invariant with respect to any transformation, the ensemble of their effects is invariant with respect to the same transformation."[2]

"The symmetry group of the cause is a subgroup of the symmetry group of the effect. Or less precisely: The effect is at least as symmetric as the cause.

Equivalent states of a cause are mapped to (*i.e.*, are correlated with) equivalent states of its effect... Also less precisely: Equivalent causes are associated with equivalent effects."[3]

---

[1] Newton, *De Gravitatione*, p. 36, Axiom 1; see also *Principia*, Book III, 2nd Rule.
[2] Paul Renaud, *Sur une généralisation du principe de symétrie de Curie*, 1935, quoted after Rosen, *Symmetry in Physics*, 1982, p. 26
[3] Rosen, *Symmetry in Science*, 1995, quoted after Rosen, *The Symmetry Principle*, 2005, pp. 308f.

'Aequat causa effectum' (the cause corresponds to its effect) is originally a scholastic rule about the relationship between reason and consequence.[4] It has been traced back to Nicholas of Autrecourt (1298 - 1369).[5] Leibniz changed it to a metaphysical law for motion.[6] Mayer exploited it to advance the energy conservation law.[7] In agreement with Helmholtz[8], I will use it rather heuristically. And I will concentrate myself on the benefit of Newton's and Euler's notions of (stationary) state, once more. Moreover, I will elucidate a less known similarity of CM and QM.

19.1. MOTIVATION

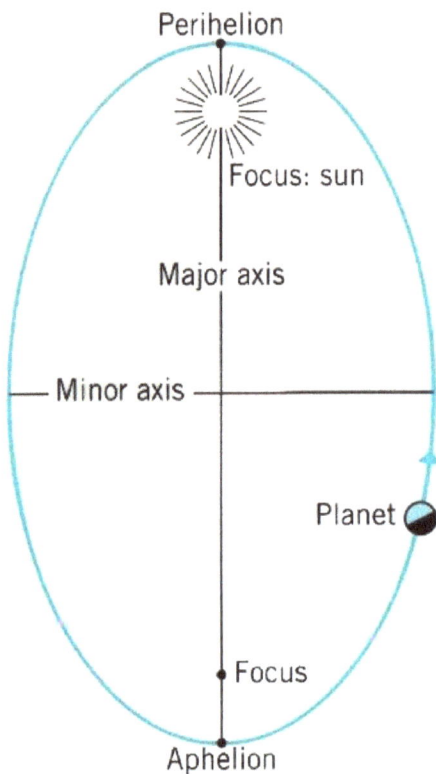

• Although the Newtonian field of gravity is *spherically* symmetric, the Kepler orbits – circles, ellipses, parabolas, hyperbolas – are *not*.

• Although the potential energy of the three-dimensional harmonic oscillator is *spherically* symmetric, its trajectories – circles, ellipses – are *not*.

• Although the Coulomb potential is *spherically* symmetric, the Bohr orbits – but *s* – are *not*.

Fig. 19.1. Plane of ecliptic

(http://www.lpl.arizona.edu/~tami/Sun/SchedulePTYS_files/lecture5-13.pdf)

---

[4] *Cf* de.wikipedia.org/wiki/Aequat_causa_effectum. – For the similar maxims "The same causes will always produce the same effects" and "like causes produce like effects", see Maxwell, *Matter and Motion*, 1877, § 19. A deviating statement is the 2nd quotation above.

[5] Hans-Ulrich Wöhler (*1950), *Dialektik in der mittelalterlichen Philosophie*, 2006, p. 163

[6] Leibniz, *Specimen Dynamicum*, 1982, pp. 22-23, 32-33 (cited after de.wikipedia.org/wiki/Aequat_causa_effectum)

[7] Julius Robert Mayer (1814 - 1878), *Bemerkungen über die Kräfte der unbelebten Natur,* 1842

[8] von Helmholtz, *Reden und Vorträge*, 1884, Vol. 1, p. 69 (cited after *Wikipedia, ibid.*)

How can it happen, that the symmetry of trajectories or wave functions is less than that of the potential field 'causing' them? Does this not contradict the rule 'aequat causa effectum'?

Shaw stresses, that the symmetry of a problem does *not* imply the symmetry of its solution.[9] For instance, the equation $x^2 = 1$ is invariant against the change $x \rightarrow -x$; its solutions: $x = \pm 1$, however, are *not*. Only the set of *all* solutions exhibits the symmetry of the problem.

Zee explains,

> "It is crucial to distinguish between the symmetry of physical laws and the symmetry imposed by a specific situation…This distinction…was one of Newton's great intellectual achievements, and it enabled physics as we know it to take shape."[10]

In his Nobel lecture, Wigner defines 'law' and 'situation' as follows.

> "The regularities in the phenomena which physical science endeavors to uncover are called the laws of nature. The name is actually very appropriate. Just as legal laws regulate actions and behavior under certain conditions, but do not try to regulate all actions and behavior, the laws of physics also determine the behavior of its objects of interest only under certain well defined conditions, but leave much freedom otherwise. The elements of the behavior which are not specified by the laws of nature are called initial conditions. These, then, together with the laws of nature, specify the behavior as far as it can be specified at all…"[11]

In view of the Titius-Bode law[12], the initial conditions of the planets appear not to be independent.[13] Einstein also noticed, that the initial conditions "are not free, but also have to obey certain laws."[14] This makes the notion 'initial condition' relative. This issue is beyond the scope of this book, however.

---

[9] Ronald "Ron" Shaw (*1929), *Symmetry, Uniqueness, and the Coulomb Law of Force*, 1965

[10] Zee, *Fearful Symmetry*, 1999, pp. 13f.

[11] Wigner, *Events, laws of nature, and invariance principles*, 1963, p. 7; see also his papers in Rosen (Ed.), *Symmetry in Physics*, 1982.

[12] Johann Daniel Titius (Johann Dietz, 1729 - 1796), *Betrachtungen über die Natur*, 1766; Johann Elert Bode (1747 - 1826), *Deutliche Anleitung zur Kentniss des Gestirnten Himmels*, 1772. – For an intriguing relationship to Pascal's triangle, see milan.milanovic.org/math/english/kepler/kepler.html.

[13] von Weizsäcker, *Über die Entstehung des Planetensystems*, 1943; Chandrasekhar, *On a New Theory of Weizsäcker on the Origin of the Solar System*, 1946

[14] Einstein, *Bietet die Feldtheorie Möglichkeiten für die Lösung des Quantenproblems?*, 1923, p. 360

I will show, that the rule 'aequat causa effectum' is validated, if one considers not only the forces/potentials, but also the initial conditions to be 'causes'.

19.2. A SHORT LOOK AT THE CLASSICAL INITIAL CONDITIONS. I

A single trajectory of a body of constant mass, m, is specified by the initial values of position, $\vec{r}_0 \equiv \vec{r}(t = 0)$, and velocity, $\vec{v}_0 \equiv \vec{v}(t = 0)$, or momentum, $\vec{p}_0 \equiv \vec{p}(t = 0)$, of the body. The developments of canonical, statistical and quantum mechanics have led to treating position and velocity/momentum on more or less equal footing. For our purpose, however, it is more appropriate to account for their different relationships to the external force. This yields one clue for the understanding of the difference in the symmetry properties of the curves $\vec{r}(t)$ and $\vec{v}(t)$, respectively $\vec{p}(t)$.

19.3. THE HODOGRAPH

In contrast to the curves $\vec{v}(t)$ and $\vec{p}(t)$, the curve $\vec{r}(t)$, the trajectory, is not immediately connected to the force, see Euler's principles in Subsection 2.3.2. For this, I examine first the curve $\vec{v}(t)$, the hodograph[15]. The literal meaning of the word 'hodograph' is 'path describer', where 'path' means not a trajectory of a particle, but the trace of the top of a parameter-dependent vector with fixed origin when the parameter is changing.

19.3.1. Central Forces

The most general central force can be written as

$$\vec{F} = -f(r)\frac{k\vec{r}}{r^3} \tag{19.1}$$

where k represents its strength. The rather arbitrary function $f(r)$ describes its deviation from Newton's force of gravity with $f(r) = 1$ and $k = GMm$, or Coulomb's force with $f(r) = 1$ and $k = -Qq/4\pi\varepsilon_0$.[16]

For the spherical harmonic oscillator, $f(r) = (r/r_0)^3$, where $r_0 = \sqrt[3]{k/k_s}$ (k_s being the 'spring constant') is a characteristic extension.

---

[15] Hamilton, *The hodograph or a new method of expressing in symbolic language the Newtonian law of attraction*, 1846; for developments till today, see José F. Cariñena, Manuel F. Rañada & Mariano Santander, *A new look at the Feynman 'hodograph' approach to the Kepler first law*, 2016.

[16] Readers wondering about the difference in the appearance of 4π as well as readers adhering to Gauss' units of measurement are advised to look at the geometric meaning of 4π as the surface of the unit sphere.

5

It is certainly interesting to consider the motion for the case that the function $f(r)$ changes its sign, so that it is attracting: $f(r) < 0$, at some distances, r, and repulsing: $f(r) > 0$, at other ones. This, however, is beyond the scope of this chapter as its results are independent of the sign of $f(r)$.

19.3.2. The Hodograph is Plane

Without referring to angular momentum conservation and the like, the fact, that the hodograph is plane, can be established as follows. Consider an arbitrary initial velocity vector not pointing towards or away from the origin of the central force.

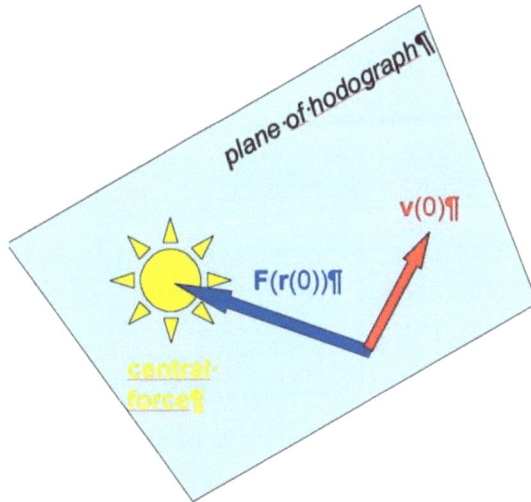

Fig. 19.2. The hodograph lies in the plane spanned by the force vector at initial position, $\vec{F}(\vec{r}(0))$, and the initial velocity vector, $\vec{v}(0)$.

There is a plane spanned by this vector and the center of the force field, see **Fig. 19.2**. This plane is a *mirror plane* of symmetry of the force field. As a consequence, there is no net force perpendicular to it, see **Fig. 19.3**. For all points of this plane, the force vector lies in this plane. Hence, the velocity vector stays in this plane for all later times.

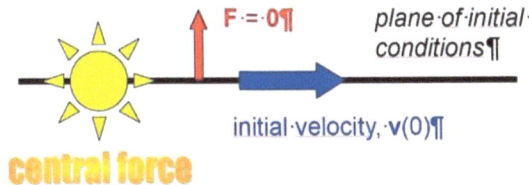

Fig. 19.3. There is no net force perpendicular to the plane of initial conditions.

19.3.3. The Hodograph is a 'Circle with Changing Radius'

The plane of the hodograph be the xy-plane. Generalizing Sommerfeld's calculations,[17] I write Newton's equation of motion in the form

$$\frac{d\dot{x}}{dt} = -\frac{kf(r)}{mr^2}\cos\varphi; \quad \frac{d\dot{y}}{dt} = -\frac{kf(r)}{mr^2}\sin\varphi \tag{19.2}$$

Here, φ is the polar angle (the centric or true anomaly of the astronomers) in the plane of motion, $z = 0$. The area law states that $\dot{\varphi} = L/mr^2$, L being the angular momentum of the orbit. Thus, dividing the equations (19.2) by $\dot{\varphi}$ yields[18]

$$\frac{d\dot{x}}{d\varphi} = -\frac{kf(r)}{L}\cos\varphi; \quad \frac{d\dot{y}}{d\varphi} = -\frac{kf(r)}{L}\sin\varphi \tag{19.3}$$

These two equations are easily integrated to

$$\dot{x} = -\frac{kf(r)}{L}\sin\varphi + \dot{x}_c; \quad \dot{y} = \frac{kf(r)}{L}\cos\varphi + \dot{y}_c \tag{19.4}$$

$\dot{x}_c$ and $\dot{y}_c$ are constants of integration, here, the coordinates of the center of the hodograph, see next equation.

The hodograph hence is given by the equation

$$[\dot{x}(t) - \dot{x}_c]^2 + [\dot{y}(t) - \dot{y}_c]^2 = \left[\frac{k}{L}f(r)\right]^2 \tag{19.5}$$

It describes a 'circle with changing radius', $v_h(t) = |f(r(t))|k/L$.

For the Kepler orbits, it is a common circle with constant radius, $v_h = GMm/L$.[19]

$$[\dot{x}(t) - \dot{x}_c]^2 + [\dot{y}(t) - \dot{y}_c]^2 = \left[\frac{GMm}{L}\right]^2 \tag{19.6}$$

The center of the hodograph is not the origin, $(0,0)$, but the center is displaced to $(\dot{x}_c, \dot{y}_c)$. It does lie in the origin: $(\dot{x}_c, \dot{y}_c) = (0,0)$, if the trajectory itself, $(x(t), y(t))$, describes a circle.

---

[17] Sommerfeld, *Mechanik*, 1994, § 6 – for purely geometrical treatments, see Cariñena, Rañada & Santander, *A new look at the Feynman 'hodograph' approach to the Kepler first law,* 2016.
[18] The case $L = 0$ is excluded as it corresponds to trajectories along a straight line through the force center.
[19] Maxwell, *Matter and Motion*, 1877, § 133; David Louis Goodstein (*1939) & Judith R. Goodstein, *Feynman's Lost Lecture. The Motion of Planets Around the Sun,* 1996

19.3.4. Spherically Symmetric Sets of Hodographs

The symmetry of a hodograph is a conjunction of the symmetries of the force field and of the initial velocity vector. The lower symmetry of the latter makes the hodograph to exhibit a non-spherical symmetry. In turn, initial conditions of higher symmetry lead to figures of higher symmetry. This can be reached through considering *sets* of hodographs belonging to certain *sets* of initial conditions.

Problem 19.1. *Consider the set of all hodographs (19.6) of constant radius, $v_h = GMm/L$, for which $\dot{x}_c$ and $\dot{y}_c$ lie on the circle $\dot{x}_c^2 + \dot{y}_c^2 = const$. Show that this set forms a circular symmetric figure around the origin!*

Problem 19.2. *Continuing the foregoing problem, build sets of hodographs (19.6) that form spherically symmetric figures!*

Problem 19.3. *Build sets of hodographs forming spherically symmetric figures for the spherical harmonic oscillator, $f(r)r^3$!*

Problem 19.4. *Build sets of hodographs forming spherically symmetric figures for general central forces ($f(r)$ being arbitrary)!*

19.4. The Trajectory

19.4.1. A Short Look at the Classical Initial Conditions. II

In contrast to the hodograph, the trajectory of a body is determined by the initial values of *both*, velocity *and* position. Hence, the symmetry of the trajectory is affected by the interplay of both. This resembles the interplay of both crystal symmetry and quasi-momentum, $\vec{k}$, in the stationary states (band structure) of Bloch electrons in non-cubic crystals.[20]

19.4.2. The Trajectory Lies in the Plane of the Hodograph

The fact, that the trajectory, $\vec{r}(t)$, of a body subject to a central force lies in the plane of the hodograph, can be proven in the same manner as the fact, that the hodograph, $\{\vec{v}(t)\}$, is plane (see Subsection 19.3.2). If the velocity vector stays

---

[20] Enders, Arthur Bärwolff, Michael Wörner & Suisky, **k.p** *theory of energy bands, wave functions, and optical selection rules in strained tetrahedral semiconductors*, 1995.

within a plane, there is no change of position off that plane. Hence, the position vector stays within that plane.

Without referring to the hodograph, one can argue as follows. The plane spanned by initial radius vector and initial velocity vector (see **Fig. 19.4**) is a mirror plane of symmetry. For this, there is no net force perpendicular to it, and thus no motion off it.

By the way, already Newton knew, that, within his mechanics, Kepler's 2nd law holds true for rather general central forces:

> "If an orbiting body is subject solely to a single attracting centripetal force, its radius vector, drawn from the body to the attracting center, sweeps out equal areas in equal times."[21]

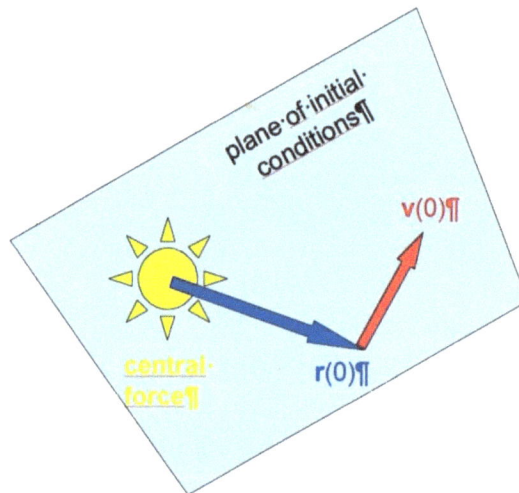

Fig. 19.4. The plane spanned by the initial position and velocity vectors.

Problem 19.5. *Examine Leibniz's (1695) statement, that the parabola is the limit case of the ellipsis, when one fixes one focus and moves the other focus into infinity! Hints: (i), consider ellipsis and parabola as conical sections; (ii) explore the equation* $(x^2/a^2 + y^2/b^2) = 1$!

Problem 19.6. (*) *Which consequences can be drawn from Newton's generalization of Kepler's 2nd law just quoted?*

---

[21] Newton, *De motu corporum in gyrum*, 1684; after en.wikipedia.org/wiki/De_motu_corporum_in_gyrum

19.4.3. Scale Symmetry, or Mechanical Similarity

"Sed res est certissima exactissimaque, quòd *proportio qua est inter binorum quorumcunque Planetarum tempora periodica, sit præ cise sesquialtera proportionis* mediarum distantiarum, id est *Orbium* ipsorum…"[22]

In this subsection, I make a sidestep to a symmetry of the trajectories that is of general interest.

The spherical symmetry of central fields and the symmetry of the orbits in them are also interrelated through a concept known as scale symmetry, or mechanical similarity[23]. Here, orbits of different size, but similar shape are compared each with another.

Given the central force (19.1) with $f(r) = (r/r_0)^b$ (r_0 – characteristic distance, b = const), Newton's equation of motion reads

$$m\frac{\mathrm{d}^2\vec{r}}{\mathrm{d}t^2} = -kr_0^{-b}r^{b-3}\vec{r} \qquad (19.7)$$

Now distances and periods be rescaled as

$$r = \alpha R, \quad t = \beta T \qquad (19.8)$$

In the new variables, $\vec{R}, T$, Newton's equation of motion (19.7) becomes

$$m\frac{\mathrm{d}^2\vec{R}}{\mathrm{d}T^2} = -\alpha^{b-3}\beta^2 kr_0^{-b}R^{b-3}\vec{R} \qquad (19.9)$$

This equation – and hence Newton's equation of motion (19.7) – is *in*dependent of the scaling transform (19.8), if

$$\alpha^{b-3}\beta^2 = 1 \qquad (19.10)$$

This invariance is called *scale symmetry*, or *mechanical similarity*.

Kepler's 3rd law quoted at the beginning of this subsection requires $b = 0$, as in Newton's force law.

---

[22] Kepler, *Harmonices mundi Libri V*, 1619, Ch. 3, § 8, p. 189 – Engl.: But it is absolutely certain and exact, that *the proportion between the periodic times of any two planets is precisely the sesquialternate proportion* [*i.e.*, the ratio of 3:2] of their mean distances, *i.e.*, of the *orbits* themselves… (*cf* https://en.wikipedia.org/wiki/Kepler's_laws_of_planetary_motion#Third_law).

[23] See, *e.g*, Roman Wladimir Jackiw (*1939), *Introducing scale symmetry*, 1972.

In bypassing I note, that the scale transform (19.8) is a special case of a Lie transform.[24] The theory of the groups of continuous transforms[25] – nowadays called Lie groups – represents a most powerful mathematical tool. It is, however, beyond the scope and the level of this book.[26]

19.4.4. Accidental, or Hidden Symmetry. Laplace-Runge-Lenz Vector

In this subsection, I make a sidestep to another symmetry of the trajectories that is of general interest.

19.4.4.1. The forces $\vec{F}_{-2} \sim \vec{r}/r^3$ (Newton, Coulomb) and $\vec{F}_1 \sim \vec{r}$ (Hooke) are distinguished

"*3. The true Theory of* Elasticity *or* Springines, *and a particular Explanation thereof in several Subjects in which it is to be found: And the way of computing the velocities of Bodies moved by them.* ceiiinossssttuu."[27]

Among the central forces (19.1), the forces $\vec{F}_{-2} \sim \vec{r}/r^3$ (Newton, Coulomb) and $\vec{F}_1 \sim \vec{r}$ (Hooke) are distinguished as follows.

Theorem 19.1. (Newton 1) *Let a body revolve in an ellipse, subject to a centripetal force tending toward the center of the ellipse. Then, the force vector is proportional to the vector from the force center to the body.*[28]

Theorem 19.2. (Newton 2) *Let a body revolve in an ellipse, subject to a centripetal force tending toward a focus of the ellipse. Then, the force vector is proportional to the vector from the force center to the body divided by the 3rd power of the length of that vector being the distance of the body from that focus.*[29]

---

[24] N. Wyn Evans, *Group theory of the Smorodinsky–Winternitz system*, 1991

[25] Marius Sophus Lie (1842 - 1899), *Theorie der Transformationsgruppen*, 1888…1893, *Vorlesungen über Differentialgleichungen mit bekannten infinitesimalen Transformationen*, 1891; *Die Grundlagen für die Theorie der unendlichen kontinuierlichen Transformationsgruppen. I*, 1891

[26] For a physically oriented introduction, see, *e.g.*, Robert Gilmore (*1941), *Lie Groups, Lie algebras, and Some of Their Applications*, 1974. The classical Kepler problem is treated in Exercises 9.VI.10…13.

[27] Robert Hooke (1635 - 1703), *A Description of Helioscopes and some other Instruments*, 1676, p. 31; http://www.e-rara.ch/zut/collections/content/pageview/731092. The anagram ceiiinossssttuu means "Ut tensio, sic vis; That is, The Power of any String is in the same proportion with the Tension thereof…" (Hooke, *Lectures de potentia restitutiva, or, of Spring*, 1678, p. 1; https://books.google.de/books?id=XeU5QzqCz6AC pg=RA1-PA117#v=onepage q f=false).

[28] After Newton, *Principia*, Book I, Sect. 2, Prop. 10. – In Corollary 2 to that proposition, Newton states, that the period of revolution is *in*dependent of the size of the ellipse.

[29] After Newton, *Principia*, Book I, Sect. 3, Prop. 11. – After that, Newton states, that the same holds true for parabolic and hyperbolic orbits.

In bypassing I note, that the pure amount of Newton's insights is overwhelming. In spite of his phenomenal mastering of geometry, it is hardly conceivable that he has obtained all of them by geometrical reasoning, *not* exploiting his equation of motion and force law.

$$m\frac{\mathrm{d}^2\vec{r}}{\mathrm{d}t^2} = \mathrm{const}\,\vec{r} \quad \mathrm{resp.} \quad -\mathrm{const}\frac{\vec{r}}{r^3} \tag{19.11}$$

Newton's insights are generalized in Bertrand's theorem[30].

Theorem 19.3. (Bertrand) *Among all central forces, there are only two ones, for which all bound orbits are closed orbits: (i), the inverse-square central force such as Newton's gravitational or Coulomb's electrostatic forces, (ii), Hooke's force of the spherical harmonic oscillator.*

Problem 19.7. (Hooke's law and exceptions*) *Write the elastic force as power series in the extension! Show, that the lowest-order term is odd! In which cases, Hooke's force law does not apply?[31] Show this behavior to stress the intimate relationship between strong electronic interactions and mechanical properties of solids![32]*

19.4.4.2. Laplace-Runge-Lenz vector

"No scientific discovery is named after its original discoverer."[33]

However, Coulomb's $(b = 0)$ and Hooke's $(b = 3)$ forces are not *per se* distinguished against all other power-law forces, $\vec{F} \sim r^{b-3}\vec{r}$. For this, the additional symmetries arising in them are sometimes referred to as *accidental*, or *hidden* symmetries.[34]

As a matter of fact, the three-dimensional rotational symmetry of the Newton-Coulomb potential is extended by the Laplace-Runge-Lenz (LRL) vector[35] to the

---

[30] Joseph Louis François Bertrand (1822 – 1900), *Théorème relatif au mouvement d'un point attiré vers un centre fixe*, C. R. Acad. Sci. 77 (1873) 849–853; F. C. Santos, V. Soares & A. C. Tort, *An English translation of Bertrand's theorem*, 2007

[31] For an example, see Elena Gati, Markus Garst, Rudra S. Manna, Ulrich Tutsch, Bernd Wolf, Lorenz Bartosch, Harald Schubert, Takahiko Sasaki, John A. Schlueter & Michael Lang, *Breakdown of Hooke's law of elasticity at the Mott critical endpoint in an organic conductor*, 2016

[32] Jörg Schmalian, *Elastisch dank einflussreicher Elektronen*, 2017, p. 25

[33] Stephen Mack Stigler (*1941), *Stigler's law of eponymy*, 1980

[34] See, *e.g*, Géza Györgyi (1930-1973) & J. Révai, *Hidden Symmetry of the Kepler Problem*, 1965.

[35] Jacob Hermann (1678 - 1733), *d'investigare l'Orbite de' Pianeti, nell'ipotesi che le forze centrali o pure le gravit'a delgi stessi Pianeti sono in ragione reciproca de' quadrati delle distanze, che i medesimi tengono dal*

four-dimensional rotational symmetry.[36] The LRL vector is originally defined for the Newton-Kepler force, $\vec{F} = -k\vec{r}/r^3$.

$$\vec{A} \underset{\text{def}}{=} \vec{p} \times \vec{L} - mk\frac{\vec{r}}{r} \qquad (19.12)$$

where $\vec{L} = \vec{r} \times \vec{p}$ is the (constant) angular momentum. It is a *constant of motion*. The symmetry extension is most simply shown as follows.

The components of the angular momentum, $L_{x,y,z}$, obey the Poisson algebra

$$\{L_i, L_j\} = \varepsilon_{ijk} L_k \qquad (19.13)$$

(always summation over repeated indices). Here, ε_{ijk} is the permutation, or Levi-Civita symbol[37] of rank 3;

$$\{L_i, L_j\} \underset{\text{def}}{=} \frac{\partial L_i}{\partial r_k}\frac{\partial L_j}{\partial p_k} - \frac{\partial L_i}{\partial p_k}\frac{\partial L_j}{\partial r_k} \qquad (19.14)$$

is the Poisson bracket of L_i and L_j.[38] Being a special Lie algebra, the Poisson algebra (19.13) is related to the rotation symmetry of free space. In turn, the rotation symmetry of the force field implies the conservation of the angular momentum, $\vec{L}$.

Now, the rescaled LRL vector $\vec{D} = (-2mE)^{-1/2}\vec{A}$ extends the Poisson algebra (19.13) for $\vec{L}$ to a Poisson algebra for $\vec{L}, \vec{D}$.[39]

$$\{D_i, L_j\} = \varepsilon_{ijk} D_k; \quad \{D_i, D_j\} = \varepsilon_{ijk} L_k \qquad (19.15)$$

Here, the total energy is gauged such that the bound states have got $E < 0$. This Poisson algebra is a special case of the Lie algebra for the *four*-dimensional rotation

---

Centro, a cui si dirigono le forze stesse, 1710; *Extrait d'une lettre de M. Herman à M. Bernoulli datée de Padoüe le 12. Juillet 1710*, 1732; Laplace, *Traité de mécanique celeste*, 1799 Tome I, Premiere Partie, Livre II, pp.165ff.; Carl David Tolmé Runge (1856 - 1927), *Praxis der Gleichungen*, 1921; *Vector Analysis*, 1923; Wilhelm Lenz (1888 - 1957), *Über den Bewegungsverlauf und die Quantenzustände der gestörten Keplerbewegung*, 1924, eq. (2) – for more sources and newer developments, see en.wikipedia.org/wiki/Laplace%E2%80%93Runge%E2%80%93Lenz_vector and links therein.

[36] Fock, *Zur Theorie des Wasserstoffatoms*, 1935; Bargmann, *Zur Theorie des Wasserstoffatoms*, 1936 – here, the Kepler problem in higher-dimensional spaces is discussed, too.

[37] Ricci-Curbastro & Levi-Civita, *Méthodes du calcul différentiel absolu et leurs applications*, 1900

[38] Poisson, *Sur la variation des constantes arbitraires dans les questions de mécanique*, 1816

[39] David M. Fradkin, *Existence of the Dynamic Symmetries O4 and SU3 for All Classical Central Potential Problems*, 1967, § 2

group $O(4)$. The generalization to rather arbitrary central forces (and special-relativistic mechanics) is also given by Fradkin (*loc. cit.*, § 4).

Problem 19.8. *Show, that the LRL vector (19.12) is invariant against the scale transform (19.8)!*

Problem 19.9. (*) *Can the LRL vector (19.12) be build through the requirements of time and scale independence?*

Problem 19.10. (*) *Continuing the foregoing problem, can analogous vectors be build for other central potentials through the requirements of time and scale independence?*

Problem 19.11. (*) *Is the LRL vector (19.12) a stationary-state function?*

19.4.4.3. Geometrical Meaning of the Laplace-Runge-Lenz Vector

The LRL vector $\vec{A}$ is parallel to the semi-major axis and points in the direction from the force center to the perihelion. Its length, $A^2 = m^2k^2 + 2mEL^2 = m^2k^2e^2$ (*e* being the eccentricity of the orbit), has no geometrical meaning. It is obvious, however, that it can be rescaled to geometrically relevant lengths. One example is the eccentricity vector, $\vec{e} = \vec{A}/mk$. Another example is the vector $\vec{A}/E$ ($E < 0$). It equals the vector from the force center to the 'empty' focus of length $2ae$.[40]

19.4.4.4. Hamilton's Binormal Vector and a Related Dyadic Tensor

If the LRL vector, $\vec{A}$, is a constant of motion, Hamilton's binormal vector[41],

$$\vec{B} = \frac{\vec{L} \times \vec{A}}{L^2} = \vec{p} - \frac{mk}{L^2 r}\vec{L} \times \vec{r}; \quad \vec{A} = \vec{B} \times \vec{L} \tag{19.16}$$

is a constant of motion, too. It is called 'binormal', because it is normal to both $\vec{A}$ and $\vec{L}$.

The two vectors, $\vec{A}$ and $\vec{B}$, are the principal axes of the symmetric dyadic tensor[42]

---

[40] Cariñena, Rañada & Santander, *A new look at the Feynman 'hodograph' approach to the Kepler first law*, 2016, § 3.2
[41] Hamilton, *Applications of Quaternions to Some Dynamical Questions*, 1845
[42] Fradkin, *Existence of the Dynamic Symmetries O4 and SU3 for All Classical Central Potential Problems*, 1967, § 3

$$\widehat{W} = \alpha \vec{A} \otimes \vec{A} + \beta \vec{B} \otimes \vec{B}; \quad W_{i,j} = \alpha A_i A_j + \beta B_i B_j \qquad (19.17)$$

where α and β are scaling constants. For the r^{-1}-potentials (Newton, Coulomb), $\widehat{W}$ is conserved for arbitrary values of α and β.

For the potential energy of the spherical harmonic oscillator: $V = (k/2)r^2$, the tensor $\widehat{W}$ is conserved, iff $\beta = mk\alpha$. If, moreover, $\alpha = 1/2m$, then,

$$\mathrm{Tr}\,\widehat{W} \equiv W_{11} + W_{22} + W_{22} = \frac{1}{2m}p^2 + \frac{k}{2}r^2 \qquad (19.18)$$

equals the Hamiltonian and total energy of the spherical harmonic oscillator. Furthermore, there are the conserved quantities[43]

$$G_1 = \frac{W_{1,2} + W_{2,1}}{2\omega}; \quad G_2 = \frac{W_{2,2} - W_{1,1}}{2\omega}; \quad \omega = \sqrt{\frac{k}{m}} \qquad (19.19)$$

Together with $G_3 = L_3/2$ they obey the Poisson algebra (19.13) as representation of the rotation symmetry.

19.4.5. Distinguished Forces – Distinguished Trajectories

Kinematically, the direction of the velocity vector represents a distinguished vector in space. For this, initial position vectors that are parallel or anti-parallel to the initial velocity vector are distinguished, too. The resulting orbits describe straight lines towards or away from the force center and exhibit a correspondingly high symmetry.

Dynamically, Newton's 2nd axiom postulates equilibrium between the field force (here, the centripetal force (19.1)) and the inertial force (here, of the moving body), $-md\vec{v}/dt$. As the velocity vector, $\vec{v}$, can be decomposed into its normal component, $\vec{v}_n$, and its tangential component, $\vec{v}_t$: $\vec{v} = \vec{v}_t + \vec{v}_n$, the inertial force can be decomposed into the centrifugal force[44]: $\vec{F}_{cf} = -m\dot{\vec{v}}_n$, which is related to the change of the direction of motion (see Fig. **19.5**[45]), and the *tangential* component: $\vec{F}_t = -m\dot{\vec{v}}_t$ being related to the change of the modulus of the velocity.

---

[43] Al-Hashimi, *Accidental Symmetry in Quantum Physics*, 2008, p. 15
[44] Descartes, *Le Monde*, 1664; Huygens, *Letter to Henry Oldenbourg, Secretary of the Royal Society, 1669*; Newton, *Principia*
[45] Wilfried Wittkowsky – own work, CC BY-SA 3.0, https://commons.wikimedia.org/w/index.php?curid=2796881.

Fig. 19.5. Action of the centrifugal force upon the seats of a swing carousel.

Accordingly, there are three distinguished cases.

1. The centrifugal force vanishes. This is the case of straight motion towards or away from the force center discussed above.

2. The centrifugal force exactly counterbalances the central force for all times, so that the distance of the revolving body from the force center is constant. Accordingly, the tangential component of the inertial force vanishes, so that the tangential velocity is constant. This is the case of circular orbits. It is possible for attracting central forces only.

3. The centrifugal force counterbalances the central force at certain points of time. This happens in the extremal points of the orbit, such as apogee and perigee.

19.4.6. Spherically Symmetric Sets of Trajectories

The set of all possible Kepler orbits fills the whole space, but the origin, and thus forms a spherically symmetric figure. For here, no single initial condition plays a role. Spherically symmetric figures are also formed for the set of all possible Kepler orbits being subject to the condition of

- given total energy, or

- given modulus of angular momentum, or

- given total energy and modulus of angular momentum.

Each of these conditions endows a classification on the set of all possible states.

Problem 19.12. (*) *Are there further subsets of spherical symmetry?* Hint*: Explore the set of all states with given Laplace-Runge-Lenz vector (19.12)!*

19.5. THE QUANTUM ANALOG

The quantum analog to the classical arguments above demonstrates the relevance of the stationary weight functions, $F_E(\vec{r})$ and $G_E(\vec{p})$, again. As in the foregoing sections, it is sufficient to treat the one-particle case.

19.5.1. Quantum Initial Conditions

Stationary quantum states are described by quantities – such as the stationary wave function, $\psi(\vec{r}, t) = \psi_E(\vec{r})\exp\{iEt/\hbar\}$ – that represent 'orbitals'. Bohr orbitals resemble a weighted set of periodic trajectories, see Fig. **19.6**.

The quantum initial conditions are thus not vectors, $\vec{r}(t_0), \vec{p}(t_0)$, in phase space, $\{\vec{r}, \vec{p}\}$, but vectors, $\psi(\vec{r}, t_0)$, in Hilbert space, $\{\psi(\vec{r}, t)\}$. Hence, averaging over initial conditions is averaging over wave functions or quantities built by means of them.

19.5.2. Accidental, or Hidden Symmetry.
Quantum Laplace-Runge-Lenz, or Pauli Vector

The concepts of Subsection 19.4.4 on accidental, or hidden symmetries of classical systems can be and has been transferred to the quantum realm. For later use I will present only the quantum Laplace-Runge-Lenz vector operator for the Coulomb potential.

$$\hat{\vec{A}} = \tfrac{1}{2}\left(\hat{\vec{p}} \times \hat{\vec{L}} - \hat{\vec{L}} \times \hat{\vec{p}}\right) - mk\hat{r}^{-1}\hat{\vec{r}} \qquad (19.20)$$

It could well be called 'Pauli operator', because Pauli[46] has invented it during his most elegant calculation of the hydrogen spectrum within matrix mechanics.[47] It is a constant of motion, because it does not explicitly depends on time and commutes with the Hamiltonian, $\hat{H} = -\hbar^2\Delta/2m - k/r$.

---

[46] Pauli, *Über das Wasserstoffspektrum vom Standpunkt der neuen Quantenmechanik*, 1926
[47] For the corresponding group-theoretical calculations see, *eg*, Gilmore, *Lie Groups, Lie algebras, and Some of Their Applications*, 1974, Exercise 9.VI.14.

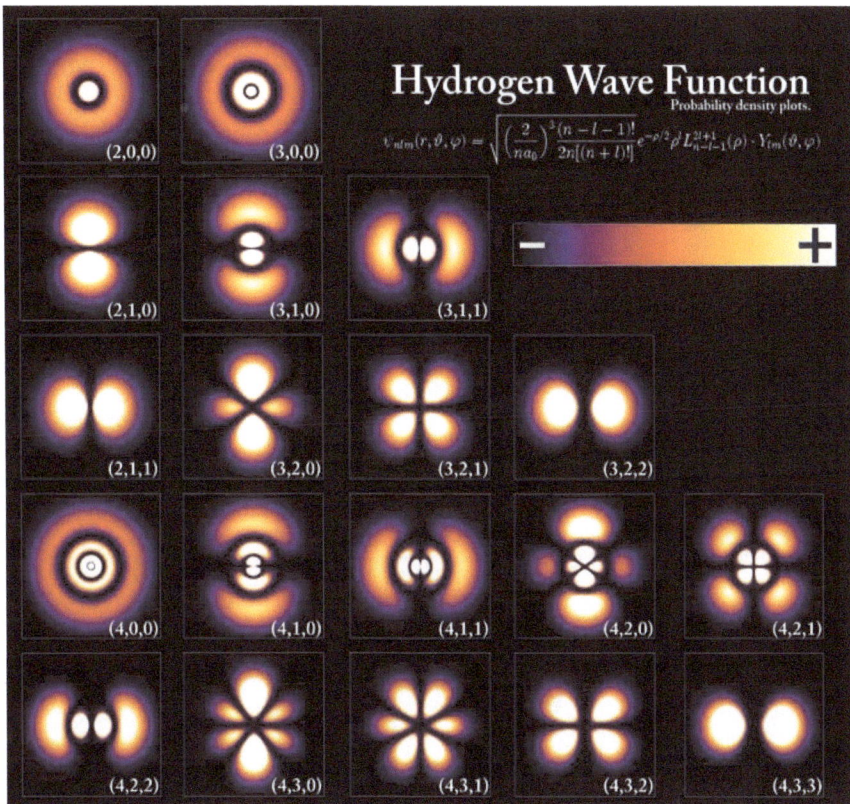

Fig. 19.6. Probability densities of Bohr orbitals, $|\psi_{nlm}(\vec{r})|^2$ (source: en.wikipedia.org/wiki/ File:Hydrogen_Density_Plots.png).

Analogously to the classical case above, the rescaling of $\hat{\vec{A}}$ to $\hat{\vec{D}} = \hat{\vec{A}}/\sqrt{-2mE}$ yields an operator, which, together with $\hat{\vec{L}}$, obeys a closed Lie algebra for $SO(4)$.[50] As common within QM, the Lie algebra is represented not by Poisson brackets (5.14), but by commutators.[51]

$$[\hat{D}_i, \hat{L}_j] = \varepsilon_{ijk}\hbar\hat{D}_k; \ [\hat{D}_i, \hat{D}_j] = \varepsilon_{ijk}\hbar\hat{L}_k \tag{19.21}$$

---

[50] Al-Hashimi, *Accidental Symmetry in Quantum Physics*, 2008, p. 17
[51] Lanczos, *The Poisson bracket*, 1972

While the spherical symmetry makes the energies of the Bohr orbitals (n, l, m) to be independent of the magnetic quantum number, m, this accidental symmetry makes them, additionally, to be independent of the azimuthal quantum number, l.

Problem 19.13. *Show, that the correspondence between Poisson bracket and commutator presupposes the existence of a universal constant of dimension action!*

19.5.3. Spherically Symmetric Sets of Bohr Orbitals

Let us consider the Bohr orbitals, $\psi_E(\vec{r}) = \psi_{nlm}(\vec{r})$. In spherical coordinates, they separate as

$$\psi_{nlm}(\vec{r}) = R_{nl}(r)\Psi_{lm}(\theta, \phi); \quad n = 1,2,3, \dots ;$$

$$l = 0,1, \dots n - 1; \quad m = -l, 1 - l, \dots l - 1, l \qquad (19.22)$$

Here we can exploit

Theorem 19.4. (Unsöld) *Due to the identity (in θ, ϕ)*

$$\sum_{m=-l}^{l} |\Psi_{lm}(\theta, \phi)|^2 \equiv \frac{2l+1}{4\pi} \qquad (19.23)$$

a filled or half-filled subshell of atomic orbitals with $l > 0$ is spherically symmetric and thus contributes an orbital angular momentum of zero.[52]

As a consequence, the weight function averaged over the direction of the angular momentum,

$$\langle F_{nlm}(\vec{r})\rangle_m \equiv \frac{\Omega_{nl}}{2l+1} \sum_{m=-l}^{l} |\psi_{nlm}(r, \theta, \phi)|^2 = \frac{\Omega_{nl}}{4\pi} R_{nl}^2(r) \qquad (19.24)$$

is spherically symmetric (Ω_{nl} being the characteristic volume). The classical analog is the set of all Kepler orbits with fixed total energy, E, and modulus of angular momentum, L.

The same (spherical symmetry) holds true for the graph of all values of $F_{nlm}(\vec{r}) = \Omega_{nlm}|\psi_{nlm}(\vec{r})|^2$ at

- given total energy, E_{nl}, or

---

[52] Albrecht Otto Johannes Unsöld (1905 - 1995), *Beiträge zur Quantenmechanik des Atoms*, 1927; S. M. Binder, *Unsöld's Theorem*, 2013

- given modulus of angular momentum, $L = \sqrt{l(l+1)}\hbar$.

The specification 'the energy equals the value E being part of the discrete spectrum'

> "characterizes a finite manifold of states in a rotationally invariant manner. This is – generally speaking – not possible within classical mechanics, because the initial conditions play too large a role…"[53]

Problem 19.14. *Are there further subsets of spherical symmetry?* Hint: *Explore all states with given quantum Laplace-Runge-Lenz vector (Pauli) operator (19.20)[54]!*

Problem 19.15. *What is the geometric interpretation of the limitation $l \le n - 1$?*

Problem 19.16. *What is the geometric interpretation of the limitation $|m| \le l$?*

19.6. WHAT METAPHYSICS WAS GOOD FOR

"There is a tremendous treasure of philosophical meaning behind the great theories of Euler and Lagrange, and of Hamilton and Jacobi…"[55]

The metaphysical postulate 'aequat causa effectum' does not solve physical problems, but it helps to explore them more profoundly.[56] Here, it leads to the examination of the role of the initial conditions as another 'cause' of a trajectory, additionally to the force. If the symmetry of the initial conditions equal that of the force field, the single effect (the symmetry of the resulting classical trajectory resp. quantum orbital) does so as well.

In both the classical and quantum cases, two different initial conditions lead to equivalent trajectories/orbitals, if the one can be obtained from the other one through a symmetry transformation of the force field/potential. The set of all such trajectories/orbitals forms a figure that exhibits the symmetry of the force field/potential.

---

[53] Wigner, *Über die elastischen Eigenschwingungen symmetrischer Systeme*, 1930, Sect. 1
[54] See also Arno Bohm (*1936), *Quantum Mechanics: Foundations and Applications*, 1986, pp. 208-222.
[55] Lanczos, *The Variational Principles of Mechanics*, 1986, p. xi
[56] For an introduction into the nature of metaphysics – its methodology, epistemology, ontology, our access to metaphysical knowledge – see Tuomas E. Tahko (*1982), *An Introduction to Metametaphysics*, 2015.

For Kepler orbits, the (spherical) symmetry of the boundary conditions is not lower than that of the force field. For this, they play no role for the symmetry of the motion. If the symmetry of the boundary conditions is lower than that of the force field, the influence of the boundary conditions on the symmetry of the motion can be treated analogously to the initial conditions.

Part V

Field Quantization as Selection Problem

CHAPTER 20

Introduction

"If quantum mechanics and relativity were revolutions in the sense of the French Revolution of 1789 or the Russian Revolution of 1917, then quantum field theory is more of the order of the Glorious Revolution of 1688: things changed only just enough so that they could stay the same."[1]

When compared with the quantization of point-mechanical systems, the quantization of fields faces two additional challenges.

1. The potential, kinetic and total energies are primarily given as energy *density fields* over space and time. Are the limiting functions to be introduced w.r.t. those densities?

2. The canonical momentum of a field is actually a momentum *density*, say, p. Hence, the kernel of the Fourier transform between the wave functions in configuration and in momentum-density configuration spaces is *not* equal to $\exp\{ipx/\hbar\}$.

Thus, one expects a commutation relation like

$$\hat{q}(x,t)\hat{p}(x',t) - \hat{p}(x',t)\hat{q}(x,t) = i\hbar\delta(x-x') \qquad (20.1)$$

and operators like

$$\hat{p}(x,t) = -i\hbar\frac{\delta}{\delta q(x,t)}; \quad \hat{q}(x,t) = i\hbar\frac{\delta}{\delta p(x,t)} \qquad (20.2)$$

(variational rather than partial derivatives), see Problem 20.1.

Again, the classical treatment turns out to be a good starting point for setting the notions, in particular, the momentum and energy *densities*.

Within the framework of this book, this part serves merely as an illustration of the power of the paradigm 'quantization as selection problem'; a concise treatment of quantum field theory (QFT) is *not* intended.

---

[1] Weinberg, *The search for unity. Notes for a history of quantum field theory*, 1977

Problem 20.1. (Variational derivatives) *Assume, that Ehrenfest's theorem[2] applies and calculate tentatively*

$$\left\langle \frac{\delta \hat{H}}{\delta \hat{p}} \right\rangle = \frac{\partial}{\partial t} \langle \hat{q} \rangle = \int \left\{ \frac{\partial \Psi^*}{\partial t} \hat{q} \Psi + \Psi^* \frac{\partial \hat{q}}{\partial t} \Psi + \Psi^* \hat{q} \frac{\partial \Psi}{\partial t} \right\} \mathrm{d}q$$

$$= \int \left\{ \frac{i}{\hbar} \hat{H} \hat{q} \Psi + \Psi^* \frac{\partial \hat{q}}{\partial t} \Psi + \Psi^* \hat{q} \frac{-i}{\hbar} \hat{H} \right\} \mathrm{d}q \qquad (20.3)$$

and

$$\left\langle -\frac{\delta \hat{H}}{\delta \hat{q}} \right\rangle = \frac{\partial}{\partial t} \langle \hat{p} \rangle = \int \left\{ \frac{\partial \Psi^*}{\partial t} \hat{p} \Psi + \Psi^* \frac{\partial \hat{p}}{\partial t} \Psi + \Psi^* \hat{p} \frac{\partial \Psi}{\partial t} \right\} \mathrm{d}q$$

$$= \int \left\{ \frac{i}{\hbar} \hat{H} \hat{p} \Psi + \Psi^* \frac{\partial \hat{p}}{\partial t} \Psi + \Psi^* \hat{p} \frac{-i}{\hbar} \hat{H} \right\} \mathrm{d}q \qquad (20.4)$$

to obtain $(\hat{H}\hat{q} - \hat{q}\hat{H})$, $(\hat{H}\hat{p} - \hat{p}\hat{H})$ and finally the guesses (20.1) and (20.2)!

---

[2] Ehrenfest, *Bemerkung über die angenäherte Gültigkeit der klassischen Mechanik innerhalb der Quantenmechanik*, 1927.

CHAPTER 21

Coupled Classical Oscillators

Abstract: Since this part *Field Quantization as Selection Problem* attempts to carry over the quantization of point-mechanical systems to that of classical fields, it begins with a chapter about coupled harmonic oscillators being the classical basis of phonons, the quantum particles of sound. After establishing the Newtonian equations of motion, the eigenmodes are calculated as a Toeplitz eigenvalue problem. Finally, the canonical momentum and the Hamiltonian are provided.

Keywords: Canonical momentum, Coupled classical oscillators, Eigenmodes, Hamiltonian, Quantization, Toeplitz eigenvalue problem.

A natural step from point-mechanical systems to spatial fields consists in considering a set of oscillators being located at different positions. As well known from crystals (see Section 16.6), the theory becomes most simple, when that set of positions forms a spatial lattice. For the purpose of this book it is sufficient to treat spatially *one*-dimensional systems.

21.1. CHAIN OF COUPLED CLASSICAL HARMONIC OSCILLATORS

Planck's (1900) resonators and – supposedly equivalently – Einstein's (1905) oscillators exist *in vacuo* and interact *via* electromagnetic radiation. In order to facilitate the treatment of those rather abstract imaginations, let me start with a quite concrete model of coupled oscillators. For the purpose of this book it is sufficient to consider the simplest case of a mono-atomic one-dimensional harmonic crystal lattice.[1] While a harmonic oscillator represents the paradigm of a single body moving in an external conservative force field, this chain represents a paradigm of interacting bodies.

Thus, I start with $N + 2$ bodies with equidistant equilibrium positions, $q_l^{(0)}$ ($l = 0,1, ... , N + 1$), $q_{l+1}^{(0)} - q_l^{(0)} = a = const.$ However, the bodies are actually not at rest, but oscillate around their respective equilibrium positions. For simplicity, I assume the two outermost bodies to be held fixed at their equilibrium positions (*fixed* boundary conditions[2]).

---

[1] For the general theory, *cf* Wigner, *Über die elastischen Eigenschwingungen symmetrischer Systeme*, 1930, and the classics William Thomson, 1st Baron Kelvin (1824 - 1907) & Peter Guthrie Tait (1831 - 1901), *Treatise on natural philosophy*, 1879; Born & Huang Kun (1919 - 2005), *Dynamical Theory of Crystal Lattices*, 1954.
[2] For the alternative of *periodic* boundary conditions, see Born & Theodore von Kármán (1881 - 1963), *Über Schwingungen in Raumgittern*, 1912; Haken, *Quantenfeldtheorie des Festkörpers*, 1973, § 7, or the books cited in the foregoing footnote.

$$q_0(t) = q_{N+1}(t) = 0 \tag{21.1}$$

The bodies are assumed to have got the mass m each and to be pairwise interconnected with massless springs of force constant κ. They oscillate not against fixed centers, but against the instantaneous positions of their next neighbors. The oscillations are assumed to be harmonic.

21.2. EQUATION OF MOTION

Under these conditions, the force upon body l due to its own elongation, $q_l(t) \equiv x_l(t) - x_l^{(0)}$, and to the elongations of its two neighbors, $q_{l\pm 1}(t)$, equals

$$K_l(t) = \kappa[q_{l+1}(t) - q_l(t)] - \kappa[q_l(t) - q_{l-1}(t)] \tag{21.2}$$

$$= \kappa[q_{l+1}(t) - 2q_l(t) + q_{l-1}(t)]; \quad l = 1(1)N \tag{21.3}$$

For harmonic oscillations, we have

$$q_l(t) = c_l\cos\omega t + s_l\sin\omega t; \quad c_l = q_l(0); \quad s_l = \dot{q}_l(0)/\omega \tag{21.4}$$

Due to $\ddot{q}_l(t) = -\omega^2 q_l(t)$, the Newtonian equations of motion,

$$m\ddot{q}_l(t) = K_l(t); \quad l = 1(1)N \tag{21.5}$$

simplify to a system of ordinary linear difference equations.

$$-m\omega^2 c_l = \kappa[c_{l+1} - 2c_l + c_{l-1}]; \quad l = 1(1)N$$

or

$$c_{l+1} - \left(2 - \frac{\omega^2}{\omega_{cl}^2}\right)c_l + c_{l-1} = 0; \quad l = 1(1)N; \quad \omega_{cl} \equiv \sqrt{\frac{\kappa}{m}} \tag{21.6}$$

The same system applies to the coefficients s_l. Here, ω_{cl} is the *cl*assical angular frequency of a single oscillator with force constant κ and mass m. The boundary conditions (21.1) become

$$c_0 = c_{N+1} = s_0 = s_{N+1} = 0 \tag{21.7}$$

21.3. TOEPLITZ EIGENVALUE PROBLEM

Together with the boundary conditions (21.7), the system (21.6) represents a spatially discrete classical eigenvalue problem. It is isomorphic with the eigenvalue problem for the matrix $(T_{ij})_{N \times N} = \delta_{j,i\pm1}$ being a special symmetric tridiagonal Toeplitz matrix[3].

$$\hat{T} \cdot \vec{r} = \mu\vec{r}; \quad \vec{r} = (x_1, \dots, x_N)^t; \quad x_0 = x_{N+1} = 0; \quad \mu \equiv 2 - \frac{\omega^2}{\omega_{cl}^2} \qquad (21.8)$$

(t denoting the transpose of a vector). The N eigenvalues are

$$\mu_j = 2 - \frac{\omega^2}{\omega_{cl}^2} = 2\cos\left(\frac{j}{N+1}\pi\right); \quad \omega_j = 2\omega_{cl}\sin\left(\frac{j}{N+1}\frac{\pi}{2}\right); \quad j = 1(1)N \qquad (21.9)$$

The corresponding N normalized eigenvectors, $\vec{r}^{(j)}$, have got the components

$$x_l^{(j)} = \sqrt{\frac{2}{N+1}}\sin\left(\frac{jl}{N+1}\pi\right); \quad j, l = 1(1)N \qquad (21.10)$$

21.4. SOLUTION TO THE EQUATION OF MOTION. EIGENMODES. PHONONS

The N eigenvectors, $\vec{r}^{(j)}$, represent N linearly independent solutions to eqs. (21.6). The actual oscillations of the bodies in the chain are

$$q_l(t) = \sum_{j=1}^{N} x_l^{(j)}\left[c_j\cos\omega_j t + s_j\sin\omega_j t\right]; \quad l = 1(1)N \qquad (21.11)$$

where the coefficients, c_j and s_j, are determined by the initial conditions.

$$q_l(0) = \sum_{j=1}^{N} c_j x_l^{(j)}; \quad l = 1(1)N \qquad (21.12a)$$

$$v_l(0) \equiv \dot{q}_l(0) = \sum_{j=1}^{N} \omega_j s_j x_l^{(j)}; \quad l = 1(1)N \qquad (21.12b)$$

In terms of the vectors $\vec{c} \equiv (c_1 \dots c_N)^t$ and $\vec{s} \equiv (s_1 \dots s_N)^t$, and of the matrices $\hat{\Omega} \equiv diag(\omega_1 \dots \omega_N)$ and

$$\hat{X} \equiv (\vec{r}^{(1)} \quad \vec{r}^{(2)} \quad \dots \quad \vec{r}^{(N)}) \qquad (21.13)$$

the initial conditions read

---

[3] Otto Toeplitz (1881 - 1940), *Zur Theorie der quadratischen and bilinearen Formen von unendlich vielen Veränderlichen. I*, 1911. My exposition follows that in Lothar Berg (1930 - 2015), *Lineare Gleichungssysteme mit Bandstruktur*, 1986, pp. 44f.

$$\vec{q}(0) = \hat{X} \cdot \vec{c}; \quad \vec{v}(0) = \hat{X} \cdot \hat{\Omega} \cdot \vec{s} \tag{21.14}$$

This immediately yields the coefficients $c^{(j)}$ and $s^{(j)}$ as ($\hat{X}^{-1} = \hat{X}^t$)

$$\vec{c} = \hat{X}^t \cdot \vec{q}(0); \quad \vec{s} = \hat{\Omega}^{-1} \cdot \hat{X}^t \cdot \vec{v}(0) \tag{21.15}$$

The eigenmodes, $\{\vec{r}^{(j)}(c_j\cos\omega_j t + s_j\sin\omega_j t)|j = 1 \dots N\}$, represent collective oscillations of the chain, which are independent each of another. They can be quantized like a single oscillator. This leads to the concept of *phonons* ('sound particles')[4].

21.5. TOTAL ENERGY OF THE LATTICE OSCILLATIONS. CANONICAL MOMENTUM. HAMILTONIAN

The force (21.2) is conservative, its potential energy being

$$V = \frac{\kappa}{2}\sum_{l=1}^{N} (q_l - q_{l-1})^2 \tag{21.16}$$

Correspondingly, the total energy of the oscillations of the chain equals

$$E = \frac{\kappa}{2}\sum_{l=1}^{N} [q_l(t) - q_{l-1}(t)]^2 + \frac{m}{2}\sum_{l=1}^{N} \dot{q}_l^2(t) \tag{21.17a}$$

$$= \frac{\kappa}{2}\sum_{l=1}^{N} [q_l(0) - q_{l-1}(0)]^2 + \frac{m}{2}\sum_{l=1}^{N} \dot{q}_l^2(0) \tag{21.17b}$$

$$= \frac{\kappa}{2}\sum_{j=1}^{N} c_j^2 + \frac{m}{2}\sum_{l=1}^{N} \omega_j^2 s_j^2 \tag{21.17c}$$

However, the coefficients c_j and s_j are *not* dynamical variables, so that no Hamiltonian is connected with this representation of E. For this, I will calculate the elongations, $q_l(t)$, the other way round.

The canonical momenta, p_l, equal the Cartesian ones: $p_l = m\dot{q}_l$; $l = 1(1)N$. Replacing in eq. (21.17a) the $\dot{q}_l$ with the p_l yields the Hamiltonian in the form $H = T + V$.

$$H = \frac{1}{2m}\sum_{l=1}^{N} p_l^2(t) + \frac{\kappa}{2}\sum_{l=1}^{N} (q_l(t) - q_{l-1}(t))^2 = E \tag{21.18}$$

Problem 21.1. *Prove the correctness of the Hamiltonian (21.18), e.g, by means of the corresponding equations of motion!*

---

[4] Frenkel, *Wave Mechanics. Elementary Theory*, 1932

CHAPTER 22

From Chains to Strings: Limit Transition to the Continuum

Abstract: The next step is the transition to the continuum being dealt with in this chapter. The Newtonian equations of motion become d'Alembert's wave equation, the energies and thus the Hamiltonian become functionals. The question, which field takes over the role of the free body, is examined by means of standing waves.

Keywords: d'Alembert's wave equation, Chain, Continuum, Energy density, Field, Free body, Hamiltonian, Hamiltonian equations of motion, Newtonian equations of motion, String.

There are at least two possibilities for the limit transition to the continuum. The one extends the oscillating bodies till they touch each another – the other one increases their number at constant length and mass of the chain. I will follow the second approach.

22.1. FROM NEWTONIAN EQUATIONS OF MOTION TO D'ALEMBERT'S WAVE EQUATION

Let us increasing the number of oscillating bodies: $N \to \infty$, at constant length, L, and constant total mass, $M = Nm$, of the chain (omitting the fixed end points, $l = 0$ and $l = N + 1$).

$$L = (N + 1)a = const \;\Rightarrow\; a = \frac{L}{N+1} \to 0 \qquad (22.1a)$$

$$M = Nm = const \;\Rightarrow\; m = \frac{M}{N} \to 0 \qquad (22.1b)$$

The set of the N elongations of the chain becomes the elongation *field* of the string.

$$\{q_l(t)\} \equiv \{q(la, t)\} \to q(x, t); \quad 0 \le x \le L \qquad (22.2)$$

with the boundary conditions

$$q(0, t) = q(L, t) = 0 \qquad (22.3)$$

The set of forces (21.2) becomes a force *field*.

Peter Enders

$$K_l(t) = \kappa[q((l+1)a,t) - 2q(la,t) + q((l-1)t)]$$

$$= \kappa\left[a^2 q''(al,t) + \frac{2a^4}{4!}q''''(al,t) + \cdots\right] \qquad \textbf{(22.4a)}$$

$$\downarrow$$

$$K(x,t) = \kappa a^2 q''(x,t); \quad q''(x,t) \equiv \frac{\partial^2 q}{\partial x^2}(x,t) \qquad \textbf{(22.4b)}$$

Here, it would be premature to set $\kappa a^2 = L^2\tilde{\kappa}$, because the inertial forces, $-mq_l(t)$, vanish, too. For this, I consider the acceleration field.

$$\ddot{q}(x,t) = \frac{1}{m}K(x,t) = \frac{N}{M}\left(\frac{L}{N}\right)^2 \kappa q''(x,t) = \frac{L^2\kappa}{MN}q''(x,t) \qquad \textbf{(22.5)}$$

It remains finite, if

$$\kappa \to \tilde{\kappa}N; \quad \tilde{\kappa} = const \qquad \textbf{(22.6)}$$

As a consequence, the oscillation frequency of the single oscillators diverges.

$$\omega_{cl} = \sqrt{\frac{\kappa}{m}} = \sqrt{\frac{\tilde{\kappa}}{M}}N \to \infty \qquad \textbf{(22.7)}$$

It looses its meaning and drops out off the theory. Its role as characteristic internal system parameter is taken over by the propagation speed, c. Indeed, Newton's equation of motion (21.5) becomes

$$M\ddot{q}(x,t) = L^2\tilde{\kappa}q''(x,t) \qquad \textbf{(22.8)}$$

that is, d'Alembert's wave equation[1].

$$\ddot{q}(x,t) = c^2 q''(x,t); \quad c^2 = \frac{L^2\tilde{\kappa}}{M} \qquad \textbf{(22.9)}$$

Its spatial content reflects the fact, that a wave represents not only a temporally, but also a spatially periodic process.

---

[1] D'Alembert, *Reflexions sur la cause generale des vents*, 1747. For the relation to Bernoulli's problem (Daniel Bernoulli, *Examen principorum mechanicae*, 1726), see Georg Friedrich Bernhard Riemann (1826 - 1866), *Über die Darstellbarkeit einer Function durch eine trigonometrische Reihe*, 1854/1867; Sydney Henry Gould (*1909), *Variational Methods for Eigenvalue Problems*, 1957, §§III.2f.

The speed of propagation, c, depends solely on intrinsic properties of the string. In contrast, frequency and wavelength represent *external* parameters of a wave, which are impressed upon it through external excitation and spatial restrictions (*e.g.*, resonators), respectively.

Problem 22.1. (Energy quanta hv*) *Comment on the fact, that, nevertheless, the field quantum has got the energy hv!*

22.2. ENERGIES AND HAMILTONIAN BECOME FUNCTIONALS

Furthermore, the potential and kinetic energies become *functionals*. For the potential energy, the limit transition is performed most simply by means of the Taylor expansion[2].

$$V(\{q(la,t)\}) = \frac{\kappa}{2}\sum_{l=0}^{N} [q((l+1)a,t) - q(la,t)]^2$$

$$= \frac{\kappa}{2}\sum_{l=0}^{N} \left[aq'(la,t) + \frac{a^2}{2!}q''(la,t) + \cdots\right]^2 \qquad \textbf{(22.10a)}$$

$$\downarrow$$

$$V[q(x,t)] = \frac{\kappa a}{2}\int_0^L \left(\frac{\partial q(x,t)}{\partial x}\right)^2 dx = \frac{\tilde{\kappa}L}{2}\int_0^L \left(\frac{\partial q(x,t)}{\partial x}\right)^2 dx \qquad \textbf{(22.10b)}$$

For the kinetic energy, the limit transition is most simple in terms of the velocities, $\dot{q}(la,t)$.

$$T(\{\dot{q}(la,t)\}) = \frac{m}{2}\sum_{l=1}^{N} \dot{q}^2(la,t) \qquad \textbf{(22.11a)}$$

$$\downarrow$$

$$T[\dot{q}(x,t)] = \frac{m}{2a}\int_0^L \dot{q}(x,t)^2 dx = \frac{M}{2L}\int_0^L \dot{q}(x,t)^2 dx \qquad \textbf{(22.11b)}$$

Now, when replacing the velocity field, $\dot{q}(x,t)$, with the Cartesian momentum field, $\dot{q}(x,t) \rightarrow p(x,t)/m$, the kinetic energy (22.11b) diverges. For this, one replaces $\dot{q}(x,t)$ with the momentum *density*, $\mathrm{p}(x,t) = p(x,t)/a$, of the Cartesian

---

[2] James Gregory (Gregorie, 1638 – 1675), *Vera circuli et hyperbolae quadratura*, 1667; *Exercitationes geometricae*, 1668; *Geometriae pars universalis*, 1668; Brook Taylor (1685 - 1731), *Methodus Incrementorum Directa et Inversa*, 1715

momentum field: $\dot{q}(x,t) \to \mathrm{p}(x,t)a/m = \mathrm{p}(x,t)L/M$. Then, the kinetic energy (22.11b) reads

$$T[\mathrm{p}(x,t)] = \frac{L}{2M}\int_0^L \mathrm{p}(x,t)^2 \mathrm{d}x \qquad (22.12)$$

The Hamiltonian thus becomes

$$H[\mathrm{p}(x,t),q(x,t)] = T[\mathrm{p}(x,t)] + V[q(x,t)]$$

$$= \frac{L}{2M}\int_0^L \mathrm{p}(x,t)^2\mathrm{d}x + \frac{\tilde{\kappa}L}{2}\int_0^L \left(\frac{\partial q(x,t)}{\partial x}\right)^2 \mathrm{d}x \qquad (22.13)$$

Here, all terms can be expressed as integrals over *densities* (set in fracture, as $\mathrm{p}(x,t)$).

$$H[\mathrm{p}(x,t),q(x,t)] = \int_0^L \mathfrak{H}(\mathrm{p},q)\mathrm{d}x = \int_0^L \{\mathfrak{T}(\mathrm{p}) + \mathfrak{V}(q)\}\mathrm{d}x \qquad (22.14)$$

with

$$\mathfrak{H}(\mathrm{p},q) = \mathfrak{T}(\mathrm{p}) + \mathfrak{V}(q) = \frac{L}{2M}\mathrm{p}(x,t)^2 + \frac{M}{2L}\left(c\frac{\partial q(x,t)}{\partial x}\right)^2 = \mathfrak{E}(x,t) \qquad (22.15)$$

The Hamiltonian equations of motion read

$$\dot{q} = \frac{\delta H[\mathrm{p},q]}{\delta \mathrm{p}} = \frac{\partial \mathfrak{H}(\mathrm{p},q,\partial q/\partial x)}{\partial \mathrm{p}} = \frac{L}{M}\mathrm{p} \qquad (22.16a)$$

$$\dot{\mathrm{p}} = -\frac{\delta H[\mathrm{p},q]}{\delta q} = -\frac{\partial \mathfrak{H}(\mathrm{p},q,\partial q/\partial x)}{\partial q} + \frac{\partial}{\partial x}\frac{\partial \mathfrak{H}(\mathrm{p},q,\partial q/\partial x)}{\partial(\partial q/\partial x)}$$

$$= \frac{M}{L}c^2\frac{\partial^2 q(x,t)}{\partial x^2} \qquad (22.16b)$$

As expected, they recover the wave equation (22.9).

22.3. STANDING WAVES

Within Newton's and Euler's representations of CM, the paradigm of the free, interaction-less body plays a distinguished role. Within QM, it allows for immediately establishing the Fourier transform as the relationship between the wave functions in configuration and momentum configuration spaces. Does the field of standing waves play that paradigmatic role?

Standing harmonic waves complying with the boundary conditions (22.3) consists of the components, or partial waves,

$$q_n(x,t) = q_n \sin(k_n x) e^{-i\omega_n t}; \quad \mathrm{p}_n(x,t) = \mathrm{p}_n \sin(k_n x) e^{-i\omega_n t};$$

$$\omega_n = ck_n = c\frac{2\pi n}{L}; \quad n = 1,2,\dots \tag{22.17}$$

The discretization of wavenumber: $k \to k_n$, and frequency: $\omega \to \omega_n$ is due to the boundary conditions. The values of the coefficients q_n and p_n follow, (i), from the initial conditions (see Problem 22.2) and, (ii), the Hamiltonian equations of motion (22.16), which yield

$$\dot{q}_n(x,t) = -i\omega_n q_n \sin(k_n x) e^{-i\omega_n t} = \frac{L}{M}\mathrm{p}_n(x,t) = \frac{L}{M}\mathrm{p}_n \sin(k_n x)e^{-i\omega_n t} \tag{22.18a}$$

$$\dot{\mathrm{p}}_n(x,t) = -i\omega_n \mathrm{p}_n \sin(k_n x) e^{-i\omega_n t} = -\frac{M}{L}c^2 k_n^2 q_n \sin(k_n x)e^{-i\omega_n t} \tag{22.18b}$$

Hence, (i), the partial amplitudes of the generalized position, q_n, and momentum densities, p_n, are interrelated as

$$\mathrm{p}_n = -i\omega_n \frac{M}{L} q_n \tag{22.19}$$

and, (ii), as expected, the angular frequency, ω_n, obeys the simplest, *viz*, linear dispersion relation.

$$\omega_n = ck_n \tag{22.20}$$

If there are *several* modes excited, the energy density is spatially and temporally *not* constant.

$$\mathfrak{E} = \frac{Mc^2}{2L}\sum_{n,n'} \{q_n^* q_{n'} k_n k_{n'}[\sin(k_n x)\sin(k_{n'}x)$$

$$+\cos(k_n x)\cos(k_{n'}x)]e^{i(\omega_{n'}-\omega_n)}\} \tag{22.21}$$

The total energy equals

$$E = \frac{Mc^2}{2}\sum_n |q_n|^2 k_n^2 \tag{22.22}$$

This looks like the total energy of a set of infinitely many single oscillators of mass M and classical turning points $\pm|q_n|$ evenly smeared out over the interval $(0, L)$.

Problem 22.2. *Calculate the coefficients q_n and p_n from the initial conditions!*

Problem 22.3 (*). *Try to simplify the mixed terms ($n \neq n'$) in formula (22.21)*

CHAPTER 23

Quantization of the String

Abstract: As an intermediate step, this chapter describes the quantization of the mechanical string. In order not to deal with the derivative in the potential energy density, a spatial Fourier transform is performed. Then, the procedure is quite analogously to the point-mechanical case, because the length of the string provides an extrinsic length parameter. Nevertheless, energy quanta, $\hbar\omega$, emerge.

Keywords: Energy quantum, Extrinsic length parameter, Quantization, String.

For pedagogical reasons, I first consider the quantization of the mechanical string of the foregoing chapter. It assumes an intermediate position between point-mechanical systems and general wave fields as

• it is described by field quantities, and

• it exhibits an extrinsic length parameter (L).

This allows for treating the two most fundamental problems of field quantization *separately*. The first one is the emergence of the field quantum, $\hbar\omega$. The second one is the dimension of the momentum *density*, $\mathfrak{p}$, in that the dimension of $q\mathfrak{p}$ equals *action/length*. It will be overcome in the next chapter.

Last but not least, the extrinsic length parameter highlights, that spatial and temporal dependencies are *not* treated on equal footing. For the normal-mode quantization of general wave fields, this deficiency has been criticized by Schleich[1]. It too will be overcome in the next chapter.

23.1. CHANGE FROM POSITION SPACE, $\{x\}$, TO ITS FOURIER SPACE, $\{k\}$

The (extrinsic) length parameter, L, allows for replacing the momentum density, $\mathfrak{p}(x,t)$, with the 'pseudo-momentum field',

$$\tilde{p}(x,t) \underset{\text{def}}{=} L\mathfrak{p}(x,t) \tag{23.1}$$

---

[1] Schleich, *Quantum Optics in Phase Space*, 2000, Ch. 10; see also Robert Bennett, Thomas M. Barlow & Almut Beige, *A physically motivated quantization of the electromagnetic field*, 2016.

Similarly, in order to deal with the derivative $\partial q(x,t)/\partial x$ in $V[q]$, one could replace it with the 'pseudo-elongation field',

$$\tilde{q}(x,t) \equiv L\,\partial q(x,t)/\partial x \qquad (23.2)$$

Then, the dimension of $\tilde{q}\tilde{p}$ equals *action*, and the wave functions in $\tilde{p}$- and $\tilde{q}$-spaces can be interrelated as ordinary Fourier transforms each to another, without introducing a new universal parameter. However, this would prevent the treatment of cases, where $q(x,t)$ itself enters the energy density. Moreover, and more important, it would prevent the natural way to the field quantum, $\hbar\omega$. In view of the new intrinsic parameter 'propagation speed', c, the natural way consists in the change from the position space, $\{x\}$, to its Fourier space, $\{k\}$, since $\hbar\omega = \hbar ck$.

In order to obey the boundary conditions (22.4) and to stick with real-valued variables, I use the Fourier sine transform.[2]

$$q(x,t) = \sum_k q_k(t)\sin(kx); \; q_k(t) = \frac{2}{L}\int_0^L q(x,t)\sin(kx)dx; \; k = \frac{\pi}{L},\frac{2\pi}{L},\ldots (23.3a)$$

$$\tilde{p}(x,t) = \sum_k \tilde{p}_k(t)\sin(kx); \; \tilde{p}_k(t) = \frac{2}{L}\int_0^L \tilde{p}(x,t)\sin(kx)dx; \; k = \frac{\pi}{L},\frac{2\pi}{L},\ldots (23.3b)$$

In terms of the Fourier sine transforms, $q_k(t)$ and $\tilde{p}_k(t)$, the Hamiltonian and the energies (3.17) read

$$H[\tilde{p}_k(t),q_k(t)] = T[\tilde{p}_k(t)] + V[q_k(t)] = \frac{1}{2M}\sum_k \tilde{p}_k^2(t) + \frac{\tilde{\kappa}L^2}{2}\sum_k k^2 q_k^2(t) \quad (23.4)$$

They are the sums of the *modal* values,

$$T_k(\tilde{p}_k(t)) = \frac{1}{2M}\tilde{p}_k^2(t) \qquad (23.5a)$$

$$V_k(q_k(t)) = \frac{\tilde{\kappa}L^2}{2}k^2 q_k^2(t) = \frac{M}{2}c^2 k^2 q_k^2(t) \qquad (23.5b)$$

$$H_k(\tilde{p}_k(t),q_k(t)) = T_k(\tilde{p}_k(t)) + V_k(q_k(t)) = E_k \qquad (23.5c)$$

Problem 23.1. *Comment on the fact, that the mixed terms (n,n') in the energy density (22.21) are absent in the formulae 23.5!*

---

[2] *Cf* formulae (3.22) and Fourier, *La Théorie Analytique de la Chaleur*, 1822, §§ 220ff.

23.2. LIMITING FUNCTIONS AND AMPLITUDES. WAVE FUNCTIONS

As in the point-mechanical case in Section 7.1, limiting functions, $F_{k,E}(q_k)$ and $G_E(\tilde{p}_k)$, can be introduced.

$$V_{k,E}(q_k) = F_{k,E}(q_k)V_k(q_k) \le E_k \qquad\qquad \textbf{(23.6a)}$$

$$T_E(\tilde{p}_k) = G_{k,E}(\tilde{p}_k)T(\tilde{p}_k) \le E_k \qquad\qquad \textbf{(23.6b)}$$

Although $T_k(\tilde{p}_k)$ does not explicitly depend on k, $G_{k,E}$ may do so *via* the characteristic length to be introduced next.

The further steps are quite analogous to that case, except that we are working in phase space $(q_k, \tilde{p}_k)$ rather than (x, p).

$F_E(q_k)$ and $G_E(\tilde{p}_k)$ are dimensionless and non-negative.

$$F_{k,E}(q_k) = F_{k,E}(q_k/q_{k,E}) \ge 0 \qquad\qquad \textbf{(23.7a)}$$

$$G_{k,E}(\tilde{p}_k) = G_{k,E}(\tilde{p}_k/\tilde{p}_{k,E}) \ge 0 \qquad\qquad \textbf{(23.7b)}$$

$q_{k,E}$ and $\tilde{p}_{k,E}$ are characteristic lengths in q_k- and $\tilde{p}_k$-space, respectively.

Without loss of generality I normalize as

$$\int_{-\infty}^{+\infty} F_{k,E}\left(\frac{q_k}{q_{k,E}}\right)\frac{dq_k}{q_{k,E}} = \int_{-\infty}^{+\infty} G_{k,E}\left(\frac{\tilde{p}_k}{\tilde{p}_{k,E}}\right)\frac{d\tilde{p}_k}{\tilde{p}_{k,E}} = 1 \qquad\qquad \textbf{(23.8)}$$

In what follows, $q_{k,E}$ and $\tilde{p}_{k,E}$ are omitted if not explicitly needed.

The corresponding limiting amplitudes and wave functions are introduced as

$$F_{k,E}(q_k) = \left|f_{k,E}(q_k)\right|^2 = q_{k,E}\left|\psi_{k,E}(q_k)\right|^2 \qquad\qquad \textbf{(23.9a)}$$

$$G_{k,E}(\tilde{p}_k) = \left|g_{k,E}(\tilde{p}_k)\right|^2 = \tilde{p}_{k,E}\left|\phi_{k,E}(\tilde{p}_k)\right|^2 \qquad\qquad \textbf{(23.9b)}$$

The almost complete analogy to the point-mechanical case suggests, that the wave functions, $\psi_{k,E}(q_k)$ and $\phi_{k,E}(\tilde{p}_k)$, are Fourier transforms each of another, where $q_{k,E}\tilde{p}_{k,E} = \hbar$.

$$\psi_{k,E}(q_k) = \sqrt{\frac{1}{2\pi\hbar}} \int_{-\infty}^{+\infty} e^{-\frac{i}{\hbar}q_k\tilde{p}_k}\phi_{k,E}(\tilde{p}_k)\mathrm{d}\tilde{p}_k;$$

$$\phi_{k,E}(\tilde{p}_k) = \sqrt{\frac{1}{2\pi\hbar}} \int_{-\infty}^{+\infty} e^{\frac{i}{\hbar}\tilde{p}_k q_k}\psi_{k,E}(q_k)\mathrm{d}q_k \qquad (23.10)$$

23.3. NON-CLASSICAL REPRESENTATION OF THE ENERGY DENSITY. STATIONARY AND TIME- DEPENDENT SCHRÖDINGER EQUATIONS

Analogously to eqs. (7.15), the non-classical representation of the modal energy becomes (in order to simplify the notation, I will mostly write E for E_k)

$$E_{k,\mathrm{ncl}} = \int_{-\infty}^{+\infty} F_{k,E}(q_k)V_k(q_k)\frac{\mathrm{d}q_k}{q_{k,E}} + \int_{-\infty}^{+\infty} G_{k,E}(\tilde{p}_k)T(\tilde{p}_k)\frac{\mathrm{d}\tilde{p}_k}{\tilde{p}_{k,E}}$$

$$= \int_{-\infty}^{+\infty} \left|\psi_{k,E}(q_k)\right|^2 V_k(q_k)\mathrm{d}q_k + \int_{-\infty}^{+\infty} \left|\phi_{k,E}(\tilde{p}_k)\right|^2 T(\tilde{p}_k)\mathrm{d}\tilde{p}_k \qquad (23.11)$$

The Fourier transform enables us to eliminate $\phi_{k,E}(\tilde{p}_k)$ and $\psi_{k,E}(q_k)$ to obtain the stationary Schrödinger equation in q_k- and $\tilde{p}_k$-space, respectively.

$$\left[-\frac{\hbar^2}{2M}\frac{\partial^2}{\partial q_k^2} + V_k(q_k)\right]\psi_{k,E}(q_k) = E_k\psi_{k,E}(q_k) \qquad (23.12\mathrm{a})$$

$$\left[\frac{\tilde{p}_k^2}{2M} + V_k\left(i\hbar\frac{\partial}{\partial\tilde{p}_k}\right)\right]\phi_{k,E}(\tilde{p}_k) = E_k\phi_{k,E}(\tilde{p}_k) \qquad (23.12\mathrm{b})$$

The time-dependence of the stationary wave functions reads

$$\psi_{k,E}(q_k,t) = \psi_{k,E}(q_k)e^{-\frac{i}{\hbar}Et}; \quad \phi_{k,E}(\tilde{p}_k,t) = \phi_{k,E}(\tilde{p}_k)e^{-\frac{i}{\hbar}Et} \qquad (23.13)$$

Finally, for each mode, k, the time-dependent Schrödinger equations in q_k- and $\tilde{p}_k$-spaces assume their standard form.

$$i\hbar\frac{\partial}{\partial t}\psi_k(q_k,t) = \left[-\frac{\hbar^2}{2M}\frac{\partial^2}{\partial q_k^2} + V_k(q_k)\right]\psi_k(q_k,t) \qquad (23.14\mathrm{a})$$

$$i\hbar\frac{\partial}{\partial t}\phi_k(\tilde{p}_k,t) = \left[\frac{\tilde{p}_k^2}{2M} + V_k\left(i\hbar\frac{\partial}{\partial\tilde{p}_k}\right)\right]\phi_k(\tilde{p}_k,t) \qquad (23.14\mathrm{b})$$

23.4. ENERGY QUANTA, $\hbar\omega$. PHONONS

Let us apply these results to the simple string we were starting from. With the modal potential energy (23.5b), the stationary Schrödinger equation (23.12a) becomes

$$\left(-\frac{\hbar^2}{2M}\frac{\partial^2}{\partial q_k^2} + \frac{Mc^2k^2}{2}q_k^2\right)\psi_{k,E}(q_k) = E_k\psi_{k,E}(q_k) \tag{23.15}$$

This is the stationary Schrödinger equation of a linear harmonic oscillator of mass M, angular frequency $\omega = \omega_k = ck$ ($k = \pi/L, 2\pi/L, 3\pi/L, ...$) and energy E_k, see eq. (9.12). The wave functions are Hermite functions, the energies are quantized as

$$E_k = E_{k;n} = \left(n + \frac{1}{2}\right)\hbar\omega_k; \; n = 0,1,2, ... \tag{23.16}$$

The energy quanta, $\hbar\omega_k$, belong to the Fourier components (modes), $q_k(t)\sin(kx)$, of the elongation field, $q(x,t)$. Like the normal modes of the oscillating chain, the Fourier components represent collective excitations of the string – *phonons* ('sound particles').

Classical Mechanics and Quantum Mechanics, 2019, 261-268

CHAPTER 24

Quantization of the Free Electromagnetic Field

Abstract: In contrast to the string, the free electromagnetic field has not got an extrinsic length parameter. However, the wave number provides an intrinsic length parameter. Energy quanta, $\hbar\omega$, emerge only for the transverse field components, for which a potential energy analogously to that of the harmonic oscillator acts. The analog to the longitudinal field modes is the free particle.

Keywords: Electromagnetic field, Energy quantum, Extrinsic length parameter, Intrinsic length parameter, Longitudinal electromagnetic field, Quantization, Transverse electromagnetic field.

Within the quantization of the string, spatial and temporal variables have not been treated on equal footing. This was justified by the existence of an external length parameter: the length, L, of the string. For a free field in space, there is no external length parameter. For this, such a length is sometimes artificially introduced (normal-mode quantization), what has been criticized by Schleich[1].

As a matter of fact, Planck (1900) has proposed Boltzmann's auxiliary concept of energy quanta, ε, as a fundamental quantity for monochromatic radiation of frequency ν in resonators: $\varepsilon = h\nu$. This provides the wavelength, $\lambda = c/\nu$, as (intrinsic) length parameter.

I will proceed quite analogously to the point-mechanical case above. For this, the discussion of the possible and impossible configurations will be omitted.

Altogether, this chapter sketches how our approach might be made fruitful for the quantization of the free electromagnetic field; it is not yet a fully fledged treatment. In contrast to the approaches, which re-interpret the field quantities as wave functions[2], wave functions will be introduced additionally to the field quantities.

24.1. CANONICAL THEORY

The energy density of the classical electromagnetic field *in vacuo* equals

---

[1] Schleich, *Quantum Optics in Phase Space*, 2000, Ch. 10; see also Bennett, Barlow & Beige, *A physically motivated quantization of the electromagnetic field*, 2016.

[2] See, *e.g*, Aleksander Ilyich Akhiezer (1911 - 2000) & Wladimir Borissowitsch Berestezki (1913 - 1977), *Quantenelektrodynamik*, 1962.

$$\mathfrak{E}(\vec{r},t) = \frac{\varepsilon_0}{2}\vec{E}(\vec{r},t)^2 + \frac{1}{2\mu_0}\vec{B}(\vec{r},t)^2 \tag{24.1}$$

being the sum of the densities of the "electrostatic" and "magnetic energies"[3]. Here, it is tempting to quantize this Hamiltonian analogously to the harmonic oscillator. Unfortunately, this is not directly possible, because $\vec{E}$ and $\vec{B}$ are not canonically conjugated each to another. Bennett *et al.*[4] have quantized $\vec{E}$ and $\vec{B}$ in terms of the creation and annihilation operators of the harmonic oscillator, postulating that the free electromagnetic field consists of photons. This approach is superior to normal-mode quantization, but not in the sense of Hertz's program. Moreover, I will include the longitudinal component of the electric field.

In terms of the vector potential, $\vec{A}(\vec{r},t)$, at vanishing scalar potential: $\Phi(\vec{r},t) \equiv 0$, the energy density (24.1) reads

$$\mathfrak{E}(\vec{r},t) = \frac{\varepsilon_0}{2}\left(\frac{\partial}{\partial t}\vec{A}(\vec{r},t)\right)^2 + \frac{1}{2\mu_0}\left(\nabla\times\vec{A}(\vec{r},t)\right)^2 \tag{24.2}$$

The two terms are considered to be the kinetic and potential energy densities, respectively.

$$\mathfrak{T}(\vec{r},t) = \frac{\varepsilon_0}{2}\left(\frac{\partial}{\partial t}\vec{A}(\vec{r},t)\right)^2; \quad \mathfrak{B}(\vec{r},t) = \frac{1}{2\mu_0}\left(\nabla\times\vec{A}(\vec{r},t)\right)^2 \tag{24.3}$$

The canonical momentum density equals

$$\vec{\mathfrak{P}}(\vec{r},t) = \frac{\partial[\mathfrak{T}(\vec{r},t)-\mathfrak{B}(\vec{r},t)]}{\partial(\partial\vec{A}(\vec{r},t)/\partial t)} = \varepsilon_0\frac{\partial}{\partial t}\vec{A}(\vec{r},t) = -\varepsilon_0\vec{E}(\vec{r},t) \tag{24.4}$$

The Hamiltonian density is the energy density (24.2) in terms of the canonical momentum density, $\vec{\mathfrak{P}}$.

$$\mathfrak{H}(\vec{r},t) = \frac{1}{2\varepsilon_0}\vec{\mathfrak{P}}(\vec{r},t)^2 + \frac{1}{2\mu_0}\left(\nabla\times\vec{A}(\vec{r},t)\right)^2 \tag{24.5}$$

Now I expand the fields into plane waves (continuous Fourier transform).

---

[3] *Cf* Maxwell, *A Dynamical Theory of the Electromagnetic Field*, 1965, Art. 71ff.; *A Treatise on Electricity and Magnetism* 1891, Art. 630ff.
[4] Bennett, Barlow & Beige, *A physically motivated quantization of the electromagnetic field*, 2016

$$\vec{\mathfrak{A}}(\vec{r}, t) = \frac{1}{(2\pi)^{3/2}} \iiint \vec{\mathfrak{A}}_{\vec{k}}(t) e^{i\vec{k}\cdot\vec{r}} d^3k \tag{24.6a}$$

$$\vec{\mathfrak{P}}(\vec{r}, t) = \frac{1}{(2\pi)^{3/2}} \iiint \vec{\mathfrak{P}}_{\vec{k}}(t) e^{i\vec{k}\cdot\vec{r}} d^3k \tag{24.6b}$$

In $\vec{k}$-space, the Hamiltonian equals

$$H = \iiint \mathfrak{H}_{\vec{k}} d^3k \tag{24.7}$$

where $\mathfrak{H}_{\vec{k}}$ is the Hamiltonian density in $\vec{k}$-space.

$$\mathfrak{H}_{\vec{k}} = \mathfrak{T}_{\vec{k}} + \mathfrak{V}_{\vec{k}} = \mathfrak{E}_{\vec{k}} = \text{const} \tag{24.8a}$$

$$\mathfrak{T}_{\vec{k}} = \frac{1}{2\varepsilon_0} \mathfrak{P}_{\vec{k}}(t)\mathfrak{P}_{-\vec{k}}(t) = \frac{1}{2\varepsilon_0} \left|\mathfrak{P}_{\vec{k}}^{\parallel}(t)\right|^2 + \frac{1}{2\varepsilon_0} \left|\mathfrak{P}_{\vec{k}}^{\perp}(t)\right|^2 \tag{24.8b}$$

$$\mathfrak{V}_{\vec{k}} = \frac{1}{2\mu_0}\left(\vec{k} \times \vec{A}_{\vec{k}}(t)\right) \cdot \left(\vec{k} \times \vec{A}_{-k}(t)\right) = \frac{1}{2\mu_0} k^2 \left|A_{\vec{k}}^{\perp}(t)\right|^2 \tag{24.8c}$$

Here, $\vec{A}_{\vec{k}} = \vec{A}_{\vec{k}}^{\parallel} + \vec{A}_{\vec{k}}^{\perp}$ and $\vec{\mathfrak{P}}_{\vec{k}} = \vec{\mathfrak{P}}_{\vec{k}}^{\parallel} + \vec{\mathfrak{P}}_{\vec{k}}^{\perp}$ have been decomposed into their components lying parallel and perpendicular to the vector $\vec{k}$, respectively. Since the plane waves ($\vec{k}$-modes) are independent each of another, the energy density in $\vec{k}$-space, $\mathfrak{E}_{\vec{k}}$, is time-independent.

Problem 24.1. *Comment on the fact, that, in eq. (24.2), the roles of kinetic and potential energy appear to be reversed when compared with eq. (24.1)!*

24.2. LIMITING FUNCTIONS AND AMPLITUDES. WAVE FUNCTIONS

For each plane wave ($\vec{k}$-mode), limiting (weight) functions and amplitudes can be introduced in a manner being quite analogous to the point-mechanical case in Section 7.1.

$$\mathfrak{V}_{\vec{k};\mathfrak{E}}\left(\vec{A}_{\vec{k}}\right) = F_{\vec{k};\mathfrak{E}}\left(\vec{A}_{\vec{k}}\right)\mathfrak{V}_{\vec{k}}\left(\vec{A}_{\vec{k}}\right) \leq \mathfrak{E}_{\vec{k}} \tag{24.9a}$$

$$\mathfrak{T}_{\vec{k};\mathfrak{E}}\left(\vec{\mathfrak{P}}_{\vec{k}}\right) = G_{\vec{k};\mathfrak{E}}\left(\vec{\mathfrak{P}}_{\vec{k}}\right)\mathfrak{T}_{\vec{k}}\left(\vec{\mathfrak{P}}_{\vec{k}}\right) \leq \mathfrak{E}_{\vec{k}} \tag{24.9b}$$

The limiting functions, $F_{\vec{k};\mathfrak{E}}$ and $G_{\vec{k};\mathfrak{E}}$ are dimensionless and non-negative.

$$F_{\vec{k};\mathfrak{E}}(\vec{A}_{\vec{k}}) = F_{\vec{k};\mathfrak{E}}\left(\frac{\vec{A}_{\vec{k}}}{A_{\vec{k};\mathfrak{E}}}\right) \geq 0 \tag{24.10a}$$

$$G_{\vec{k};\mathfrak{E}}(\vec{\mathfrak{P}}_{\vec{k}}) = G_{\vec{k};\mathfrak{E}}\left(\frac{\vec{\mathfrak{P}}_{\vec{k}}}{\mathfrak{P}_{\vec{k};\mathfrak{E}}}\right) \geq 0 \tag{24.10b}$$

Here, $A_{\vec{k};\mathfrak{E}}$ and $\mathfrak{P}_{\vec{k};\mathfrak{E}}$ are characteristic lengths in $\vec{A}_{\vec{k}}$- and $\vec{\mathfrak{P}}_{\vec{k}}$-space, respectively. Because we are working with energy densities in $\vec{k}$-space, the product of them has got the dimension action $\times$ length^3. In what follows, I will omit the subscript $\vec{k}$, if not needed.

Without loss of generality I normalize as before, *i.e.*, as

$$\iiint F_{\mathfrak{E}}\left(\frac{\vec{A}}{A_{\mathfrak{E}}}\right)\frac{d^3 A}{A_{\mathfrak{E}}^3} = \iiint G_{\mathfrak{E}}\left(\frac{\vec{\mathfrak{P}}}{\mathfrak{P}_{\mathfrak{E}}}\right)\frac{d^3\mathfrak{P}}{\mathfrak{P}_{\mathfrak{E}}^3} = 1 \tag{24.11}$$

The corresponding limiting (weight) amplitudes and wave functions are introduced as

$$F_{\mathfrak{E}}\left(\frac{\vec{A}}{A_{\mathfrak{E}}}\right) = \left|f_{\mathfrak{E}}\left(\frac{\vec{A}}{A_{\mathfrak{E}}}\right)\right|^2; \quad \psi(\vec{A}) = \frac{1}{A_{\mathfrak{E}}^{3/2}}f_{\mathfrak{E}}\left(\frac{\vec{A}}{A_{\mathfrak{E}}}\right) \tag{24.12a}$$

$$G_{\mathfrak{E}}\left(\frac{\vec{\mathfrak{P}}}{\mathfrak{P}_{\mathfrak{E}}}\right) = \left|g_{\mathfrak{E}}\left(\frac{\vec{\mathfrak{P}}}{\mathfrak{P}_{\mathfrak{E}}}\right)\right|^2; \quad \phi(\vec{\mathfrak{P}}) = \frac{1}{\mathfrak{P}_{\mathfrak{E}}^{3/2}}g_{\mathfrak{E}}\left(\frac{\vec{\mathfrak{P}}}{\mathfrak{P}_{\mathfrak{E}}}\right) \tag{24.12b}$$

In what follows, I will omit $A_{\vec{k};\mathfrak{E}}$ and $\mathfrak{P}_{\vec{k};\mathfrak{E}}$, if not explicitly needed.

24.3. NON-CLASSICAL REPRESENTATION OF THE FIELD ENERGY DENSITY

Within $\vec{k}$-space, the non-classical representation of the field energy density equals

$$\mathfrak{E}_{ncl} = \iiint |\psi_{\mathfrak{E}}(\vec{A})|^2 \mathfrak{B}(\vec{A})d^3 A + \iiint |\phi_{\mathfrak{E}}(\vec{\mathfrak{P}})|^2 \mathfrak{T}(\vec{\mathfrak{P}})d^3\mathfrak{P} \tag{24.13a}$$

$$= \iiint |\psi_{\mathfrak{E}}^{\perp}(\vec{A}^{\perp})|^2 \frac{k^2}{2\mu_0}|A^{\perp}|^2 d^3 A^{\perp} + \iiint |\phi_{\mathfrak{E}}^{\perp}(\vec{\mathfrak{P}}^{\perp})|^2 \frac{1}{2\varepsilon_0}|\mathfrak{P}^{\perp}|^2 d^3\mathfrak{P}^{\perp}$$

$$+ \iiint |\phi_{\mathfrak{E}}^{\parallel}(\vec{\mathfrak{P}}^{\parallel})|^2 \frac{1}{2\varepsilon_0}|\mathfrak{P}^{\parallel}|^2 d^3\mathfrak{P}^{\parallel} \tag{24.13b}$$

$$\equiv \mathfrak{E}^{\perp}_{\text{ncl}} + \mathfrak{E}^{\parallel}_{\text{ncl}} \tag{24.13c}$$

The formula for the transverse modes, $\mathfrak{E}^{\perp}_{\text{ncl}}$, is quite different from that for the longitudinal modes, $\mathfrak{E}^{\parallel}_{\text{ncl}}$. This will result in quite different quantizations: energy quanta (photons) emerge solely for the transverse field, while the longitudinal field is analogous to the case of free particles.

24.4. THE TRANSVERSE FIELD. PHOTONS

> "I therefore take the liberty of proposing for this hypothetical new atom, which is not light but plays [as the carrier of radiant energy] an essential part in every process of radiation, the name *photon*."[5]

The non-classical energy density in $\vec{k}$-space of the transverse field components,

$$\mathfrak{E}^{\perp} = \iiint \left|\psi_{\mathfrak{E}^{\perp}}(\vec{A}^{\perp})\right|^2 \frac{k^2}{2\mu_0} |A^{\perp}|^2 d^3 A^{\perp} + \iiint \left|\phi_{\mathfrak{E}^{\perp}}(\vec{\mathfrak{P}}^{\perp})\right|^2 \frac{1}{2\varepsilon_0} |\mathfrak{P}^{\perp}|^2 d^3 \mathfrak{P}^{\perp} \tag{24.14}$$

is analogous to the non-classical energy of a two-dimensional harmonic oscillator. For this, is it quantized as

$$\mathfrak{E}^{\perp} = \left(n_1 + \frac{1}{2}\right) A^{\perp}_{\mathfrak{E}^{\perp},1} \mathfrak{P}^{\perp}_{\mathfrak{E}^{\perp},1} c_0 k + \left(n_2 + \frac{1}{2}\right) A^{\perp}_{\mathfrak{E}^{\perp},2} \mathfrak{P}^{\perp}_{\mathfrak{E}^{\perp},2} c_0 k \tag{24.15}$$

where $c_0 = 1/\sqrt{(\varepsilon_0 \mu_0)}$ is the speed of transverse electromagnetic waves in vacuo[6].

$$n_{1,2} = 0, 1, \dots \tag{24.16}$$

are the corresponding quantum numbers.

As mentioned above, Planck's (1900) "energy elements" and Einstein's (1905) "light quanta" are defined for monochromatic radiation of angular frequency ω. In this case,

$$A^{\perp}_{\vec{k};\mathfrak{E};1(2)} \mathfrak{P}^{\perp}_{\vec{k};\mathfrak{E};\vec{k};1(2)} = \frac{\tilde{h}_{\omega;1(2)}}{4\pi k^2} \delta(k - \omega/c_0) \tag{24.17}$$

---

[5] Gilbert Newton Lewis (1875 - 1946), *The conservation of Photons*, 1926
[6] *Cf* Maxwell, *A Dynamical Theory of the Electromagnetic Field*, 1864, eq. (71)

where $\tilde{h}_{\omega;1(2)}$ is a quantity of dimension action that may depend on angular frequency, ω, and polarization direction (1,2). Then, the total energy of the transverse field,

$$E^\perp = \iiint \left[\left(n_1 + \frac{1}{2}\right) A^\perp_{\vec{k};\mathfrak{E}^\perp_{\vec{k}};1} \mathfrak{B}^\perp_{\vec{k};\mathfrak{E}^\perp_{\vec{k}};1} + \left(n_2 + \frac{1}{2}\right) A^\perp_{\vec{k};\mathfrak{E}^\perp_{\vec{k}};2} \mathfrak{B}^\perp_{\vec{k};\mathfrak{E}^\perp_{\vec{k}};2} \right] c_0 k\, d^3 k$$

$$= \left(n_1 + \frac{1}{2}\right) \tilde{h}_{\omega;1}\omega + \left(n_2 + \frac{1}{2}\right) \tilde{h}_{\omega;2}\omega \tag{24.18}$$

is composed of energy quanta $\tilde{h}_{\omega;1(2)}\omega$. Compton scattering[7] and other experiments reveal, that $\tilde{h}_{\omega;1(2)} = \hbar$. This indicates, again, that there is one and only one quantum of action (see also Problem 24.2).

The time-dependence of the wave functions (24.12) and their equations of motion are not dealt with in this book, because this requires the inclusion of spin, *cf* Problem 24.3.

Problem 24.2. (Unique natural constants*) *Comment on the fact, that there is one and only one*

- quantum of action, h, within quantum physics;

- vacuum permittivity, ε_0, within electrostatics;

- vacuum permeability, μ_0, within magnetostatics;

- vacuum speed of light, c_0, within special relativity;

- gravitational constant, G, within theories of gravity!

Which branch(es) of physics do not have got such a characteristic constant? What does this mean for the relationships of the branches of physics each among another?[8]

Problem 24.3. (Spin*) *According to eq. (27.6) below,*

---

[7] Arthur Holly Compton (1892 – 1962, Nobel Award 1927), *A Quantum Theory of the Scattering of X-Rays by Light Elements*, 1923

[8] Hint: *Cf* Rompe & Treder, *Was sind und was bedeuten die Elementarkonstanten?*, 1985, § 3; *Grundfragen der Physik*, 1980, p. 36; *Über Physik*, 1979.

$$\vec{S} \equiv \iiint \varepsilon_0 \vec{E}_{rot}(\vec{r},t) \times \vec{A}_{rot}(\vec{r},t) d^3 r \qquad (24.19a)$$

$$= \iiint \varepsilon_0 \vec{E}^{\perp}_{-\vec{k}}(t) \times \vec{A}^{\perp}_{\vec{k}}(t) d^3 k \qquad (24.19b)$$

$$= - \iiint \overrightarrow{\mathfrak{P}}^{\perp}_{-\vec{k}}(t) \times \vec{A}^{\perp}_{\vec{k}}(t) d^3 k \qquad (24.19c)$$

is the total classical spin of the electromagnetic field. Can the fact, that the spin of a single photon equals ħ, be derived from that formula?

Problem 24.4. (Energy densities*) *For a free classical electromagnetic field in vacuo, the electric energy density equals the magnetic one:*

$$\frac{\varepsilon_0}{2} \vec{E}(\vec{r},t)^2 = \frac{1}{2\mu_0} \vec{B}(\vec{r},t)^2 \qquad (24.20)$$

Can this relationship be exploited for the text and problems above?

24.5. THE LONGITUDINAL FIELD

The non-classical energy density in $\vec{k}$-space of the longitudinal field components,

$$\mathfrak{E}^{\|}_{\vec{k}} = \iiint \left| \phi_{\mathfrak{E}^{\|}_{\vec{k}}}(\overrightarrow{\mathfrak{P}}^{\|}_{\vec{k}}) \right|^2 \frac{1}{2\varepsilon_0} \left| \mathfrak{P}^{\|}_{\vec{k}} \right|^2 d^3 \mathfrak{P}^{\|}_{\vec{k}} \qquad (24.21)$$

is analogous to the non-classical energy of a free particle. I do not elaborate this parallel, because it is, perhaps, not meaningful to treat the longitudinal field as a free field. For the classical longitudinal field component, $\vec{E}^{\|}(\vec{r},t)$, always goes from one charge to another charge.[9] Here, a Lagrangian constraint mentioned in Subsection 4.1.3 occurs that need special treatment. In contrast, the transverse components of the electrical and magnetic fields propagates through mutually embracing each another, see Fig. **24.1**[10]. It is also well known, that the longitudinal and scalar photons corresponding to $\vec{E}_L$ are quite different from the transverse photons corresponding to $\vec{E}_T$ and $\vec{B}_T = \vec{B}$.

For the wave functions and their equations of motion, the remarks at the end of the foregoing section apply.

---

[9] Gustav Adolf Ludwig Mie (1868 - 1957), *Lehrbuch der Elektrizität und des Magnetismus*, 1941, No. 84
[10] After Matthew Norton Wise (*1940), *The Mutual Embrace of Electricity and Magnetism*, 1979, Fig. 5

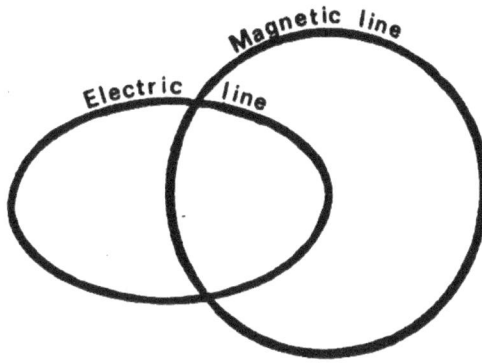

Fig. 24.1. Faraday's symbol of oneness of electrical and magnetic field lines embracing each another.

Part VI

Prospects

"Gegenüber den Rätseln der Körperwelt ist der Naturforscher längst gewöhnt, mit männlicher Entsagung sein ‚Ignoramus' auszusprechen. Im Rückblick auf die durchlaufene siegreiche Bahn trägt ihn dabei das stille Bewußtsein, daß, wo er jetzt nicht weiß, er wenigstens unter Umständen wissen könnte, und dereinst vielleicht wissen wird. Gegenüber dem Rätsel aber, was Materie und Kraft seien, und wie sie zu denken vermögen, muß er ein für allemal zu dem viel schwerer abzugebenden Wahrspruch sich entschließen: ‚Ignorabimus'."[11]

Hilbert opposed du Bois-Reymond's pessimistic 'ignoramus et ignorabimus' with his optimistic view "We must know - we will know" written – in German – on his tomb.[12]

"Wir dürfen nicht denen glauben, die heute mit philosophischer Miene und überlegenem Tone den Kulturuntergang prophezeien und sich in dem Ignorabimus gefallen. Für uns gibt es kein Ignorabimus, und meiner Meinung nach auch für die Naturwissenschaft überhaupt nicht. Statt des törichten Ignorabimus heisse im Gegenteil unsere Losung:
Wir müssen wissen,
Wir werden wissen."[13]

---

[11] Emil Heinrich du Bois-Reymond (1818 - 1896), *Über die Grenzen des Naturerkennens*, 1872 – shortly: we do not know and will not know.

[12] Picture by Kassandro - Own work, CC BY-SA 3.0,
https://commons.wikimedia.org/w/index.php?curid=4219496.

[13] Hilbert, *Naturerkennen und Logik*, 1930 – Engl.: "We must not believe those, who today with philosophical bearing and a tone of superiority prophesy the downfall of culture and accept the *ignorabimus*. For us there is no *ignorabimus*, and in my opinion even none whatever in natural science. In place of the foolish *ignorabimus* let stand our slogan: We must know, We will know." (transl. by James T. Smith); similar Hilbert' famous 1900 speech *Mathematische Probleme*.

CHAPTER 25

Summary and Conclusions

Abstract: This chapter summarizes the main results of this book and draws some conclusions. Special attention is paid to the axiomatic basis, *i.e.*, to Schrödinger's requirements and Hertz's program, which lead to the novel paradigm of quantization as selection problem (rather than eigenvalue problem). Both ones are realized by means of limiting (Schrödingerian weight) functions, which guarantee the validity of the energy conservation law in the classically forbidden regions of configuration and momentum configuration spaces (*e.g*, beyond the turning points of classical oscillators). Further topics are the separation of external and internal parameters as initiated by Euler, the differences between classical and quantum wave equations, causality and determinism, hidden variables, symmetry and field quantization.

Keywords: Axiomatic, Causality, Configuration, Determinism, Euler, External parameters, Field quantization, Hertz's program, Hidden variables, Internal parameters, Limiting functions, Momentum configuration, Quantization, Schrödinger's requirements, Symmetry, Trajectory, Wave equation, Weight functions.

25.1. SCHRÖDINGER'S REQUIREMENTS

This ebook presents an axiomatic transition from CM to QM, which obeys *all four* requirements posed by Schrödinger in 1926.

1. The "quantum equation" should "carry the quantum conditions in itself" (*2nd Commun.*, p. 511);

2. There should be a special mathematical method for solving the stationary Schrödinger equation, which accounts for the *non*-classical character of the quantization problem, *i.e.*, which is different from the classical methods for calculating the eigenmodes of strings, resonators *etc.* (*cf ibid.*, pp. 511ff.);

3. The derivation should uniquely decide, that the energy rather than the frequency values are discretized, since frequency discretization is a *classical* phenomenon (*ibid.*, pp. 511, 519);

4. The use of the classical expressions for the potential and kinetic energies should be justified (*4th Commun.*, p. 113).

These requirements challenge all approaches to QM that claim to be independent of CM. Even more important, they doubt the claim, that there is *no smooth* way from CM to QM.

25.2. AXIOMATIC BASIS

The fulfillment of all Schrödinger's requirements is possible, when – following Euler – Newton's axiomatic is reduced to his 1st and 3rd axioms concerning the *conservation* of *stationary* states. Despite of the change in the representation of the momentum when going from CM to QM, these two axioms hold true also within QM. In contrast, Newton's 2nd axiom, which implies the motion along trajectories and thus *prevents* the smooth transition to QM, is *discarded*.

Another step apart from trajectories is Helmholtz's introduction of conservative systems *not via* Newton's equation of motion, but *via* Leibniz's theorem on the conservation of kinetic energy. In order to treat space and momentum variables, a dual theorem on the conservation of potential energy has been added. Both theorems are largely *in*dependent of the time-dependence of position and velocity or momentum, *i.e.*, on the motion along *trajectories*. This allows for considering various relationships between potential, kinetic and total energies, which are systematized as *selection* problems.

The result, that generalizations of CM can be axiomatically derived from Euler's rather than from Hamilton's and Jacobi's representations of CM, does not contradict the fact, that the latter ones are considered to formally complete CM. On the contrary, novel forms are known to emerge not from the most developed existing forms, but from earlier ones.

25.3. QUANTIZATION AS SELECTION PROBLEM

It is often assumed or presupposed, that the difference between classical and quantum systems is caused by the existence of the quantum of action. As a matter of fact, this assumption is not necessary. It is sufficient to *release* certain assumptions about the set of (momentum) configurations a mechanical system can assume.

Following Planck (1900) and, in particular, Einstein (1907), the stationary Schrödinger equation is solved as the problem to *select* the quantum states out off the set of classical states. Within CM, the physically relevant spectrum of the total energy, E, is – desite of the dimension 'energy' – the continuum of non-negative real-valued numbers: $0 \leq E < \infty$ (0 being the energy of the ground state). Within

QM, this continuum is excluded even for *mathematical* reasons. As a consequence, the physical selection does not deal with the continuum, so that a question like 'what about the physically discarded continuous energy values?' does not arise at all. The physically discarded values, $E = -\frac{1}{2}\hbar\omega, -\frac{3}{2}\hbar\omega, \cdots$ for the linear harmonic oscillator, are even not part of the classical spectrum ($E \geq 0$).

Moreover, the solvability of the Schrödinger equation *without* using boundary conditions as presented here shows, that it exhibits "maximum strength" in the sense of Einstein.[1]

25.4. UNITY OF PHYSICS: HERTZ'S PROGRAM

Hertz's program of representing CM such, that all other branches of physics can be derived from it *without* additional assumptions, has been realized for non-relativistic QM.

Special-relativistic mechanics had been obtaining through relaxing Euler's implicit assumption (*Anleitung*, § § 52ff.), that the change of the velocity, v, of a body per time unit depends solely on its mass and on the external force, but not on the velocity itself[2]. Compatibility with non-relativistic mechanics implies, that the units of all quantities involved remains the same. As a consequence, that influence of the velocity is described by a dimension*less* function: $f = f(v/v_{ref})$, whereby a reference velocity emerges, analogously to x_E in the weight function $F_E(x/x_E)$, see Subsection 7.3.1.

The Maxwell-Lorentz equations as well as the Yukawa and Proca equations[3] had been obtained through extending Helmholtz's explorations on the relationships between forces and energies, see Sections 1.3 and 17.2. Generalizing them to QM has led us to the gauge invariance of the Schrödinger equation.

The derivation of QM requires much deeper changes than those two ones, because the motion of bodies (particles) proceeds no longer along trajectories, see also Section 26.1 below. Following Helmholtz, the energy law has been formulated in terms of configurations and momentum configurations, including the impossibility

---

[1] Einstein, *Grundzüge der Relativitätstheorie* [*The Meaning of Relativity*], 2014, Appendix II.
[2] Suisky & Enders, *Dynamische Begründung der Lorentz-Transformation*, 2005.
[3] Yukawa, *On the interaction of elementary particles I*, 1935; Proca, *Sur la théorie ondulatoire des électrons positifs et négatifs*, 1936.

of a *perpetuum mobile* and accounting for the relation between energy and extension.

This way, the relationship between CM and non-CM becomes well defined, and the *physical* content of non-CM is formulated *on equal footing* with the *mathematical* method (and *vice versa*). Moreover, *ad-hoc* assumptions, which may be suggested by experimental results, but are *not* supported by the axiomatic of CM, can be avoided. The wave and particle aspects emerge from the time-dependent Schrödinger equation and its solutions. The classical path in phase space is replaced with the wave functions in position and momentum representations.

25.5. SETS OF CONFIGURATIONS AND MOMENTUM CONFIGURA-TIONS REPLACE THE TRAJECTORIES

Most difficulties in understanding QM arise from the trial to assign *classical* properties to *quantum* particles or systems, notably, position and momentum as *points* in phase space. This trap is avoided here.

The release of the impenetrability of (classical) bodies does *not* lead to an absolute, indifferent, quality-less penetrability, but to a relative, quantitatively sophisticated space occupation by more than one quantum particle. It is governed by the spatial properties of the wave function, where even statistical effects enter, *cf* Figs. (**18.1** and **18.2**).

Due to the normalization $\int |\psi(x,t)|^2 dx = 1$, $|\psi|^2 dx$ can be interpreted as the relative occupation of the interval $[x, x + dx]$; analogously in p- space; for the phase space occupation, see Subsection 10.3.4.

The dynamics in space and in momentum space are treated on equal footing. As a consequence, the Schrödinger equation in momentum representation is obtained at once with that in position representation. This, too, enables one to keep maximum contact to CM and to explain, why QM is a non-classical mechanics of conservative systems, where the *classical* potential and kinetic energy functions and, consequently, the *classical* Lagrangians and Hamiltonians can still be used. The interpretation of the Hamiltonian as a Newtonian (stationary-)state function provides a natural explanation of "the peculiar significance of the energy in quantum mechanics."[4]

---

[4] Weyl, *Group theory and quantum mechanics*, 1950, p. 80.

Transitions from CM to QM starting with the requirement to abandon classical notions, in particular, the notion 'trajectory', lack that explanation. The reason for that consists in the fact that 'trajectory' is *not* such a *fundamental* notion of CM as the notion 'state'. Moreover, it remains open, why the classical functions $V(x)$ and $T(p)$ are still used, although they emerge from that notion 'trajectory'.

Transitions from CM to QM starting with an equation of motion also lack that explanation. For equations of motion do not contain conserved quantities, while the conservation of state plays a central role in QM. As a consequence, that role must be *additionally* postulated, so that jumps occur, or *ad-hoc* assumptions become necessary.

In quantum systems, the contributions of the configuration, x, and of the momentum configuration, p, to the total energy, E, are no longer the classical values $V(x)$ and $T(p)$, but $x_E|\psi_E(x)|^2 V(x)$ and $p_E|\phi_E(p)|^2 T(p,t)$, respectively. Here, x_E and p_E are characteristic lengths (extensions) that are at once exploited as normalization factors. $\psi_E(x)$ and $\phi_E(p)$ are the wave functions in the stationary state E (usually, they bear additional labels or quantum numbers). This means, that a quantum system has got a characteristic extension even then, if the corresponding classical system does not. In particular, there is no characteristic length assigned to a free body with momentum p. QM assigns to a free quantum particle the de Broglie wavelength, $\lambda = h/p$.

Problem 25.1. *Compare Newton-Euler's (Subsection 4.1.1) and Pauli's (p. 203) exclusion principles!*[5]

25.6. BENEFITS FROM THE LIMITING (WEIGHT) FUNCTIONS

The building of QM primarily on the base of the limiting functions, $F_E(x) = |f_E(x)|^2 = x_E|\psi_E(x)|^2$ and $G_E(p) = |f_E(p)|^2 = p_E|\phi_E(p)|^2$, rather than on the stationary, $\psi_E(x)$ and $\phi_E(p)$, or non-stationary wave functions, $\psi(x,t)$ and $\phi(p,t)$, exhibits the following advantages.

- The superposition principle needs not to be postulated, but is a consequence of the homogeneity of the time-dependent Schrödinger equation in $\psi(x,t)$ and

---

[5] Hint: Pauli, *Exclusion principle and quantum mechanics*, 1946; Sir Nevill Francis Mott (1905 - 1996, Nobel Award 1977), *The exclusion principle and aperiodic systems*, 1929; Mehra & Rechenberg, *The Historical Development of Quantum Theory*, Vol. 6, *The Completion of Quantum Mechanics 1926-1941*, 2000, Pt. I, Ch. II, Sect. II.7.c, pp. 274-280.

$\phi(p,t)$, respectively.

- The boundary conditions for the wave function at infinity or at infinitely high potential walls are obtained without resorting to disputed interpretations of the wave function.

- Normalizable (square integrable) wave functions for classically *un*bound systems are obtained without the construction of 'eigendifferentials' ('Weyl packets').

- The 'tunnel effect' is demystified: A quantum particle does not 'tunnel' *through* the *classical* hill: $V(x) > E$, but 'crawls' *over* the *quantum* hill, $V_E(x) = x_E|\psi_E(x)|^2 V(x) < E$.

- The energy conservation law is fulfilled *beyond* the classically allowed (momentum) configurations. In this sense,

  QM is the mechanics of oscillators without classical turning points.[6]

- Fundamental properties of many-body quantum systems are *derived* rather than postulated. The permutation-symmetrical properties of the wave functions are a consequence of the permutation invariance of the *classical* stationary states after Newton and Euler.

- Gauge invariance emerges from a natural generalization of Helmholtz's explorations about the relationships between force fields and conserved quantities.

25.7. EXTERNAL AND INTERNAL PARAMETERS

Euler's explorations about internal and external system parameters are fruitful not only within CM, but also for the relationship between CM and QM.

Within CM, they have guided us to Huygens' principle for the linear harmonic oscillator in Subsection 9.1.3. Here, novel combinations of classical dynamical variables (position and momentum) appear, which become even more important within QM, *viz*, as annihilation and creation operators, see Section 10.2.

---

[6] This adds another interpretation of QM to the eight ones sketched by Dimitris K. Lazarou, *Interpretation of Quantum Theory. An overview*, 2009.

With respect to the relationship between CM and QM, let me remark the following (being partially at variance with earlier texts).

*   By its very definition through the classical equation $E = H(x_0, p_0)$, the total energy, E, is a function of *external* parameters, *viz*, the initial values x_0 and p_0. In contrast, the stationary Schrödinger equation, $\hat{H}\psi_E = E\psi_E$, looks like an *implicit* formula for E. However, the stationary state E is not assumed by chance, but by explicit preparation.

*   In Newton's equation of motion and in the inhomogeneous wave equation, the external excitation is represented by the inhomogeneous term of the differential equation. In contrast, the time-dependent Schrödinger equation is a homogeneous differential equation; the excitation proceeds parametrically. In his discussion of the particularities of the quantum-mechanical reaction to external influences, Schrödinger stresses this as a "lack of analogy" ("*Mangel an Analogie*", 4th *Commun.*, p. 115).

The relationship between external and internal parameters is useful for other branches, too. For instance, Sommerfeld (*Elektrodynamik*, 2001, p. 290) emphasized, that is was the *internal* properties of a surface, which led Gauß[7] to the line element and to the invariant measure of curvature, which were exploited by Riemann[8] and eventually by Einstein.

25.8. CLASSICAL *VERSUS* QUANTUM WAVE EQUATION

Classical wave equations stem from the combination of, (i), continuity equations expressing the conservation of substances like electrical charge and, (ii), constitutive equations describing the reaction of the fluxes to the forces.[9] In contrast, within this (and Schrödinger's) approach, the *quantum* wave equation, *i.e*, the time-dependent Schrödinger equation, results from the extension of the time-*in*dependent Schrödinger equation. The latter one is rather an "oscillation or amplitude equation" (Schrödinger, 4th *Commun.*, p. 110). Euler's principles of stationary-state change enable us to replace Schrödinger's ingenious, though tentative approach with a sound foundation.

---

[7] Johann Carl Friedrich Gauß (1777 - 1855), *Disquisitiones generales circa superficies curvas*, 1827.
[8] Riemann, *Üeber die Hypothesen, welche der Geometrie zugrunde liegen*, 1854.
[9] Enders, *Physical, metaphysical and logical thoughts about the wave equation and the symmetry of space-time*, 2011.

25.9. CAUSALITY AND DETERMINISM

The Schrödinger equation is as causal and deterministic as Newton's equation of motion.[10] Contrary statements about QM base on the fact, that it does not yield (classical) trajectories. The 'trajectories' of quantum particles are determined only as much as they are limited by energy and momentum conservation. (Born, *ibid.*) For this, it is not appropriate to connect the issue of causality and determinism with the motion along trajectories. Indeterminacy occurs not with respect to the quantum states, but with respect to certain observations.[11]

However, the uncertainty relation[12] allows for the discrimination of conserved and not conserved quantities. The not conserved quantities – such as, usually, position and momentum – exhibit a finite uncertainty. The conserved quantities have got determined values, while the values of the corresponding conjugated quantities are completely indefinite. For a free quantum particle, this are momentum and position, respectively. The latter complies with Newton's, but not with the nowadays notion of stationary state.

25.10. HIDDEN VARIABLES

Our approach provides not any hint for additional, 'hidden' variables, which would allow for a 'more accurate' description of quantum motion.[13] New quantities have been introduced if and only if being inevitable, and all of them are well defined. For sure, there are no additional parameters that affect the energetic spectrum, else, it would not be discrete in the known manner.

Of course, like any other one, this claim has to be decided by experiment. Here, it should be emphasized, that the 'theoretically arbitrarily accurate' *simultaneous* measurement of position and momentum of classical bodies is carried out by means of methods that are *beyond* CM, notably, by means of light. So far, this is not possible within QM, *i.e*, no experiment reaches the interior of a phase space cell.

Despite of its misleading title, Bohm's famous paper[14] has not really introduced "hidden variables". Instead, in the spirit of de Broglie's pilot wave approach, to each wave function, $\psi(\vec{r}, t) = R(\vec{r}, t)\exp\{(i/\hbar)S(\vec{r}, t)\}$, a velocity $\vec{v} = (1/$

---

[10] *Cf* Born, *Quantenmechanik der Stoß vorgänge*, 1926, Introduction.

[11] *Cf* also the witty description about the change from 'uncertainty *relation*' to 'uncertainty *principle*' in Hans Graß mann (*1960), *Das Top Quark, Picasso und Mercedes-Benz – oder Was ist Physik?*, 1997, pp. 123ff.

[12] Heisenberg, *Über den anschaulichen Inhalt der quantentheoretischen Kinematik und Mechanik*, 1927.

[13] *Cf* Born, *Quantenmechanik der Stoß vorgänge*, 1926, § 9.

[14] Bohm, *A Suggested Interpretation of the Quantum Theory in Terms of 'Hidden' Variables, I and II*, 1952.

$m)\nabla S(\vec{r}, t)$ is assigned. So far, this has not yet led to results being different from those of ordinary QM. Nevertheless, it may be heuristically useful.

25.11. SYMMETRY

Our approach yields immediately the fact, that the stationary-state quantities $|\psi_E(\vec{r})|^2$ and $\langle \psi_E | \hat{H}_0 | \psi_E \rangle = E$ have got the same symmetries as the classical Hamiltonian, $H_0 = E$. (Spin effects are not included, of course.) This yields an *axiomatic* foundation of all related symmetry properties and corresponding effects in quantum systems.

Similarly, many further quantum symmetry effects emerge in an axiomatic manner, *i.e.*, *without* posing additional requirements. Examples are gauge invariance, the Bloch wave and the Ehrenberg-Siday-Aharonov-Bohm effect.

The metaphysical rule 'aequat causa effectum' is shown to hold true, if one

- properly accounts for the initial conditions, or

- includes appropriate *sets* of 'effects' (trajectories, orbitals).

If the energy of a system represents its ability to deliver work to its environment, its scale is defined by the ground state. This makes the claim of global gauge invariance to be admissible at most for pedagogical reasons.

Weyl's reinterpretation of his earlier speculative proposal on gauge invariance had actually been suggested before by London and Fock, but it was Weyl who emphasized its role as a *symmetry principle* from which electromagnetism can be *derived*.

> "We all know that the gauge symmetries of the Standard Model are very hidden and it is, therefore, not astonishing that progress was very slow indeed."[15]

From an axiomatic point of view, the latency of the gauge symmetries makes local gauge invariance to be a heuristic approach rather than a general principle.

---

[15] O'Raifeartaigh & Straumann, *Early History of Gauge Theories and Kaluza-Klein Theories, with a Glance at Recent Developments*, 1999, p. 4.

25.12. FIELD QUANTIZATION

Within the method of normal-mode expansion, only the temporal, but not the spatial part of the field variables is concerned, because the latter is fixed by the boundary conditions. Schleich's criticism of that circumstance[16] encourages the separation of the quantization of a field from its spatial and temporal distributions.[17]

The attempt to apply the concept 'quantization as selection problem' to fields sketched here looks quite promising as it yields, *without* any additional assumptions,

- the energy quanta, $h\nu$,

- a crucial difference between the transverse and the longitudinal components of the free electromagnetic field: the transverse components are quantized like two-dimensional harmonic oscillators, while the longitudinal component is quantized like free particles.

---

[16] Schleich, *Quantum Optics in Phase Space*, 2000, Ch. 10; see also Bennett, Barlow & Beige, *A physically motivated quantization of the electromagnetic field*, 2016.
[17] *Cf* Einstein, *Über einen die Erzeugung und Verwandlung des Lichtes betreffenden heuristischen Gesichtspunkt*, 1905, Preamble.

CHAPTER 26

Open Questions – Suggestions

Abstract: This chapter summarizes open issues and suggests further explorations. For example, how do quantum particles propagate, and how to deal with time as a classical parameter?

Keywords: Bohmian mechanics, Euler's principles of stationary-state change, Guiding wave, Motion, Pilot wave, Time.

"Gerade in Zeiten, in denen die Physik nicht einmal fünf Prozent des Universums zu beschreiben scheint, während der Rest im Dunklen liegt, kann es lohnen, die Grundlagen einer Wissenschaft auf den Prüfstand zu stellen und zu revidieren."[1]

"Even the fundamental equations have to be questioned as long as it is not known, *why* they work."[2]

If the reader has got the impression, that neither QM in general, nor its multitudinous applications are widely 'grazed fields', in which no severe problems are left unsolved and hence spectacular discoveries are no longer to be expected, then this book has reached one of its goals. Several proposals for further explorations have been placed in the *-problems. Few other ones will be mentioned in this and the next chapter.

26.1. HOW DO QUANTUM PARTICLES PROPAGATE?

Quitting Newton's 2nd axiom, Heisenberg (1925) and Schrödinger (1926) have abandoned not only the relationship between force and trajectory, but also the classical notions of force and trajectory themselves. As a result, QM makes not any statement about the *single* configurations, x, and momentum configurations, p, of a system.

---

[1] Meinhard Kuhlmann (*1967), *Sein oder Nichtsein? Felder, Teilchen, Tropen – die Quantenfeldtheorie im Dialog zwischen Philosophie und Physik* [To be or not to be? Fields, particles, tropes – quantum field theory in dialog between philosophy and physics], 2016, p. 35; see also *What is Real?*, Scientific American, August 2013.
[2] Hans Graßmann (*1960), *Das Top Quark, Picasso und Mercedes-Benz*, 1997, p. 121

Many authors still consider the expression $-\nabla V(x)$ to represent a force – without connecting it with an acceleration, $d^2x(t)/dt^2$, of course. On the other hand, the particle threads in cloud chambers resembles classical trajectories.

As a matter of fact, we *are* forced to give up the connection between force and trajectory according to Newton's 2nd axiom. Possibly, the concepts 'force' and 'trajectory' persist in modified forms? Is it possible to extract out of the set of all configurations (C_{all}) *non*-classical orbital parameters?

Initially, Born did *not* exclude the persistence of some kind of classical orbits:

> "Keine der beiden Auffassungen [Born, Heisenberg & Jordan, 1926 – matrix mechanics; Schrödinger, 1926 – wave mechanics (PE)] scheint mir befriedigend zu sein. Ich möchte versuchen, hier eine dritte Interpretation zu geben... Dabei knüpfe ich an eine Bemerkung Einsteins über das Verhältnis von Wellenfeld und Lichtquanten an; er sagte etwa, daß die Wellen nur dazu da seien, um den korpuskularen Lichtquanten den Weg zu weisen, und er sprach in diesem Sinne von einem "Gespensterfeld".
>
> Das Führungsfeld ... breitet sich nach der Schrödingerschen Differentialgleichung aus. Impuls und Energie aber werden so übertragen, als wenn Korpuskeln (Elektronen) tatsächlich herumfliegen."[3]

The concept of a guiding, or pilot wave was followed for decades by de Broglie.[4] Its practical realization is due to Bohm.[5] Within his extension of Schrödinger's wave mechanics – called 'Bohmian mechanics' or 'de Broglie-Bohmian mechanics' – the position of a particle changes as

---

[3] Born, *Quantenmechanik der Stoßvorgänge*, 1926, Introduction; – Engl.: For me, none of both conceptions [Born, Heisenberg & Jordan, 1926 – matrix mechanics; Schrödinger, 1926 – wave mechanics (PE)] appears to be satisfying. I wish to try to give a third interpretation... Here, I begin with a remark by Einstein about the relation between wave field and light quanta. Approximately, he said, that the waves exist only to show the light quanta the way, and, in this sense, spoke about a "ghost field". ...The guiding field...propagates according to Schrödinger's differential equation. However, momentum and energy are transferred as if corpuscles (electrons) actually fly around. – *Cf* also Martin Daumer (*1968), *Quantenmechanik und Determinismus*, 1997, pp. 173f.; *Streutheorie aus der Sicht Bohmscher Mechanik.*
[4] de Broglie, *Ondes et mouvements*, 1926; *Étude critique des bases de l'interprétation actuelle de la mécanique ondulatoire*, 1963; *cf* Henning Sievers, *Louis de Broglie und die Quantenmechanik*, 1998.
[5] Bohm, *A Suggested Interpretation of the Quantum Theory in Terms of 'Hidden' Variabls*, 1952

$$\frac{dx}{dt}(t) = \frac{\hbar}{m}\,\Im\left\{\frac{\nabla_{x(t)}\psi(x(t),t)}{\psi(x(t),t)}\right\} \tag{26.1}$$

This unifies the classical wave and particle aspects in explicit rather than merely interpretative form. If the initial values, $x(0)$, are suitably chosen, one obtains the conventional results for measured quantities.[6]

Obviously, this works well for the free particle, for which $\psi(x,t)\sim\exp(ipx/\hbar)$. On the other hand, it yields a vanishing velocity: $dx(t)/dt = 0$, for all cases, in which the wave function is real-valued, notably, for the linear oscillator.

On the other hand, Schrödinger's wave mechanics deals more symmetrically with position and momentum. This suggests, (i), to complete the association (26.1) through

$$\frac{dp}{dt}(t) = -\hbar\Im\left\{\frac{\nabla_{p(t)}\phi(p(t),t)}{\phi(p(t),t)}\right\} \tag{26.2}$$

However, this does not hold true already for dimensional reasons.

(ii), the association (26.1) could be replaced with

$$p(t) = -\hbar\Im\left\{\frac{\nabla_{x(t)}\psi(x(t),t)}{\psi(x(t),t)}\right\};\quad x(t) = -\hbar\Im\left\{\frac{\nabla_{p(t)}\phi(p(t),t)}{\phi(p(t),t)}\right\} \tag{26.3}$$

The exploration of these ansatzes is left to the reader in Problems 26.1 and 26.2.

Problem 26.1. *Examine eqs. (26.1) and (26.2) for the ground state of the H-atom!*

Problem 26.2. *Examine eq. (26.3) for the ground state of the H-atom!*

26.2. TIME AS CLASSICAL PARAMETER

"In all books which develop quantum mechanics from the Schrödinger equation it is considered that the time-dependent equation (TDSE) to be more fundamental than the time-independent equation (TISE). In the TDSE,

---

[6] For a review and critical analysis, see Oliver Passon (*1969), *Bohmsche Mechanik – eine elementare Einführung in die deterministische Interpretation der Quantenmechanik*, 2010.

$$\left(H_S(t) - i\hbar \frac{\partial}{\partial t}\right)\psi_S(t) = 0 \qquad\qquad \textbf{(BR-1)}$$

for some quantum system S, it is usually pointed out, that time is a parameter, not enjoying the status of a quantum operator. However, it is almost never pointed out, that the time occurring in $H_S(t)$ always is a classical time arising from the classical time development...of external fields or material particles interacting with the quantum system. In this sense the TDSE is, from the outset, a mixed quantum-classical equation."[7]

Within our approach, these and related difficulties do *not* occur, because – in agreement with Schrödinger (*4th Commun.*, p. 109) – the TISE is more fundamental than the TDSE. The time is not related to the quantization of the stationary states and thus not needs to become an operator.

Recently, Briggs & Rost[8] have generalized Mott's forgotten derivation of the TDSE from the TISE.[9] A sufficiently large quantum system – described by a TISE – be divided into two subsystems, the one being much smaller than the other one. If the large subsystem is treated classically, the small one gets to be described by a TDSE.

In contrast, I have derived the TDSE from the TISE through a straightforward generalization of Euler's principles of stationary-state change (Section 11.3). Admittedly, in Euler's principles as well as in Newton's 2nd axiom and equation of motion, it is implicitly presupposed that the external force is not affected by the body it acts upon. The desirable unification of both approaches is left to the reader (Problem 26.3).

Problem 26.3. (*) *Can Euler's principles of stationary-state change be derived by means of a classical-mechanical version of the treatment by Briggs & Rost (loc. cit.)? If not, can, in turn, Briggs & Rost's treatment benefit from Euler's principles?*

---

[7] John Stuart Briggs (*1942) & Jan-Michael Rost, *On the derivation of the time-dependent equation of Schrödinger*, 2001, p. 693
[8] Briggs & Rost, *Time Dependence in Quantum Mechanics*, 1999
[9] Mott, *Theory of excitation by collision with heavy particles*, 1931; the principle had been formulated in the 'Three-men paper' Born, Heisenberg & Jordan, *Über Quantenmechanik II*, 1926, Sect. 5.

CHAPTER 27

Points of Extension or Generalization

Abstract: Accordingly to the purpose of this book, it concludes with a prospect on extensions and generalizations. Explicitly mentioned will be special relativity, spin, quantum electrodynamics and dissipative systems.

Keywords: Dissipative systems, Quantum electrodynamics, Special relativity, Spin.

27.1. SPECIAL RELATIVITY

Starting from non-relativistic CM, we have obtained non-relativistic QM. For this, it is tempting to obtain special-relativistic QM from special-relativistic CM. The latter can be derived from non-relativistic CM, too, see Problem 27.1. The result should be Newton-Lorentz's equation of motion.

$$\frac{\mathrm{d}}{\mathrm{d}t}\left(\frac{m\vec{v}}{\sqrt{1-v^2/c^2}}\right) = \vec{F}_{\text{ext}} \qquad (27.1)$$

The corresponding special-relativistic quantum wave equation is expected to read somethink like

$$[-\hbar^2\nabla^2 + m^2c^4]\psi_E(\vec{r}) = [E - U(\vec{r}, -i\hbar\nabla)]^2\psi_E(\vec{r}) \qquad (27.2)$$

where $U(\vec{r}, \vec{p}) \equiv \tilde{U}(\vec{r}, \vec{v}(\vec{p}))$, see Problem 27.2. For free particles ($U \equiv 0$), this becomes the time-independent Klein-(Fock-)Gordon equation[1].

$$[-\hbar^2\nabla^2 + m^2c^4]\psi_E(\vec{r}) = E^2\psi_E(\vec{r}) \qquad (27.3)$$

On the other hand, one may start from the special-relativistic energy-momentum 4-vector for a free body: $p = (iE/c, p_x, p_y, p_z)$, and its conservation law: $p^2 = p_x^2 + p_y^2 + p_z^2 - E^2/c^2 = -m^2c^2$, and replace p with $p - qA$, where q is the 'coupling constant' like the electrical charge or the gravitating mass of the body under consideration, and $A(x)$ the 4-vector potential representing the external force field.

---

[1] Fock, *Zur Schrödingerschen Wellenmechanik*, 1926; Walter Gordon (1893 - 1939), *Der Comptoneffekt nach der Schrödingerschen Theorie*, 1926; Oskar Benjamin Klein (1894 - 1977), *Die Reflexion von Elektronen an einem Potentialsprung nach der relativistischen Dynamik von Dirac*, 1929; for similar claims at this time, see en.wikipedia.org/wiki/Klein%E2%80%93Gordon_equation.

This 'minimum coupling' assumption has been justified by the requirement of local gauge invariance. Within our approach, local gauge invariance is not required, but a natural symmetry of the stationary-state functions, see Subsection 17.3.1.

Within the basic special-relativistic theories, time is *both* a coordinate (that can be changed, *i.e.*, transformed rather arbitrarily), and the development parameter (that is bound to increase monotonously). In order to resolve this contradiction, Fock has proposed to separate these two functions by means of an additional development variable[2]. His idea has found broad resonance.[3] For this, the reader is encouraged to explore this approach, too.

Problem 27.1 *Generalize Euler's derivation of Newton's equation equation of motion (see Subsection 2.4.1) to the case, that the change of $\vec{v}$, $d\vec{v}$, does depend on $\vec{v}$.[4] Show, that the result, eq. (27.1), is Lorentz-covariant, i.e, the equation of motion of special-relativistic mechanics, if $\vec{F}_{ext}$ Lorentz-transforms like the (Maxwell) Lorentz force, $q(\vec{E} + \vec{v} \times \vec{B})$[5]! How Euler's principles of stationary-state change have to be modified?*

Problem 27.2 *Multiply both sides of eq. (27.1) from the left with $\vec{v}$ and assume that there exists a potential function, $\tilde{U}(\vec{r}, \vec{v})$ such, that $\vec{v} \cdot \vec{F}_{ext} = -d\tilde{U}/dt$, to obtain*

$$\frac{d}{dt}\left[\frac{mc^2}{\sqrt{1-v^2/c^2}} + \tilde{U}(\vec{r}, \vec{v})\right] = 0 \qquad (27.4)$$

27.2. SPIN

Besides of entanglement (Schrödinger) and interference (Feynman), spin is an enigmatic feature of QM, too. In order to see, what can be said about an 'angular momentum proper' within CM, let me resort to the

Theorem 27.1 (Huygens-Steiner[6]) Suppose a body of mass m has got the moment of inertia I_{cm} with respect to an axis passing through its center of mass. Let d be the distance of another axis being parallel to that axis. Then, the moment of inertia with respect to the latter axis, $I(d)$, equals

---

[2] Fock, *Die Eigenzeit in der Klassischen und in der Quantenmechanik*, 1937
[3] See, *e.g*, Oliver Davis Johns, *Analytical Mechanics for Relativity and Quantum Mechanics*, 2011.
[4] Suisky & Enders, *Dynamische Begründung der Lorentz-Transformation*, 2005
[5] Planck, *Das Prinzip der Relativität und die Grundgleichungen der Mechanik*, 1906
[6] Named to the honor of Huygens and Jakob Steiner (1796 - 1863), see en.wikipedia.org/wiki/Parallel_axis_theorem.

$$I(d) = I_{cm} + md^2 \qquad (27.5)$$

This formula represents the total moment of inertia, $I(d)$, as the sum of a coordinate-*in*dependent, hence, *in*trinsic contribution, I_{cm}, and a coordinate-*dependent*, hence, *ex*trinsic contribution, md^2.

This suggests the angular momentum to be the sum of a coordinate dependent and a coordinate *in*dependent terms; the latter one being termed *spin*.

Such a division exists for an electromagnetic field, too. The quantity

$$\vec{\mathbb{S}}_{\text{EM,spin}}(\vec{r}, t) \equiv \varepsilon_0 \vec{E}_{rot}(\vec{r}, t) \times \vec{A}_{rot}(\vec{r}, t) \qquad (27.6)$$

is the origin-*in*dependent part of its angular momentum density, where $\vec{E}_{rot}$ and $\vec{A}_{rot}$ are the rotational components of the vector potential and electrical field strength, respectively.[7]

It is hoped that this way the spin can be more axiomatically included into QM, although – despite of the common commutator rules and the common conservation law – the properties of the quantum spin angular momentum are rather different from those of the quantum orbital angular momentum.

27.3. QUANTIZATION OF THE ELECTROMAGNETIC FIELD

The gravito-electromagnetic force and field equations can be obtained through continuing Helmholtz's explorations on the relations between forces and energies. They are isomorphic with the (Maxwell-)Lorentz force and the Maxwell-Lorentz equations for electrical charges and currents *in vacuo*. For this, it is hoped that some basic equations of quantum electrodynamics can be obtained through extending our approach accordingly. The following two points should be accounted for.

1. For electrons and positrons (each having got spin $1/2$), the wave functions have got four components.[8] For this, the wave functions (24.2) are expected to have more than one component, too. This will perhaps modify the calculations made in Chapter 24.

---

[7] Léon Rosenfeld (1904 - 1974), *Sur la définition du spin d'un champ de rayonnement*, 1942; for a recent critical review, see Kirk T. McDonald, *Orbital and Spin Angular Momentum of Electromagnetic Fields*, 2015.
[8] Dirac, *The Quantum Theory of the Electron*, 1928

2. Within wave mechanics, the time-dependent dynamical variables position, $\vec{r}(t)$, and momentum, $\vec{p}(t)$, become time-*in*dependent operators, $\hat{\vec{r}}$ and $\hat{\vec{p}}$, respectively. Shall field quantities like the electrical field strength, $\vec{E}(\vec{r}, t)$, become space- and time-*in*dependent operators, $\hat{\vec{E}}$?[9]

27.4. DISSIPATIVE SYSTEMS

The paramount role of the Hamiltonian within QM makes the application to dissipative systems rather intricate.[10]

For instance, for the equation of motion

$$m\ddot{x} = F(x)(1 - \alpha\dot{x}^2) \tag{27.7}$$

one can construct two different Hamiltonians, which describe one and the same classical system, but two different quantum systems.[11]

Our approach may be applicable when there is a suitable conserved quantity.

Problem 27.3 (*) *Consider the linear damped oscillator with the Newtonian equation of motion*

$$m\ddot{x} = -kx - c\dot{x} \tag{27.8}$$

where k is the spring constant and c the damping parameter. Show, that

$$B = \frac{m}{2}\exp(2\beta t)(\dot{x} - \gamma x)(\dot{x} - \gamma^* x) \tag{27.9}$$

is a constant of motion, where $\beta = c/2m$ and $\gamma = -\beta + i\omega$, $\gamma^* = -\beta - i\omega$, with $\omega = \sqrt{k/m - \beta^2}$![12] Can Bohlin's constant of motion, B, serve as Newtonian state function and starting point for our approach to quantization?

---

[9] See also Dirac's pioneering paper *The quantum theory of emission and absorption of radiation*, 1927.

[10] For a recent review, in particular, concerning the inclusion of non-equilibrium thermodynamics, see Gyula Vincze & Andras Szasz, *Nonequilibrium Thermodynamic and Quantum Model of a Damped Oscillator*, 2015.

[11] Gustavo (Velazques) López, X. E. López & Hector (Hernandez-)González (*1981), *Ambiguities on the Quantization of a One-Dimensional Dissipative System with Position Depending Dissipative Coefficient*, 2007

[12] K. Bohlin, *Note sur le Problème des Deux Corps et sur une Intégration Nouvelle dans le Problème des Trois Corps*, Bull. Astr. 28 (1911) 113-119

27.5. MECHANICS IN PHASE SPACE

Hamiltonian mechanics is classical mechanics in a $2f$-dimensional phase space, $\Gamma = \{x_1, \dots, x_f, p_1, \dots, p_f\}$, where f is the number of degrees of freedom of the mechanical system under consideration. Correspondingly, we have explored the limitation of the motions in configuration and momentum configuration spaces imposed by the energy conservation law for motions along trajectories in parallel. Eventually, it turned out, that the quantum-mechanical motion in momentum configuration space, *i.e*, the wave function $\varphi(p, t)$, is completely determined by the quantum-mechanical motion in configuration space, *i.e*, by the wave function $\psi(x, t)$, and *vice versa*. How this dubbing is related to the discretization of the quantum phase space into cells of size h^f?

Thus, there are constructions of distributions in phase space. Best known is Wigner's distribution (6.24). Can our approach be modified such, that we are led to the latter one instead to Schrödinger's wave functions?

Problem 27.4 (*) Consider the Wigner function[13]

$$P(x, p) = \frac{1}{\pi\hbar} \int \psi(x + y)\psi^*(x - y) e^{-\frac{2iyp}{\hbar}} \, dy$$

It allows for building average values, *e.g*, for the energy, as (*cf* Wigner, *loc. cit.*, p. 750)

$$E = \iint H(x, p)P(x, p) \, dxdp$$

Furthermore, we obtain the size of a quantum phase space cell as

$$h \iint P(x, p) \, dxdp = h$$

In turn, can we ask for a distribution in phase space with these properties and derive the stationary Schrödinger equation from it?

27.6. MECHANICS IN HYDRODYNAMICAL FORM

In Subsection 2.5.3 I have introduced force fields for describing the force acting upon a (*whole*) body as depending on the relative position of two interacting bodies.

---

[13] *Cf* Wigner, *On the Quantum Correction for Thermodynamic Equilibrium*, 1932, eq. (5), where Wigner refers to the coauthorship of Leó Szilárd (1898 – 1964) (p. 750, fn. 2).

Another possibility to introduce force fields consists in accounting for the finite extension of real bodies and to consider the forces acting upon the *parts* of a body. To do so, I neglect its atomic constitution and assume its mass, m, to be smoothly distributed over its volume, V.

$$m = \int_V dm(\vec{r}) = \int_V \rho(\vec{r})d^3r \qquad (27.10)$$

$\rho(\vec{r})$, the mass density, describes that mass distribution.

Now, the non-relativistic mass conservation implies the following. The change of total mass, $m_V(t)$, within a fixed volume, V, equals the of mass flow, $I_{\partial V}(t)$, through its boundary, ∂V.

$$\frac{dm_V(t)}{dt} = \int_V \frac{\partial\rho(\vec{r},t)}{\partial t} d^3r = -I_{\partial V}(t) = -\int_{\partial V} \vec{j}(\vec{r},t) \cdot d^2\vec{r} \qquad (27.11)$$

Here, $\vec{j}(\vec{r},t)$ is the (mass) current density, *i.e*, the flow per area unit; the vector $d^2\vec{r}$ points toward outside the volume V. Using the integral theorem[14]

$$\int_V \nabla \cdot \vec{j}(\vec{r},t)d^3r = \int_{\partial V} \vec{j}(\vec{r},t) \cdot d^2\vec{r} \qquad (27.12)$$

we obtain the equation of *local* mass conservation.

$$\frac{\partial\rho(\vec{r},t)}{\partial t} + \nabla \cdot \vec{j}(\vec{r},t) = 0 \qquad (27.13)$$

Equations of that type:

'temporal change of density' + 'spatial change of flux' = 0

are called *continuity equation*. The conservation of extensive quantities are often expressed in this form (see also Subsection 27.1).

Often, the flux exhibits the hydrodynamic form

$$\vec{j}(\vec{r},t) = \rho(\vec{r},t)\vec{v}(\vec{r},t) \qquad (27.14)$$

---

[14] Lagrange, *Nouvelles recherches sur la nature et la propagation du son*, 1762 (Lagrange transforms triple integrals into double integrals using integration by parts; after https://en.wikipedia.org/wiki/Divergence_theorem#cite_note-9); M. Ostrogradsky, *Démonstration d'un thééorème du calcul intégral*, Paris 1826; *Première note sur la théorie de la chaleur*, 1828/1831; V. J. Katz, *The History of Stokes' Theorem*, 1979

where $\vec{v}(\vec{r}, t)$ is the velocity of the 'hydrodynamic particle' at position $\vec{r}$ and time t. Then, the continuity equation (27.13) assumes the form

$$\frac{D\rho(\vec{r},t)}{Dt} \equiv \frac{\partial\rho(\vec{r},t)}{\partial t} + \vec{v}(\vec{r}, t) \cdot \nabla\rho(\vec{r}, t) = -\rho(\vec{r}, t)\nabla \cdot \vec{v}(\vec{r}, t) \qquad \textbf{(27.15)}$$

In order to determine density and flux, one needs a second relationship between them.[15] One of the simplest cases is that the flux results from local density variations.

$$\vec{j}(\vec{r}, t) = -D\nabla\rho(\vec{r}, t) \qquad \textbf{(27.16)}$$

(Fourier's 1st law[16]) Such equations between density and flux are called *constitutive equation*, because the describe properties of the mattter under consideration.

Inserting the constitutive equation (27.16) into the continuity equation (27.13) results in the *equation of motion*,

$$\frac{\partial\rho(\vec{r},t)}{\partial t} = D\nabla^2\rho(\vec{r}, t) \qquad \textbf{(27.17)}$$

(diffusion equation, Fourier's 2nd law[17]).

The scalar single-particle Schrödinger field as described by the wave function $\psi(\vec{r}, t)$ obeys the continuity equation[18]

$$\frac{\partial}{\partial t}(\psi\overline{\psi}) = \frac{\hbar}{2mi}\nabla\{\psi\nabla\overline{\psi} - \overline{\psi}\nabla\psi\} \qquad \textbf{(27.18)}$$

(see also Subsection 14.1.2). For this, Schrödinger has proposed

$$\vec{j}_\psi = \frac{\hbar}{2mi}\{\overline{\psi}\nabla\psi - \psi\nabla\overline{\psi}\} \qquad \textbf{(27.19)}$$

to be "the flux of of the weight function $[|\psi|^2]$ in configuration space" (*ibid.*, p. 137).

---

[15] For more details, see Enders, *Physical, metaphysical and logical thoughts about the wave equation and the symmetry of space-time*, 2011.
[16] *Cf* Fourier, *La Théorie Analytique de la Chaleur*, 1822, Ch. I, Sect. IV, Art. 68.
[17] *Cf* Fourier, *La Théorie Analytique de la Chaleur*, 1822, Ch. II, Sect. I, Art. 105.
[18] *Cf* Schrödinger, *Quantisierung als Eigenwertproblem. Vierte Mitteilung*, 1926, § 7, eq. (41) with $\rho = 1$ and eq. (42).

Dividing the flux j_ψ (27.19) by the 'density'

$$\rho_\psi = \psi\overline{\psi} \equiv |\psi|^2 \tag{27.20}$$

one obtains the 'hydromechanical velocity'

$$\vec{v}_\psi = \frac{\hbar}{2mi|\psi|^2}\{\overline{\psi}\nabla\psi - \psi\nabla\overline{\psi}\} = \frac{\hbar}{m}\Im(\nabla\ln\psi) \tag{27.21}$$

This is just Bohm's velocity (2.22).

This suggests to consider the Schrödinger field like a fluid with density $\rho_\psi = |\psi|^2$. This becomes obvious, if one sets[19] $\psi(\vec{r}, t) = \alpha(\vec{r}, t)e^{i\beta(\vec{r}, t)}$. Density and velocity are $\rho_\psi = \alpha^2$ and $\vec{v}_\psi \sim \nabla\beta$, respectively. Recently, this picture has regained considerable popularity.[20]

In turn, one could consider a 'fluid', for which the equation of continuity in the form (27.15) holds true and the density can be written in the form $\rho = |\psi|^2$. The reader is encouraged to search for reasonable constitutive relations and derive the corresponding equations of motion. Here, she should account for the possibility, that ψ is complex-valued and non-scalar. The cases found should include the Bohmian velocity field (2.22) and the time-dependent Schrödinger equation. Which fluxes lead to the Klein-(Fock-)Gordon and Dirac equations, respectively?

---

[19] Erwin Madelung (1881 - 1972), *Quantentheorie in hydrodynamischer Form*, 1927.
[20] See, *e.g*, J. V. Lill, Michael I. Haftel & G. H. Herling, *Mixed state quantum mechanics in hydrodynamic form*, 1989; Pierre-Henri Chavanis & Tonatiuh Matos, *Covariant theory of Bose-Einstein condensates in curved spacetimes with electromagnetic interactions: the hydrodynamic approach*, 2016.

REFERENCES AND FURTHER READING

M. Abramowitz & I. A. Stegun (Eds.), *Handbook of Mathematical Functions*, Washington: NBS 1964; New York: Dover 1972; abridged reprint: Danos & Rafelski, *Pocketbook of mathematical functions*, 1984, http://people.math.sfu.ca/ cbm/aands/index.htm (1.10.2015), http://people.maths.ox.ac.uk/macdonald/aands/abramowitz_and_stegun.pdf (1.10.2015)

A. I. Achieser & W. B. Berestezki, *Quantenelektrodynamik*, Leipzig: Teubner 1962

Y. Aharonov & D. Bohm, *Significance of Electromagnetic Potentials in the Quantum Theory*, Phys. Rev. 115 (1959) 485-491; reprint in: Shapere & Wilczek, 1989, paper [2.6]

Y. Aharonov & D. Bohm, *Further considerations of electromagnetic potential in the quantum theory*, Phys. Rev. 123 (1961) 1511-1524

H. Aichmann & G. Nimtz, *On the Traversal Time of Barriers*, Found. Phys. 44 (2014) 678-688

M. Al-Hashimi, *Accidental Symmetry in Quantum Physics*, Thesis, Bern 2008; http://www.wiese.itp.unibe.ch/theses/al-hashimi_phd.pdf (1.9.2015)

H. Albert, *Traktat über kritische Vernunft*, Tübingen: Mohr 1991

J. d'Alembert, *Traité de Dynamique*, 1743; German: *Abhandlung zur Dynamik*, Thun · Frankfurt/Main: Deutsch 1997 (Ostwalds Klassiker 106)

J. Le Rond d'Alembert, *Reflexions sur la cause generale des vents*, Paris: David l'Aine 1747

R. F. S. Andrade, A. M. C. Souza, E. M. F. Curado & F. D. Nobre, *A thermodynamical formalism describing mechanical interactions*, EPL 108 (2014) 20001

M. Arndt, O. Nairz, J. Vos-Andreae, C. Keller, G. van der Zouw & A. Zeilinger, *Wave-particle duality of C_{60} molecules*, Nature 401 (1999) 680-682

V. I. Arnol'd, *Huygens & Barrow, Newton & Hooke*, Basel: Birkhäuser 1990

J. Audretsch, *Verschränkte Welt*, Weinheim: Wiley-VCH 2002

Augustinus Aurelius, *Confessiones*, http://leaderu.com/cyber/books/augconfessions/bk11.html (1.9.2015)

M. Ya. Azbel', *Time, tunneling and turbulence*, Physics-Uspekhi 41 (1998) 543-552; Russ.: М. Я. Азбель, *Время, туннелирование и турбулентность*, УФН 168 (1998) 613-623, http://ufn.ru/ufn98/ufn98_6/Russian/r986b.pdf (1.9.2012)

A. Bach, *Indistinguishable Classical Particles*, Berlin *etc.*: Springer 1997 (Lecture Notes in Physics m44)

R. Bach, D. Pope, S.-H. Liou & H. Batelaan, *Controlled double-slit electron diffraction*, New J. Phys. 15 (2013) 033018 (7pp), http://iopscience.iop.org/article/10.1088/1367-2630/15/3/033018/pdf (1.10.2014)

E. Baird, *Newton's aether model*, xxx.lanl.gov/abs/physics/0011003 (1.10.2013)

D. Baird, R. I. G. Hughes & A. Nordmann (Eds.), *Heinrich Hertz: Classical Physicist, Modern Philosopher*, Dordrecht: Springer 1998 (BSPS 198), https://books.google.de/books?id=9iEy BwAAQBAJ (30.07.2017)

P. Ball, *Focus: New Crystal Type is Always in Motion*, Physics 9, 4, Jan. 8, 2016, http://physics.aps.org/articles/v9/4 (10.4.2016)

S. Bargmann, *Ideas and Opinions by Albert Einstein*, New York: Crown 1954, http://namnews.files.wordpress.com/2012/04/29289146-ideas-and-opinions-by-albert-einstein.pdf (13.9.2014)

V. Bargmann, *Zur Theorie des Wasserstoffatoms. Bemerkungen zur gleichnamigen Arbeit von V. Fock*, Z. Physik 99 (1936) 7, 576-582

V. Bargmann, *Note on Wigner's Theorem on Symmetry Operations*, J. Math. Phys. 5 (1964) 862-868; reprint in: Rosen, *Symmetry in Physics*, 1982, pp. 147-153, staff.science.uu.nl/ henri105/ Teaching/CFTclass-Bar64.pdf (7.2.2014; retriev. 22.9.2015)

A. O. Barut, *Dynamical Symmetry Group Based on Dirac Equation and Its Generalization to Elementary Particles*, Phys. Rev. 135 (1964) 839-842

K. Baumann & R. U. Sexl, *Die Deutung der Quantentheorie*, Braunschweig: Vieweg 1987

C. Baumgarten, *Minkowski Spacetime and QED from Ontology of Time*, 2015, arXiv:1409.5338v5 [physics.hist-ph] 30 Nov 2015 (24.4.2017)

C. Baumgarten, *Old Game, New Rules: Rethinking The Form of Physics*, 2016, http://arxiv.org/ abs/1604.03060v1 (29.5.2016)

F. Beck & K. Neitmann (Eds.), *Brandenburgische Landesgeschichte und Archivwissenschaft. Festschrift für Lieselott Enders zum 70. Geburtstag*, Weimar: Böhlau 1997 (Veröffentlichungen des Brandenburgischen Landeshauptarchivs 34)

J. S. Bell, *On the Einstein-Podolsky-Rosen Paradox*, Physics 1 (1964) 195-200

J. S. Bell, *Bertlmann's socks and the nature of reality*, J. Phys. Coll. 42 (1981) C2-41-C2-62; reprint in: Bell, *Speakable and Unspeakable in Quantum Mechanics*, 2004, cds.cern.ch/record/142461/files/198009299.pdf (19.9.2013)

J. S. Bell, *Speakable and Unspeakable in Quantum Mechanics*, Cambridge: Cambridge Univ. Press [2]2004

M. Le Bellac, *Quantum and Statistical Field Theory*, Oxford: Clarendon 1991

Ch. H. Bennett, G. Brassard & A. K. Ekert, *Quanten-Kryptographie*, Spektrum d. Wiss. Digest-ND 3/2003, 90-98

I. Bengtsson & K. Zyczkowski, *Geometry of Quantum States. An Introduction to Quantum Entanglement*, Cambridge: Cambridge Univ. Press 2006

R. Bennett, Th. M. Barlow & A. Beige, *A physically motivated quantization of the electromagnetic field*, Eur. J. Phys. 37 (2016) 014001 (11 pp), http://iopscience.iop.org/0143-0807/37/1/014001 (19.4.2016)

F. A. Berezin, *The Method of Second Quantization*, New York: Academic Press 1965

L. Berg, *Lineare Gleichungssysteme mit Bandstruktur*, Berlin: Dtsch. Verlag d. Wiss. 1986

K. Berndl, M. Daumer & D. Dürr, *Bohmsche Mechanik und die Quantenphysik*, Einsichten 1 (1996); complete text: mathematik.uni-muenchen.de/ bohmmech/BohmHome/files/einsichten.pdf (28.04.2015)

D. Bernoulli, *Examen principorum mechanicae et demonstrationes geometricae de compositione et resolutione virium*, Comment. Acad. Petrop. (1726) Febr., 126-142

D. Bernoulli, *Hydrodynamica, sive de viribus et motibus fluidorum commentarii*, Argentorati (Straßburg) 1738

M. V. Berry, *Quantal phase factors accompanying adiabatic changes*, Proc. R. Soc. Lond. A 392 (1984) 45-57

I. B. Bersuker (Ed.), *The Jahn-Teller-Effect. A Bibliographic Review*, New York *etc.*: Plenum 1984

J. Bertrand, *Théorème relatif au mouvement d'un point attiré vers un centre fixe*, C. R. Acad. Sci. 77 (1873) 849–853; http://gallica.bnf.fr/ark:/12148/bpt6k3034n (22.3.2016)

S. M. Binder, *Unsöld's Theorem*, demonstrations.wolfram.com/UnsoeldsTheorem/ (20.06.2013)

G. Birkhoff & J. von Neumann, *The logic of quantum mechanics*, Ann. of Math. [2] 37 (1936) 823-843; fulviofrisone.com/attachments/article/451/the%20logic%20of%20quantum%20mechanics%201936.pdf (1.9.2012)

F. Bloch, *Über die Quantenmechanik der Elektronen im Kristallgitter*, Z. Physik 52 (1928) 555-600

S. C. Bloch, *Introduction to Classical and Quantum Harmonic Oscillators*, New York: Wiley ^{2}2013

D. I. Blokhintsev, *Principles of Quantum Mechanics*, Moscow: "Nauka", 5th, rev. ed. 1976 (in Russian)

J. E. Bode, *Deutliche Anleitung zur Kentniss des Gestirnten Himmels* [*Clear Instruction for the Knowledge of the Starry Heavens*], Hamburg: Harmsen 1772

A. N. Bogoljubow (Ed.), *Mechanics and Physics of the 2nd Half of 19th Century*, Moscow: "Nauka" 1978 (in Russian)

A. Bohm, *Quantum Mechanics: Foundations and Applications*, New York: Springer ^{2}1986

A. Bohm, A. Mostafazadeh, H. Koizumi, Q. Niu & J. Zwanziger, *The Geometric Phase in Quantum Systems. Foundations, Mathematical Concepts, and Applications in Molecular and Condensed Matter Physics*, Berlin *etc.*: Springer 2003

D. Bohm, *Quantum Theory*, Englewood Cliffs: Prentice-Hall 1951; reprint: New York: Dover 1989

D. J. Bohm, *A Suggested Interpretation of the Quantum Theory in Terms of 'Hidden' Variables, I and II*, Phys. Rev. 85 (1952) 166-193; reprint in: Wheeler & Zurek, *Quantum Theory and Measurement*, 1983; German transl. of I in: Baumann & Sexl, *Die Deutung der Quantentheorie*, 1987

N. Bohr, *On the Constitution of Atoms and Molecules*, Phil. Mag. 26 (1913) 1-25; home.tiscali.nl/physis/HistoricPaper/HistoricPapers.html (25.6.2011)

N. Bohr, *On the Spectrum of Hydrogen*, Fysisk Tidsskrift 12 (1914) 97ff.; home.tiscali.nl/physis/HistoricPaper/HistoricPapers.html; Engl. in: Udden, *The Theory of Spectra and Atomic Constitution–Three Essays*, 1922; ia902604.us.archive.org/22/items/theoryofspectraa00bohriala/theoryofspectraa00bohriala.pdf (25.6.2011); s. also F. R. Moulton & J. J. Schifferes (Eds.), *Autobiography of Science*, New York: Doubleday 1950; web.lemoyne.edu/giunta/bohr.html (25.6.2011)

N. Bohr, *On the Quantum Theory of Line-Spectra*, Kgl. Danske Vid. Selsk. Skr., Nat. Math. Afd. 8. Ræ kke IV.1,1-3 (1918); reprint in: van der Waerden, *Sources of Quantum Mechanics*, 1967, paper 3; home.tiscali.nl/physis/HistoricPaper/HistoricPapers.html (25.6.2011); strangepaths.com/wp-content/uploads/2008/01/bohr1918.pdf (25.6.2011)

N. Bohr, *The Structure of the Atom and the Physical and Chemical Properties of the Elements*, Fysisk Tidsskrift, 19 (1921) 153-154; Engl. in: Udden, *The Theory of Spectra and Atomic Constitution–Three Essays*, 1922; s. also F. R. Moulton & J. J. Schifferes (Eds.), *Autobiography of Science*, New York: Doubleday 1950; home.tiscali.nl/physis/HistoricPaper/HistoricPapers.html (25.6.2011)

N. Bohr, *The structure of the atom*, Nobel Lecture, Dec. 11, 1922; in: *Nobel Lectures, Physics 1922-1941*, Amsterdam: Elsevier 1965; nobelprize.org/nobel_prizes/physics/laureates/1922/bohr-lecture.html (25.6.2011)

N. Bohr, *Über die Quantentheorie der Linienspektren*, Braunschweig: Vieweg 1923

N. Bohr, *Das Quantenpostulat und die neuere Entwicklung der Atomistik*, Volta Congress, Como, Sept. 1927; Naturwiss. 16 (1928) 245-257; in: *Atomtheorie und Naturbeschreibung*, 1931, pp. 34-59

N. Bohr, *Atomtheorie und Naturbeschreibung*, Berlin: Springer 1931

N. Bohr, *Can Quantum-Mechanical Description of Physical Reality be Considered Complete?*, Phys. Rev. 48 (1935) 696-702

N. Bohr, H. A. Kramers & J. C. Slater, *The Quantum Theory of Radiation*, Phil. Mag. [6] 47 (1924) 785-802; reprint in: van der Waerden, 1968, paper 5

E. du Bois-Reymond, *Über die Grenzen des Naturerkennens*, 45. Versammlung deutscher Naturforscher und Ärzte, Leipzig, 14.08.1872, Leipzig: Veit 1872; http://www.deutschestextarchiv.de/book/show/dubois_naturerkennen_1872; in: *Reden von Emil du Bois-Reymond in zwei Bänden* (Estelle du Bois-Reymond, Ed.), Vol. 1, Leipzig: Veit 1912, pp. 441-473; http://vlp.mpiwg-berlin.mpg.de/library/data/lit28636; in: *Vorträge über Philosophie und Gesellschaft*, Hamburg: Meiner 1974

A. Bokulich, *Open or Closed? Dirac, Heisenberg, and the Relation between Classical and Quantum Mechanics*, Stud. Hist. Philos. Mod. Phys. 35 (2004) 377-396; philsci-archive.pitt.edu/1614/1/Open_or_Closed-preprint.pdf (11.05.2015)

L. Boltzmann, *Über die Beziehung zwischen dem zweiten Hauptsatze der mechanischen Wärmetheorie und der Wahrscheinlichkeitsrechnung respektive den Sätzen über das Wärmegleichgewicht*, Sitz.ber. Kais. Akad. Wiss. Math.-Naturwiss. Classe, Abt. II, LXXVI (1877) 373-435; reprint in: Wiss. Abhandlungen, Vol. II, Leipzig: Barth 1909/New York: Chelsea 1968, reprint 42, pp. 164-223; Engl.: *On the Relationship between the Second Fundamental Theorem of the Mechanical Theory of Heat and Probability Calculations Regarding the Conditions for Thermal Equilibrium* (transl. by K. Sharp & F. Matschinsky), Entropy 17 (2015) 1971-2009; http://www.mdpi.com/1099-4300/17/4/1971 (01.08.2015)

L. Boltzmann, *Wissenschaftliche Abhandlungen* (F. Hasenöhrl, Ed.), Leipzig: Barth 1909 (3 Vols.); s. a. C. Weber, *Boltzmann's Bibliography*, faculty.washington.edu/vienna/boltzmann/boltzmannbib.htm (18.02.2006)

P. Boonserm & M. Visser, *Transmission probabilities and the Miller-Good transformation*, J. Phys. A: Math. Theor. 42 (2009) 1-10

H. Lawrence Bond (Ed.), *Nicholas of Cusa: Selected Spiritual Writings*, New York: Paulist Press 1997 (Classics of Western Spirituality)

F. Bopp, *Grundlagen der klassischen Physik in gegenwärtiger Sicht*, Sitz.ber. Bayer. Akad. Wiss., Math.-Naturwiss. Kl. (1971) Sonderdruck 6

H.-H. von Borczeskowski & R. Wahsner, *Newton und Voltaire. Zur Begründung und Interpretation der klassischen Mechanik*, Berlin: Akademie-Verlag 1980 (WTB 123)

M. Born, *Die Dynamik der Kristallgitter*, Leipzig: Teubner 1915

M. Born, *Die Relativitätstheorie Einsteins*, Berlin: Springer 1920, [7]2003 (with comments and additions by J. Ehlers & M. Pössel); Engl.: *Einstein's theory of relativity*, New York: Dutton 1922; https://archive.org/details/einsteinstheoryo00born (18.05.2006)

M. Born, *Über Quantenmechanik*, Z. Phys. 26 (1924) 379-395; home.tiscali.nl/physis/HistoricPaper/HistoricPapers.html (18.05.2006)

M. Born, *Zur Quantenmechanik der Stoßvorgänge*, Z. Phys. 37 (1926) 863-867; home.tiscali.nl/physis/HistoricPaper/HistoricPapers.html (18.05.2006)

M. Born, *Quantenmechanik der Stoßvorgänge*, Z. Phys. 38 (1926) 803-827; reprint in: G. Ludwig, *Wellenmechanik*, 1970, pp. 237-259

M. Born, *Zur Wellenmechanik der Stoßvorgänge*, Nachr. Ges. d. Wiss. Göttingen, Math.-Phys. Kl. 1926, 146-160; Berlin: Weidmannsche Buchh. 1927; gdz.sub.uni-goettingen.de/dms/load/img/?IDDOC=63981 (18.05.2006)

M. Born, *The statistical interpretation of quantum mechanics* (Nobel Lecture 1954); in: G. Ekspong (Ed.), *Nobel Lectures in Physics, Vol 3, 1942-1962*, 1998, pp. 256-267

M. Born, *Ausgewählte Abhandlungen*, Göttingen: Vandenhoek & Ruprecht 1963 (2 Vols.)

M. Born & M. Goeppert-Mayer, *Dynamische Gittertheorie der Kristalle*, in: H. Geiger & K. Scheel (Eds.), *Handbuch der Physik*, Vol. 24, Berlin: Springer 1933, pp. 623-794

M. Born & W. Heisenberg, *La méchanique des quanta*, in: *Électrons et Photons*, 5th Solvay Congress, Brussels, Oct. 24.-29, 1927, Institut International de Physique Solvay 1928, pp. 143-184

M. Born, W. Heisenberg & P. Jordan, *Über Quantenmechanik II*, Z. Phys. 35 (1926) 557-615 ("Drei-Männer-Arbeit"); reprint in: W. Heisenberg, *Gesammelte Werke / Collected Works* (W. Blum, H. P. Dürr & H. Rechenberg, Eds.), Heidelberg: Springer 1985, Ser. A, Part I, pp. 387ff.; Engl.: *On Quantum Mechanics II*, in: van der Waerden, *Sources of Quantum Mechanics*, 1968, pp. 321-385 ("Three-men paper"), web.archive.org/web/20080420214519/, web.ihep.su/owa/dbserv/hw.part2?s_c=BORN+1926 (18.05.2006)

M. Born & K. Huang, *Dynamical Theory of Crystal Lattices*, Oxford: Oxford Univ. Press 1954

M. Born & L. Infeld, *Foundations of the New Field Theory*, Proc. Roy. Soc. A 144 (1934) 425-451; home.tiscali.nl/physis/HistoricPaper/Historic%20Papers.html (18.05.2006)

M. Born & P. Jordan, *Zur Quantenmechanik*, Z. Phys. 34 (1925) 858-888; reprint in: G. Ludwig, *Wellenmechanik*, 1970, pp. 210-237; Engl.: *On Quantum Mechanics*, in: van der Waerden, *Sources of Quantum Mechanics*, 1968, pp. 277-306; web.archive.org/web/20080420214921/ (18.05.2006)[1]

M. Born & Th. v. Kármán, *Über Schwingungen in Raumgittern*, Phys. Zs. 13 (1912) 297-309; reprint in: M. Born, *Ausgewählte Abhandlungen 1*, 1963, pp. 244-248

H.-H. v. Borzeszkowski & R. Wahsner, *Newton und Voltaire – Zur Begründung und Interpretation der klassischen Mechanik*, Berlin: Akademie-Verlag 1980 (WTB 123)

N. Bose, *Plancks Gesetz und Lichtquantenhypothese*, Z. Phys. 26 (1924) 178-181

---

[1] The missing section on field quantization is available at users.phys.psu.edu/ collins/qm-foundations/ (8.5.2015).

D. Bouwmeester, A. K. Ekert & A. Zeilinger (Eds.), *The Physics of Quantum Information. Quantum Cryptography, Quantum Teleportation, Quantum Computation*, Heidelberg *etc.*: Springer 2000

L. Boyle, J. Y. Khoo & K. Smith, *Symmetric Satellite Swarms and Choreographic Crystals*, 2015, Phys. Rev. Lett. 116 (2016) 015503; arXiv:1407.5876v2 (18.06.2016)

K. A. Brading, *Which symmetry? Noether, Weyl, and conservation of electric charge*, Stud. Hist. Phil. Modern Physics 33 (2002) 3-22, phys.cts.ntu.edu.tw/workshop/2012/1010806center/PDF/Brading%20Which%20symmetry_%20Noether,%20Weyl,%20and%20conservation%20of%20electric%20charge.pdf (01.05.2014)

K. Bräuer, *Die fundamentalen Phänomene der Quantenmechanik und ihre Bedeutung für unser Weltbild. Physikalische Grundlagen der Phänomene und Gedanken berühmter Physiker, Biologen und Psychologen*, Berlin: Logos 2000

S. L. Braunstein, C. M. Caves & G. J. Milburn, *Generalized uncertainty relations: Theory, examples, and Lorentz invariance*, arXiv:quant-ph/9507004v1 (24.05.2012)

A. Bravais, *Mémoire sur les systèmes formés par les points distribués régulièrement sur un plan ou dans l'espace*, J. Ecole Polytech. 19 (1850) 1–128; reprint in: *Études Cristallographiques*, Paris: Gauthier-Villars 1866, pp. 1-98, 103, http://gallica.bnf.fr/ark:/12148/bpt6k96124j/f5.image; English: *Memoir 1*, Crystall. Soc. Am. 1949; German: *Abhandlung über die Systeme von regelmässig auf einer Ebene oder im Raum vertheilten Punkten* (transl. and ed. by C. & E. Blasius), Leipzig: Engelmann 1897 (Ostwalds Klassiker 90), https://archive.org/details/abhandlungberdi01bravgoog

R. Brenneke, *Die Verdienste Leonard Eulers um den Potentialbegriff*, Z. Phys. 25 (1924) 1-6

J. S. Briggs & J. M. Rost, *Time Dependence in Quantum Mechanics*, 1999, arxiv.org/pdf/quant-ph/9902035.pdf (24.04.2002)

J. S. Briggs & J. M. Rost, *On the derivation of the time-dependent equation of Schrödinger*, Found. Physics 31 (2001) 693-712; mpipks-dresden.mpg.de/ rost/jmr-reprints/brro01.pdf (24.04.2002)

L. Brillouin, *Les électrons dans les métaux et le classement des ondes de de Broglie correspondantes*, Compt. Rend. [4] 191 (1930) 292-294

L. de Broglie, *Waves and Quanta*, Compt. rend. 177 (1923) 507-510; home.tiscali.nl/physis/HistoricPaper/Historic%20Papers.html (04.04.2012); Engl.: *Radiation – Waves and Quanta*, strangepaths.com/wp-content/uploads/2008/01/rendus-e.pdf (04.04.2012)

L. de Broglie, *Les quanta, la théorie cinétique des gaz et le principe de Fermat*, Comptes rendus 177 (1923) 630-632, home.tiscali.nl/physis/HistoricPaper/Historic%20Papers.html (04.04.2012)

L. de Broglie, *Waves and quanta*, Nature 112 (1923) 540

L. de Broglie, *Recherches sur la théorie des quanta*, Thésis, Paris 1924; Ann. Physique [10] III (1925) 22-128; tel.archives-ouvertes.fr/tel-00006807 (04.04.2012); reprint: Ann. Fond. Louis de Broglie 17 (1992) 1, 1-109; German in: G. Ludwig, *Wellenmechanik*, 1970, pp. 85-108 (Chs. I-III)

L. de Broglie, *Ondes et mouvements*, Paris: Gauthier-Villars 1926

L. de Broglie, *Sur la fréquence propre de l'électron*, Comptes rendus 180, (1925) 498-500; home.tiscali.nl/physis/HistoricPaper/Historic%20Papers.html (04.04.2012)

L. de Broglie, *La structure atomique de la matière et du rayonnement et la Mécanique ondulatoire*, C. R. Acad. Sci. Paris 183 (1926) 447-448

L. de Broglie, *The wave nature of the electron* (Nobel Lecture 1929); in: G. Ekspong (Ed.), *Nobel Lectures in Physics, Vol 2, 1922-1941*, 1998, pp. 244-256; nobelprize.org/nobel_prizes/physics/laureates/1929/broglie-lecture.pdf (04.04.2012)

L. de Broglie, *A New Conception of Light* (transl. by D. H. Delphenich), Paris: Herman & Co.1934; neo-classical-physics.info/uploads/3/0/6/5/3065888/de_broglie_-_new_conception_of_light.pdf (04.04.2012)

Louis de Broglie: Physicien et penseur, Paris: Albin Michel 1953

L. de Broglie, *The Theory of Measurement in Wave Mechanics (Usual Interpretation and Causal Interpretation)* (transl. by D. H. Delphenich), Paris: Gauthier-Villars 1957 (The great problems in science VII); neo-classical-physics.info/uploads/3/0/6/5/3065888/de_broglie_-_measurement_in_wave_mechanics.pdf (04.04.2012)

L. de Broglie, *Étude critique des bases de l'interprétation actuelle de la mécanique ondulatoire*, Paris: Gauthier-Villars 1963; Engl.: *The Current Interpretation of Wave Mechanics: A Critical Study*, Amsterdam: Elsevier 1964

L. de Broglie, *Certitude et incertitude de la science*, Paris: Gauthier-Villars 1966

P. Bruno, *Comment on "Space-Time Crystals of Trapped Ions": And Yet it Moves* Not!, https://arxiv.org/abs/1211.4792v1

D. Bruß, *Quanteninformation*, Frankfurt am Main: Fischer 2003 (Fischer Kompakt)

S. G. Brush, *Kinetische Theorie. Band I. Die Natur der Gase und der Wärme, Band II. Irreversible Prozesse*, Berlin: Akademie-Verlag / Oxford: Pergamon / Braunschweig: Vieweg 1970 (WTB 65, 67); new Engl. ed.: *Kinetic Theory of Gases: An Anthology of Classic Papers With Historical Commentary* (N. S. Hall, Ed.), London: Imperial College Press 2003 (History Mod. Phys. Sci. 1)

A. Budó, *Theoretische Mechanik*, Berlin: Dtsch. Verlag d. Wiss. 121990 (Hochschulbücher Physik 25)

J. B. Calvert, *The Hodograph*, 2003, mysite.du.edu/126jcalvert/phys/hodo.htm (09.08.2015)

T. Cao (Ed.), *Conceptual Foundations of Quantum Field Theory*, Cambridge: Cambridge Univ. Press 1999

A. Caprez, B. Barwick, H. Batelaan, *A macroscopic test of the Aharonov-Bohm effect*, arXiv:0708.2428v1 (04.10.2012)

D. Carfi, *The Pointwise Hellmann-Feynman Theorem*, AAPP | Atti Accad. Peloritana Pericolanti, Sci. Fis., Mat. Nat. LXXXVIII (2010) 1, C1A1001004-1...14

J. F. Cariñena, M. F. Rañada & M. Santander, *A new look at the Feynman 'hodograph' approach to the Kepler first law*, Eur. J. Phys. 37 (2016) 025004 (19 pp.)

S. Carlson, *Why not energy conservation?*, Eur. J. Phys. 37 (2016) 015801 (12 pp.)

S. Carlson, *A novel formalism of classical mechanics*, to be publ.

P. Carruthers & M. M. Nieto, *Coherent States and the Forced Quantum Oscillator*, Am. J. Phys. 33 (1965) 537-544

E. Castellani & P. Mittelstaedt, *Leibniz's Principle, Physics, and the Language of Physics*, Found. Phys. 30 (2000) 1587-1604

S. Chandrasekhar, *On a New Theory of Weizsäcker on the Origin of the Solar System*, Rev. Mod. Phys. 18 (1946) 94-102

S. Chandrasekhar, *Newton's Principia for the Common Reader*, Oxford: Oxford Univ. Press 1995

S. Chapman & T. G. Cowling, *The Mathematical Theory of Non-Uniform Gases. An account of the kinetic theory of viscosity, thermal conduction, and diffusion in gases*, Cambridge: Cambridge Univ. Press 1939

G.-E. Du Châtelet, *Institutions de Physique*, Paris: Prault 1740

F. Chatelin, *Valeurs propres de matrices*, Masson 1988; Engl.: *Eigenvalues of Matrices*, transl. with add. material by Walter Lederman, Chichester *etc*: Wiley 1995

P.-H. Chavanis & T. Matos, *Covariant theory of Bose-Einstein condensates in curved spacetimes with electromagnetic interactions: the hydrodynamic approach*, Eur. Phys. J. Plus (2016) June, http://arxiv.org/abs/1606.07041v1, https://www.researchgate.net/publication/304268938

P. L. Chebyshev, *Sur le développement des fonctions à une seule variable*, Bull. Acad. Sci. St. Petersb. I (1859) 193-200; *Oeuvres* I, pp. 501-508

P. Chen & H. Kleinert *Deficiencies of Bohmian Trajectories in View of Basic Quantum Principles*, EJTP 13 (2016) 35, 1-12, http://www.ejtp.com/articles/ejtpv13i35p1.pdf (16.08.2017)

J. Clauser, M. Horne, A. Shimony & R. Holt, *Proposed Experiment to Test Local Hidden-Variable Theories*, Phys. Rev. Lett. 23 (1969) 15, 880–884

R. Clausius, *Über einen auf die Wärme anwendbaren mechanischen Satz*, Sitz.-Ber. Niederrhein. Ges. Bonn (1870) 114-119; in: Brush, *Kinetische Theorie I*, 1970, pp. 245-253

I. B. Cohen (Ed.), *Isaac Newton's Papers and Letters on Natural Philosophy*, Cambridge (MA): Harvard Univ. Press 1958

E. Colomés, Z. Zhan & X. Oriols, *Comparing Wigner, Husimi and Bohmian distributions: Which one is a true probability distribution in phase space?*, J. Comput. Electron. (2015) July; https://www.researchgate.net/publication/280590425 (14.04.2016)

G. P. Collins, *Das kälteste Gas im Universum*, Spektrum d. Wiss. Dossier 1/2003, 32-39

S. Collins, David Lowe & J. R. Barker, *The quantum mechanical tunnelling time problem– revisited*, J. Phys. C: Solid State Phys. 20 (1987) 6213-6232

A. H. Compton, *A Quantum Theory of the Scattering of X-Rays by Light Elements*, Phys. Rev. [2] 21 (1923) 5, 483–502, http://www.gsjournal.net/Science-Journals/Historical%20Papers-Relativity%20Theory/Download/2456 (04.05.2016)

E. U. Condon, *60 Years of Quantum Physics*, Physics Today 15 (1962) 37-49

J. H. Conway, H. Burgiel & Ch. Goodman-Strauss, *The Symmetries of Things*, A K Peters/CRC Press 2008

G.-G. Coriolis, *Sur les équations du mouvement relatif des systèmes de corps*, J. l'Ecole royale polytechnique 15 (1835) 144-154

E. A. Cornell & C. E. Wieman, *Bose-Einstein Condensation in a Dilute Gas; The First 70 Years and Some Recent Experiments* (Nobel Lecture 2001), nobel.se/physics/laureates/2001/cornellwieman-lecture.html (27.11.2011)

Coulomb, *Premier mémoire sur l'électricité et le magnétisme*, Hist. l'Acad. Roy. Sci. (1785) 569-577; books.google.de/books?id=by5EAAAAcAAJ pg=PA569 hl=de#v=onepage q f=false (27.10.2013)

R. Courant & D. Hilbert, *Methoden der mathematischen Physik*, Berlin: Springer 1924/1935 (2 Vols.)

A. P. Cracknell, *Applied Group Theory*, Oxford: Pergamon 1968 (Selected Readings in Physics); German: *Angewandte Gruppentheorie. Einführung und Originaltexte* (transl. by F. Cap), Braunschweig: Vieweg / Berlin: Akademie-Verlag 1971 (WTB 84)

P. Curie, *Sur la symétrie dans les phénomènes physiques, symétrie d'un champ électrique et d'un champ magnétique*, J. Physique (Paris) [3] 3 (1894) 393-415; hal.archives-ouvertes.fr/docs/00/23/98/14/PDF/ajp-jphystap_1894_3_393_0.pdf (04.01.2014); Engl.: *On Symmetry in Physical Phenomena, Symmetry of an Electric Field and of a Magnetic Field*, in: J. Rosen (Ed.), *Symmetry in Physics*, 1982, pp. 17-25

M. Danos & J. Rafelski, *Pocketbook of mathematical functions. Abridged edition of 'Handbook of Mathematical Functions', Milton Abramowitz and Irene A. Stegun (Eds.)*, Thun: Deutsch 1984

M. Daumer, *Quantenmechanik und Determinismus*, in: G. Schaefer, *Das Elementare im Komplexen*, 1997, 4. Bereich: Physik und Philosophie, pp. 167-192; mathematik.uni-muenchen.de/~bohmmech/BohmHome/files/daumer_qm_det.pdf (04.04.2004)

M. Daumer, *Streutheorie aus der Sicht Bohmscher Mechanik*, PhD theses, mathematik.uni-muenchen.de/ bohmmech/theses/Daumer_Martin_PhD.pdf (14.08.2015)

P. C. W. Davies, *Quantum tunneling time*, Am. J. Phys. 73 (2005) 1, 23-27; cosmos.asu.edu/publications/papers/%27Quantum%20Tunelling%20Time%27%20AJP000023.pdf (01.04.2012)

O. Derzhko, J. Richter & M. Maksymenko, *Strongly correlated ?at-band systems: The route from Heisenberg spins to Hubbard electrons*, Int. J. Mod. Phys. B 29 (2015) 12, 1530007 (72 p.)

J. A. DeSanto, *Scalar Wave Theory. Green's Functions and Applications*, Berlin *etc.*: 1992 (Springer Series in Wave Phenomena 12)

R. Descartes, *Principia philosophiae*, Amsterdam 1644; German: *Die Prinzipien der Philosophie*, Hamburg: Meiner 1922, [7]1965

R. Descartes, *Le Monde*, 1664; German: *Le Monde ou Traité de la Lumière – Die Welt oder Abhandlung über das Licht*, Weinheim: VCH / Berlin: Akademie-Verlag 1989; Engl.: *The World or Treatise on Light*, transl. by M. S. Mahoney, http://www.princeton.edu/hos/mike/texts/descartes/world/worldfr.htm (29.06.2013)

D. Deutsch, *Pitch Circularity*, Acoustics Today, July 2010, 8-15; http://deutsch.ucsd.edu/psychology/pages.php?i=213 (29.05.2017)

F. J. Dijksterhuis, *Lenses and Waves. Christiaan Huygens and the Mathematical Science of Optics in the Seventeenth Century*, New York *etc.*: Kluwer 2005 (Archimedes 9)

P. A. M. Dirac, *Quantum Mechanics and a Preliminary Investigation of the Hydrogen Atom*, Proc. Roy. Soc. A 110 (1926) 561-579; reprint in: B. L. van der Waerden, *Sources of Quantum Mechanics*, 1968, pp. 417-427

P. A. M. Dirac, *On the theory of quantum mechanics*, Proc. Roy. Soc. L. A 112 (1926) 661-677; jstor.org/discover/10.2307/94692?uid=3737864 uid=2 uid=4 sid=21105957412141 (01.04.2012)

P. A. M. Dirac, *The physical interpretation of the quantum mechanics*, Proc. Phys. Soc. A 113 (1927) 621-641

P. A. M. Dirac, *The quantum theory of emission and absorption of radiation*, Proc. Roy. Soc. L. A114 (1927) 243-265; reprint in: Schwinger (Ed.), *Selected Papers on Quantum Electrodynamics*, 1958, No. 1; http://hermes.ffn.ub.es/luisnavarro/nuevo_maletin/Dirac_QED_1927.pdf, http://rspa.royalsocietypublishing.org/content/royprsa/114/767/243.full.pdf (03.04.2012)

P. A. M. Dirac, *The Quantum Theory of the Electron*, Proc. Roy. Soc. L. A 117 (1928) 778, 610-624; http://www.math.ucsd.edu/ nwallach/Dirac1928.pdf (03.04.2012)

P. A. M. Dirac, *Quantum mechanics of many-electron systems*, Proc. Royal Soc. L. A 123 (1929) 714-733

P. A. M. Dirac, *The basis of statistical mechanics*, Proc. Cambr. Phil. Soc. 25 (1929) 62-66

P. A. M. Dirac, *Note on the exchange phenomena in the Thomas atom*, Proc. Cambr. Phil. Soc. 26 (1930) 376-385

P. A. M. Dirac, *The Principles of Quantum Mechanics*, Oxford: Clarendon 1930, 41958

P. A. M. Dirac, *Quantized singularities in the electromagnetic field*, Proc. Roy. Soc. (L.) A 133 (1931) 60-72

P. A. M. Dirac, *The Lagrangian in Quantum Mechanics*, Phys. Zs. Sowjetunion 3 (1933) No. 1; reprint in: J. Schwinger (Ed.), *Selected Papers on Quantum Electrodynamics*, 1958, No. 26, http://www.ifi.unicamp.br/ cabrera/teaching/aula%2015%202010s1.pdf (04.04.2012)

P. A. M. Dirac, *Theory of electrons and positrons* (Nobel Lecture 1933); in: G. Ekspong (Ed.), *Nobel Lectures in Physics, Vol 2, 1922-1941*, 1998, pp. 320-325

P. A. M. Dirac, *On the Analogy Between Classical and Quantum Mechanics*, Rev. Mod. Phys. 17 (1945) 195ff.

P. A. M. Dirac, *Is There an Aether?*, Nature 168 (1951) 906-907

P. A. M. Dirac, *Lectures on Quantum Mechanics*, New York: Yeshiva University 1964

P. A. M. Dirac, *The Relativistic Electron Wave Equation*, Eur. Conf. Particle Phys., Budapest 1977; Hungarian Acad. Sci. Central Res. Inst. Phys., preprint K.FKI-1977-62; Russ.: П.А.М.Дирак, *Релятивисткое Волновое Уравнение Электрона*, УФН 129 (1979) 4, 681-691, https://ufn.ru/ufn79/ufn79_8/Russian/r798e.pdf

P. A. M. Dirac, V. A. Fock & B. Podolsky, *On Quantum Electrodynamics*, Sow. Phys. 2 (1932) 468-479; reprint in: Fock, *Papers on Quantum Field Theory* (2007), pp. 70-82 (in Russian)

E. A. Donley, B. P. Anderson & C. E. Wieman, *New Twists in Bose-Einstein Condensation*, Optics & Photonics News 12 (2001) H.10, 34-37

A. Douglas Stone, *Einstein and the Quantum: The Quest of the Valiant Swabian*, Princeton: Princeton Univ. Press 2013

I. Duck, *Discovering Quantum Mechanics Once Again*, arXiv:quant-ph/0307121 (03.04.2012)

I. Duck & E. C. G. Sudarshan (Eds.), *Pauli and the Spin-Statistics Theorem*, Singapore: World Scientific 1998

D. Dürr, *Bohmsche Mechanik als Grundlage der Quantenmechanik*, Berlin: Springer 2001

D. Dürr, S. Goldstein, R. Tumulka & N. Zanghi, *Bohmian Mechanics and Quantum Field Theory*, Phys. Rev. Lett. 93 (2004); arXiv:quant-ph/0303156 (05.04.2004)

R. Dugas, *L'histoire de la mécanique*, Neuchâtel: Griffon 1955; Engl.: *A History of Mechanics*, New York: Dover 1988

P. Duhem, *The Origins of Statics*, Dordrecht: Kluwer 1991

P. L. Dulong & A. T. Petit, *Recherches sur quelque points important de la Théorie de la Chaleur*, Ann. Chim. et Phys. 10 (1819) 395-413; Engl.: Ann. Phil. 14 (1819) 189-198; web.lemoyne.edu/giunta/PETIT.html (11.02.2006)

Jю Duoandikoetxea, *Fourier Analysis*, American Mathematical Society, 2001

W. Durant, *The Story of Philosophy. The Lives and Opinions of the Greater Philosophers*, New York: Time 1926; https://archive.org/details/THESTORYOFPHILOSOPHY1TheLivesAndOpinionswillDurant1926; revis. ed.: New York: Garden City ; http://www.rosenfels.org/Durant.pdf (12.03.2006)

F. J. Dyson, *The radiation theories of Tomonaga, Schwinger, and Feynman*, Phys. Rev. 75 (1949) 486-502; reprint in: J. Schwinger (Ed.), *Selected Papers on Quantum Electrodynamics*, 1958, No. 24; imotiro.org/repositorio/howto/artigoshistoricosordemcronologica/1949b%20-DYSON%201949B%20Covariant%20quantum%20electrodynamics%20b.pdf (07.02.2006)

F. J. Dyson, *The S-matrix in quantum electrodynamics*, Phys. Rev. 75 (1949) 1736-1753; reprint in: J. Schwinger (Ed.), *Selected Papers on Quantum Electrodynamics*, 1958, No. 25

F. J. Dyson, *Advanced Quantum Mechanics*, 1951 (lectures at the Cornell University, autumn 1951); 2nd ed. (M. J. Moravcsik, Ed.): arXiv:quant-ph/0608140 (07.02.2006)

F. J. Dyson, *George Green and physics*, Physics World (1993) Aug., 33-38

C. Eckart, *The penetration of a potential barrier by electrons*, Phys. Rev. 35 (1930) 11, 1303-1309

W. Ehrenberg & R. E. Siday, *The Refractive Index in Electron Optics and the Principles of Dynamics*, Proc. Phys. Soc. B 62 (1949) 8-21

P. Ehrenfest, *Welche Züge der Lichtquantenhypothese spielen in der Theorie der Wärmestrahlung eine wesentliche Rolle?*, Ann. Phys. [4] 36 (1911) 91-118

P. Ehrenfest, *Bemerkung über die angenäherte Gültigkeit der klassischen Mechanik innerhalb der Quantenmechanik*, Z. Phys. 45 (1927) 455-457

A. Einstein, *Über einen die Erzeugung und Verwandlung des Lichtes betreffenden heuristischen Gesichtspunkt*, Ann. Phys. 17 (1905) 132-148; reprint in: J. Renn (Ed.), *Einstein's Annalen Papers*, 2005, pp. 164-181; strangepaths.com/files/1905.pdf; Engl.: *On a Heuristic Point of View about the Creation and Conversion of Light*, in: D. ter Haar, *The Old Quantum Theory*, 1967, pp. 91-107; http://users.physik.fu-berlin.de/~kleinert/files/eins_lq.pdf (all retriev. 07.02.2006)

A. Einstein, *Zur Elektrodynamik bewegter Körper*, Ann. Phys. 17 (1905) 891–921, reprint in: J. Renn (Ed.), *Einstein's Annalen Papers*, 2005, pp. 196-224; http://users.physik.fu-berlin.de/~kleinert/files/1905_17_891-921.pdf (07.02.2006)

A. Einstein, *Zur Theorie der Lichterzeugung und Lichtabsorption*, Ann. Phys. 20 (1906) 199-206, reprint: J. Renn (Ed.), *Einstein's Annalen Papers*, 2005, pp. 259-267

A. Einstein, *Die Plancksche Theorie der Strahlung und die Theorie der spezifischen Wärme*, Ann. Phys. 22 (1907) 180-190, correction: *ibid.*, p. 800; reprint in: J. Renn (Ed.), *Einstein's Annalen Papers*, 2005, pp. 280-291 resp. 296; physik.uni-augsburg.de/annalen/history/einstein-papers/1907_22_180-190.pdf, physik.uni-augsburg.de/annalen/history/einstein-papers/1907_22_800.pdf (07.02.2006)

A. Einstein, *Zum gegenwärtigen Strand des Strahlungsproblems*, Phys. Zs. 10 (1909) 185-193

A. Einstein, *Über die Entwicklung unserer Anschauungen über das Wesen und die Konstitution der Strahlung*, Phys. Zs. 10 (1909) 817-826; http://ekkehard-friebe.de/EINSTEIN-1909-P.pdf (07.02.2006) – Engl.: *The Development of Our Views on the Composition and Essence of Radiation*, astrofind.net/documents/the-composition-and-essence-of-radiation.php (30.04.2015)

A. Einstein, *Eine Beziehung zwischen dem elastischen Verhalten und der spezifischen Wärme bei festen Körpern mit einatomigem Molekül*, Ann. Phys. [4] 38 (1912) 881-884

A. Einstein, *Zum gegenwärtigen Stande des Problems der spezifischen Wärme*, Abhandl. Dtsch. Bunsengesell. 3 (1914) 330-364

A. Einstein, *Strahlungs-Emission und -Absorption nach der Quantentheorie*, Verh. Dtsch. Phys. Ges. Berlin [2] 18 (1916) 318-323

A. Einstein, *Zur Quantentheorie der Strahlung*, Mitt. Phys. Ges. (Zürich) 18 (1916) 47-62; reprint: Phys. Zs. 18 (1917) 121-128; Engl.: *On the Quantum Theory of Radiation*, in: van der Waerden, 1968, Paper 1

A. Einstein, *Zum Quantensatz von Sommerfeld und Epstein*, Verh. Dtsch. Phys. Ges. Berlin [2] 19 (1917) 82-92

A. Einstein, *Bietet die Feldtheorie Möglichkeiten für die Lösung des Quantenproblems?*, Sitzungsber. Preuss. Ak. Wiss. phys.-math. Kl., 13. Dez., XXXIII (1923) 359-364

A. Einstein, *Anmerkung zu S. N. Boses Abhandlung Plancks Gesetz und Lichtquantenhypothese*, Z. Phys. 26 (1924) 181ff.

A. Einstein, *Quantentheorie des einatomigen idealen Gases*, Sitzungsber. Preuss. Akad. Wiss. Phys.-math. Kl. 22 (1924) 261-267; reprint in: *Akademie-Vorträge*, 1978, No. 27; see also library.oregonstate.edu/specialcollections/coll/pauling/bond/notes/sci3.001.23.html (07.02.2006)

A. Einstein, *Quantentheorie des einatomigen idealen Gases. Zweite Abhandlung*, Sitzungsber. Preuss. Akad. Wiss. Phys.-math. Kl. 23 (1925) 3-14; reprint in: *Akademie-Vorträge*, 1978, No. 28

A. Einstein, *Zur Quantentheorie des idealen Gases*, Sitzungsber. Preuss. Akad. Wiss. Phys.-math. Kl. 23 (1925) 18-25; reprint in: *Akademie-Vorträge*, 1978, No. 29

A. Einstein, *Maxwell's Einfluss auf die Entwicklung der Auffassung des Physikalisch-Realen*, 1931; in: *Mein Weltbild*, Amsterdam 1934; Engl.: *Maxwell's Influence of the Development of the Conception of Physical Reality*; in: Maxwell, *Dynamical Theory of the electromagnetic field*, 1996, pp. 29-32; photontheory.com/Einstein/Einstein09.html; mountainman.com.au/aether_2.html[2] (07.02.2006)

A. Einstein, *Einleitende Bemerkungen über Grundbegriffe*, in: *Louis de Broglie, Physicien et penseur*, 1953

A. Einstein, *Letters on Wave Mechanics: Correspondence with H. A. Lorentz, Max Planck, and Erwin Schrödinger* (K. Przibram, Ed.; M. J. Klein, transl. and Introduction), New York: Philos. Libr. 1967, Kindle ed. 2011

A. Einstein, *Grundzüge der Relativitätstheorie*, Berlin: Akademie-Verlag [5]1977 (WTB 58), Anhang II. *Relativistische Theorie des nichtsymmetrischen Feldes*; Engl.: *The Meaning of Relativity*, Appendix II. *Relativistic Theory of the Non-Symmetric Field*, Princeton: Princeton Univ. Press [5]2014

A. Einstein, *Akademie-Vorträge*, Berlin: Akademie-Verlag 1978 / Weinheim: Wiley-VCH 2012

A. Einstein, *The Collected Papers of Albert Einstein – Gesammelte Schriften*, Princeton: Princeton Univ. Press 1987ff.

A. Einstein, H. A. Lorentz, H. Minkowski & H. Weyl, *The Principle of Relativity. A Collection of Original Memoirs on the Special and General Theory of Relativity* (with notes by A. Sommerfeld), New York: Dover 1923

---

[2] References to reprints are solely for providing easy access to original texts, not for promoting their viewpoints.

A. Einstein, B. Podolsky & N. Rosen, *Can Quantum-Mechanical Description of Physical Reality Be Considered Complete?*, Phys. Rev. 47 (1935) 777-780; strangepaths.com/wp-content/uploads/2008/01/epr.pdf (07.02.2006)

A. Einstein, *Lettres à Maurice Solovine. Reproduites en Facsimilé et traduites en français*, Paris: Gauthier-Villars 1956 / *Briefe an Maurice Solovine. Faksimile-Wiedergabe von Briefen aus den Jahren 1906 bis 1955, mit französischer Übersetzung, einer Einführung und drei Fotos*, Berlin: Deutscher Verlag d. Wiss. 1960

G. Ekspong (Ed.), *Nobel Lectures in Physics 1981-1990 / 1991-1995 / 1996-2000*, Singapore: World Scientific 1993 / 1997 / 2002

G. Ekspong (Ed.), *Nobel Lectures in Physics Vol 1 1901-1921, Vol 2 1922-1941, Vol 3 1942-1962, Vol 4 1963-1980*, Singapore: World Scientific 1998

R. S. Elliott, *Electromagnetics: History, Theory, and Applications*, Piscataway (NJ): IEEE 1993

A. Enders & G. Nimtz, *On superluminal barrier traversal*, J. Physique I. 2 (1992) 1693–1698

A. Enders & G. Nimtz, *Evanescent-mode propagation and quantum tunneling*, Phys. Rev. E 48 (1993) 632-634

G. Enders, *Zur Kassation von Akten statistischer Dienststellen*, Archivmitteilungen 1 (1954) 10-13

G. Enders, *Archivverwaltungslehre*, Berlin: Rütten & Loening 1962; Leipzig: Univ.-Verlag 2004 (reprint of the 3rd ed., with prefaces by Eckart Henning, Gerald Wiemers, and Lieselott Enders)

P. Enders, *Schrödinger Equation and Wave-Function Matching Conditions for Spatially Varying Effective Mass*, phys. stat. sol. (b) 139 (1987) K113-K116

P. Enders, *Equality and Identity and (In)distinguishability in Classical and Quantum Mechanics from the Point of View of Newton's Notion of State*, 6th Int. Symp. *Frontiers of Fundamental and Computational Physics*, Udine 2004; in: Sidharth, Honsell & De Angelis (Eds.), *Frontiers of Fundamental Physics*, 2006, pp. 239-245

P. Enders, *Von der klassischen Physik zur Quantenphysik. Eine historisch-kritische deduktive Ableitung mit Anwendungsbeispielen aus der Festkörperphysik*, Berlin · Heidelberg: Springer 2006; springer.com/de/book/9783540393955 (07.06.2016)

P. Enders, *Is Classical Statistical Mechanics Self-Consistent? (A paper of honour of C. F. von Weizsäcker, 1912-2007)*, Progr. Phys. 3 (2007) 85-87; allbusiness.com/science-technology/physics/ 5518225-1.html (07.02.2009)

P. Enders, *Equality and Identity and (In)distinguishability in Classical and Quantum Mechanics from the Point of View of Newton's Notion of State*, Icfai Univ. J. Phys. I (2008) 71-78; iupindia.org/108/IJP_Classical_and_Quantum_Mechanics_71.html (07.02.2009)

P. Enders, *Towards the Unity of Classical Physics*, Apeiron 16 (2009) 22-44; redshift.vif.com/JournalFiles/V16NO1PDF/V16N1END.pdf

P. Enders, *Huygens' principle as universal model of propagation*, Latin Am. J. Phys. Educ. 3 (2009) 19-32; dialnet.unirioja.es/servlet/articulo?codigo=3688899

P. Enders, *Gibbs' Paradox in the Light of Newton's Notion of State*, Entropy 11 (2009) 454-456; http://mdpi.com/1099-4300/11/3/454

P. Enders, *State, Statistics and Quantization in Einstein's 1907 Paper, 'Planck's Theory of Radiation and the Theory of Specific Heat of Solids'*, Icfai Univ. J. Phys. II (2009) 176-195; iupindia.org/709/IJP_Einsteins_1907_Paper_176.html (10.05.2009)

P. Enders, *Are there physical systems obeying the Maxwell-Boltzmann statistics?*, Apeiron 16 (2009) 542-554; redshift.vif.com/JournalFiles/V16NO4PDF/V16N4END.pdf (10.10.2009)

P. Enders, *Underdeterminacy and Redundance in Maxwell's Equations. Origin of Gauge Freedom – Transversality of Free Electromagnetic Waves – Gaugefree Canonical Treatment without Constraints*, EJTP 6 (2009) 135-166; ejtp.com/articles/ejtpv6i22p135.pdf (10.10.2009)

P. Enders, *Precursors of force fields in Newton's 'Principia'*, Apeiron 17 (2010) 22-27; redshift.vif.com/JournalFiles/V17NO1PDF/V17N1END.PDF (10.10.2010)

P. Enders, *Bose-Einstein versus Maxwell-Boltzmann distributions*, Apeiron 18 (2011) 15-17; redshift.vif.com/JournalFiles/V18NO1PDF/V18N1EN1.pdf (10.02.2011)

P. Enders, *Physical, metaphysical and logical thoughts about the wave equation and the symmetry of space-time*, Apeiron 18 (2011) 203-221; redshift.vif.com/JournalFiles/V18NO2PDF/V18N2END.pdf (10.05.2011)

P. Enders, *Veni – Vidi – Cassavi. Methodologische Gespräche zwischen Historikerin und Physiker*, Gedenkcolloquium für Lieselott Enders 2011, in: Brandenburgische Archive 29 (2012) 43-46; blha.de/filepool/brbgarchive_29_web.pdf (10.02.2014)

P. Enders, *Huygens' Principle for Linguistics*, Abai' Inst. Khab. 6 (12) (2011) 70-74; abai-inst.kz/pdf/Abai_habarshysy_6(12)2011.pdf (10.02.2014)

P. Enders, *Electromagnetic Momentum Balance in Maxwell's and Hertz's Works*, Galilean Electrodynamics 23 (2012) 5, 83-94

P. Enders, *Quantization as Selection rather than Eigenvalue Problem*, in: P. Bracken (Ed.), *Advances in Quantum Mechanics*, Rijeka: InTech 2013, Ch. 23, pp. 543-564; http://cdn.intechopen.com/pdfs/41542/InTech-Quantization_as_selection_rather_than_eigenvalue_problem.pdf (10.02.2014)

P. Enders, *The divergence between the historical and the logical developments of physics– Forgotten old insights can serve modern physics*, Asian J. of Physics 23 (2014) 1 & 2, 265-286

P. Enders, *Gerhart Enders als Wissenschaftler*, Brandenburgische Archive 32 (2015) 77-79; opus4.kobv.de/opus4-slbp/files/8026/Brandenburgische+Archive+32.pdf (10.05.2015)

P. Enders, *Historical prospective: Boltzmann's versus Planck's state counting – Why Boltzmann did not arrive at Planck's distribution law*, J. Thermodyn. 2016 (2016) Art. ID 9137926, 13 p., http://www.hindawi.com/journals/jther/2016/9137926/ (11.05.2016); Proc. Conf. 'Radiation-thermal phenomena and innovative technologies', ded. to A. I. Kupchishin, Nov. 10-11, 2015, Abai University, Almaty, 2015, 43-61

P. Enders, A. Bärwolff, M. Wörner & D. Suisky, **k.p** *theory of energy bands, wave functions, and optical selection rules in strained tetrahedral semiconductors*, Phys. Rev. B 51 (1995) 16695-16704

P. Enders & D. Suisky, *Quantization as selection problem*, Int. J. Theor. Phys. 44 (2005) 161-194

W. Engelmann, *Ankündigung*, in: *Ostwalds Klassiker der exakten Wissenschaften*, No. 121, Leipzig 1911; gutenberg.org/files/40854/40854-h/40854-h.htm (17.08.2013)

Ch. P. Enz & K. v. Meyenn (Eds.), *Wolfgang Pauli. Das Gewissen der Physik*, Wiesbaden: Vieweg+Teubner 1988

H. M. Enzensberger, *Die Elixiere der Wissenschaft*, Frankfurt a. M.: Suhrkamp 2002

L. Euler, *Mechanica sive motus scientia analytice exposita*, Petropoli 1736, in: *Opera Omnia* II, 1 & 2 (E015f.)[3]

L. Euler, *De motu projectorum in medio non resistente, per Methodum maximorum ac minimorum determinando*, 1743, Additamentum [Appendix] II to *Methodus inveniendi lineas curvas maximi minimive proprietate gaudentes ...*, 1744

L. Euler, *De Curvis Elasticis*, Additamentum [Appendix] I to *Methodus inveniendi lineas curvas maximi minimive proprietate gaudentes ...*, 1744; Engl.: *Leonhard Euler's Elastic Curves* (transl. and annot. by W. A. Oldfather, C. A. Ellis & D. M. Brown), Isis 20 (1933) 1, 72-160; http://www.jstor.org/stable/224885, https://www.princeton.edu/ssp/joseph-henry-project/euler-buckling/Trans-lation-of-1744-Euler.pdf (10.02.2015)

L. Euler, *Methodus inveniendi lineas curvas maximi minimive proprietate gaudentes, sive solutio problematis isoperimetrici latissimo sensu accepti*, Lausanne · Genf: Bousquet 1744; in: *Opera Omnia* I, 24; German: *Eine Methode sich der Eigenschaft des Maximums oder Minimums erfreuender Kurven zu finden, oder die Lösung des im weitesten Sinn aufgefassten isoperimetrischen Problems*, http://download.uni-mainz.de/mathematik/Algebraische%20Geometrie/Euler-Kreis%20Mainz/65.pdf (10.02.2015) (E065)

L. Euler, *Gedancken von den Elementen der Cörper : in welchen das Lehr-Gebäude von den einfachen Dingen und Monaden geprüfet, und das wahre Wesen der Cörper entdecket wird*, Berlin: Haude & Spener, 1746; digital.bibliothek.uni-halle.de/hd/content/pageview/764562 (27.11.2013); Engl.: *Thoughts on the Elements of Bodies, in which the Theory of the Simple Things and Monads is examined and the true essence of bodies is discovered* (E. Hirsch, transl.), 17centurymaths.com/contents/euler/e842/E81tr.pdf (E081) (10.02.2015)

L. Euler, *Réflexions sur l'espace et le temps*, Mém. ac. sci. Berlin 4 (1748) 324-333; in: *Opera Omnia*, Ser.II, Vol.5; Engl.: S. Uchii (Ed.), *Euler on Space and Time, Physics and Metaphysics*, bun.kyoto-u.ac.jp/phisci/Newsletters/newslet_41.html (14.06.2011) (E149)

L. Euler, *Recherches sur l'origine des forces*, Mém. ac. sci. Berlin 6 (1750); Berlin: Haude et Spener 1752) 419-447; in: *Opera Omnia*, II, 5 (E181)

L. Euler, *Harmonie entre les principes généraux de repos et de mouvement de M. de Maupertuis*, Berlin: Haude et Spener 1751; *Opera Omnia* II 5, pp. 152ff. (E197)

L. Euler, *Découverte d'un nouveaux principe de mécanique*, Mém. acad. Berlin 6 (1752) 185-217; in: *Opera Omnia* II, 5, pp. 81-108 (E177)

L. Euler, *Anleitung zur Naturlehre worin die Gründe zur Erklärung aller in der Natur sich ereignenden Begebenheiten und Veränderungen festgesetzet werden*, ca. 1750[4]; in: *Opera Omnia*, III, 1, pp. 17-178; *Opera posthuma* 2, 1862, pp. 449-560; Engl.: *An Introduction to Natural Science, Establishing the Fundamentals for the Explanation of the Events and Changes that occur in Nature* (transl. by E. Hirsch), 17centurymaths.com/contents/contentse842.html (27.11.2013) (E842)

---

[3] All papers of the Eneström index (Gustaf Hjalmar Eneström (1852 - 1923), *Verzeichnis der Schriften Leonhard Eulers*, Ergänzungsband 4 zum Jahresbericht der DMV, Leipzig: Teubner 1910 (erste Lieferung), 1913 (zweite Lieferung); http://eulerarchive.maa.org//index/enestrom.html (10.10.2016)) are available on math.dartmouth.edu/ euler/; see also eulerarchive.maa.com/; for transl. into English by I. Bruce and others, see 17centurymaths.com/ (all retriev. 10.02.2015).

[4] Truesdell dates it on 1756.

L. Euler, *Elementa doctrinae solidorum*, Novi Comm. acad. sci. Petropolitanae 4 (1752/3) 1758, pp. 14-17, 109-140; German: : uni-koeln.de/math-nat-fak/didaktiken/mathe/volkert/euler-230.pdf (transl. by R. Krömer); : http://download.uni-mainz.de/mathematik/Algebraische%20Geometrie/Euler-Kreis%20Mainz/E230.pdf (transl. by A. Aycock) (all retriev. 14.06.2014) (E230)

L. Euler, *Dissertation sur le principe de la moindre action, avec l'examen des objections de M. le Professeur Koenig faites contre ce principe*, Berlin 1753[5] (14.08.2012) (E186=E198+E199)

L. Euler, *Theoria motus corporum solidorum seu rigidorum*, 1765 (E289)

L. Euler, *Lettres à une princesse d'Allemagne*, Vol. 1, St. Petersburg 1768; in: *Opera Omnia*, III, 11; German: *Briefe an eine deutsche Prinzessin*, Leipzig: Junius 1769, Leipzig: Reclam 1965 (philosophical selection; RUB 239), Braunschweig/Wiesbaden: Vieweg 1986, https://books.google.de/books?id=_gfLXsGHIX4C (E343)

L. Euler, *Leonardi Euleri Opera Omnia sub auspiciis Societatis Scientarium Helveticae*, Zürich · Basel 1911-1986

Leonhard Euler 1707-1783. Beiträge zu Leben und Werk, Basel *etc.*: Birkhäuser 1983

N. W. Evans, *Group theory of the Smorodinsky–Winternitz system*, J. Math. Phys. 32 (1991) 3369–3375

P. P. Ewald, *Zur Theorie der Interferenzen der Röntgenstrahlen in Kristallen*, Phys. Z. 14 (1913) 465-472

P. P. Ewald, *Das reziproke Gitter in der Strukturtheorie*, Z. Kristallographie 56 (1921) 129-156

W. Fahrner (Ed.), *Nanotechnology and Nanoelectronics. Materials, Devices, Measurement Techniques*, Heidelberg *etc.*: Springer 2005

A. E. Faraggi & M. Matone, *The Equivalence Postulate of Quantum Mechanics*, https://arxiv.org/pdf/hep-th/9809127v2.pdf (14.06.2001)

W. A. Fedak & J. J. Prentis, *The 1925 Born and Jordan paper "On quantum mechanics"*, Am. J. Phys. 77 (2009) 2, 128-139; http://people.isy.liu.se/icg/jalar/kurser/QF/references/onBornJordan1925.pdf (13.06.2014)

E. A. Fellmann, *Leonhard Euler*, Hamburg: Rowohlt 1995

E. Fermi, *Zur Quantelung des idealen einatomigen Gases*, Z. Phys. 36 (1926) 902-912; Engl. in: *Collected Papers I*, 1962, pp. 186-195

E. Fermi, *Quantum theory of radiation*, Rev. Mod. Phys. 4 (1932) 87-132

E. Fermi, *Collected Papers I*, Chicago: Univ. Chicago Press 1962

R. P. Feynman, *Forces in molecules*, Phys. Rev. 56 (1939) 340-343

R. P. Feynman, *Space-Time Approach to Non-Relativistic Quantum Mechanics*, Rev. Mod. Phys. 20 (1948) 367-387; reprint in: J. Schwinger (Ed.), *Selected Papers on Quantum Electrodynamics*, 1958, No. 27; ffn.ub.es/luisnavarro/nuevo_maletin/Feynman_Approach_1948.pdf (14.06.2011)

R. P. Feynman, *Space-Time Approach to Quantum Electrodynamics*, Phys. Rev. 76 (1949) 769-789; reprint in: J. Schwinger (Ed.), *Selected Papers on Quantum Electrodynamics*, 1958, No. 22

R. P. Feynman, *An operator calculus having applications in quantum electrodynamics*, Phys. Rev. 84 (1951) 108-128

---

[5] See also https://www.bibnum.education.fr/sites/default/files/analyse-euler.pdf

R. P. Feynman, *QED: The Strange Theory of Light and Matter*, Princeton: Princeton Univ. Press 1986

R. P. Feynman & A. R. Hibbs, *Quantum Mechanics and Path Integrals* (emended by D. F. Styer), New York: McGraw-Hill 2005, Dover 2010

R. P. Feynman, R. B. Leighton & M. L. Sands, *The Feynman Lectures on Physics*, San Francisco: Pearson/Addison-Wesley 2006

H. Fischer & H. Kaul, *Mathematik für Physiker*, Vol. 2: *Gewöhnliche und partielle Differentialgleichungen, mathematische Grundlagen der Quantenmechanik*, Wiesbaden: Teubner ²2004

G. Floquet, *Sur les équations différentielles linéaires à coefficients périodiques*, Ann. ENS [2] 12 (1883) 47-88; http://www.numdam.org/item?id=ASENS_1883_2_12__47_0 (19.06.2003)

W. A. Fock, *Zur Schrödingerschen Wellenmechanik*, Z. Phys. 38 (1926) 242-250; Russ.: *О вольновой механике Шредингера*, http//:web.ihep.su/dbserv/compas/src/fock26/rus.pdf (14.04.2015)

W. A. Fock, *Über die invariante Form der Wellen- und Bewegungsgleichungen für einen geladenen Massenpunkt*, Zs. Physik 39 (1926) 226-232

W. A. Fock, *Näherungsmethoden zur Lösung des quantenmechanischen Mehrkörperproblems*, Z. Phys. 61 (1930) 126-148

V. A. Fock, *Konfigurationsraum und zweite Quantelung*, Z. Phys. 75 (1932) 622-647; corr. reprint in: Fock, *Papers on Quantum Field Theory*, 2007, pp. 25-51 (in Russian)

W. A. Fock, *Zur Theorie des Wasserstoffatoms*, Z. Physik 98 (1935) 3, 145-154

V. Fock, *Die Eigenzeit in der Klassischen und in der Quantenmechanik*, Sow. Phys. 12 (1937) 404-425

V. A. Fock, *Papers on Quantum Field Theory*, Moscow: LKI ²2007 (in Russian)

V. A. Fock & B. Podolsky, *On the quantization of electromagnetic waves and the interaction of charges in Dirac's theory*, Sow. Phys. 1 (1932) 801-817; corr. reprint in: Fock, *Papers on Quantum Field Theory*, 2007, pp. 55-69 (in Russian)

R. Folk, *Die fluide Welt des Leonhard Euler*, in: P. M. Schuster & F. Pichler (Eds.), *Georg v. Peuerbach Symposium 2010. Modelle der Wirklichkeit vom späten Mittelalter bis zur Zeit der Aufklärung in Astronomie, Mathematik, Physik*, Living Edition Pöllauberg Hainault Atascadero 2012

P. Forrest, *The Identity of Indiscernibles*, Stanford Enc. Phil. (E. N. Zalta, ed.), plato.stanford.edu/archives/win2012/entries/identity-indiscernible/ (25.07.2013)

M. Foucault, *Les mots et les choses: Une archéologie des sciences humaines*, Paris: Gallimard 1966, 1998

J. Fourier, *Mémoire sur la propagation de la chaleur dans les corps solides*, 21.12.1807, l'Institut national, in: Nouv. Bull. sci. Soc. philomatique de Paris I (1808) 6 (March), pp. 112–116; reprint in: *Œuvres complètes*, t. 2, Paris: Gauthier-Villars 1890, pp. 215–22; http://gallica.bnf.fr/ark:/12148/bpt6k33707/f220n7.capture[6]

[6] See fn. 1 on p. 215 about the authorship of this paper. In note 3, the author of https://en.wikipcdia.org/wiki/Fourier_series correctly observes "the consistent use of the third person to refer to him [Fourier]".

J. Fourier, *La Théorie Analytique de la Chaleur*, Paris: Firmin Didot 1822; https://books.google.de/books?id=TDQJAAAAIAAJ redir_esc=y; Engl.: *The Analytical Theory of Heat* (transl. with notes by A. Freeman), New York: Dover 1955

M. Fowler, *Multiparticle Wavefunctions and Symmetry*, galileo.phys.virginia.edu/classes/252/symmetry/Symmetry.pdf (15.03.2015)

D. M. Fradkin, *Existence of the Dynamic Symmetries O4 and SU3 for All Classical Central Potential Problems*, Progr. Theor. Phys. 37 (1967) 798–812; http://ptp.oxfordjournals.org/content/37/5/798.full.pdf

Ch. Francis, *Quantum Logic*, rqgravity.net/FoundationsOfQuantumTheory#QuantumLogic, 2010

J. Franck & G. Hertz, *Über Zusammenstöße zwischen langsamen Elektronen und den Molekülen des Quecksilberdampfes und die Ionisierungsspannung derselben*, Verh. dtsch. phys. Ges. [2] 16 (1914) 457-467; excerpts: leifiphysik.de/themenbereiche/atomarer-energieaustausch/geschichte#Franck-Hertz%20-%20Originalarbeit (14.06.2011)

J. Franck & G. Hertz, *Über die Erregung der 2536-Å-Quecksilberresonanzlinie durch Elektronenstöße und die Ionisierungsspannung derselben*, Verh. dtsch. phys. Ges. [2] 16 (1914) 512-517; reprint in: D. ter Haar, *Quantentheorie*, 1970, pp. 201-208

J. Franck & P. Jordan, *Anregung von Quantensprüngen durch Stöße*, Berlin: Springer 1926 (Struktur der Materie in Einzeldarstellungen III)

J. Frenkel, *On the Transformation of Light into Heat in Solids*, Phys. Rev. 37 (1931) 17-44, 1276-1294

Y. I. Frenkel *Wave Mechanics. Elementary Theory*, Oxford: Clarendon 1932

J. Frenkel, *On the Absorption of Light and the Trapping of Electrons and Positive Holes in Crystalline Dielectrics*, Phys. Zs. Sowjetunion 9 (1936) 533-536

L. Fritsche & M. Haugk, *A new look at the derivation of the Schrödinger equation from Newtonian mechanics*, Ann. Phys. (L.) 12 (2003) 371-403; http://lothar-fritsche.de/pdf/schroedinger_equation.pdf (19.06.2004)

G. Galilei, *Dialogo sopra i due massimi sistemi del Mondo, Tolemaico, e Copernicano*, Florenz 1632; Engl.: *Dialogue Concerning Two New Sciences*, New York: Macmillan 1914, galileoandeinstein.physics.virginia.edu/tns_draft/index.html (19.06.2004)

G. Galilei, *Discorsi e dimostrazioni matematiche, intorno à due nuove scienze attenenti alla mecanica e i movimenti locali*, Leiden 1638; Engl.: *Dialogues Concerning Two New Sciences [1638]* (transl. by A. de Salvio & H. Crew, introd. by A. Favaro), New York: Macmillan 1914; http://oll.libertyfund.org/titles/753 (19.06.2004)

E. Galindo-Linares, E. Navarro-Morale, G. Silva-Ortigoza, R. Suárez-Xique, M. Marciano-Melchor, R. Silva-Ortigoza & E. Román-Hernández, *Any Hamiltonian System Is Locally Equivalent to a Free Particle*, World J. Mechanics 2 (2012) 5, 246-252; http://www.ljemail.org/journal/PaperInformation.aspx?PaperID=23892 (19.06.2044)

G. Gamow, *Zur Quantentheorie des Atomkernes*, Z. Phys. 51 (1928) 204-212

G. Gamow, *The Great Physicists from Galileo to Einstein*, New York: Harper & Brothers 1961 (Harper Modern Science Series); reprint: New York: Dover 1988

F. R. Gantmacher, *Matrizentheorie*, Berlin: Dtsch. Verlag d. Wissensch. 1986

S. V. Gantsevich & V. L. Gurevich, *Correlation, Entanglement and Locality of Quantum Theory*, Theor. Phys. 2 (2017) 2, 63-69, http://www.isaacpub.org/images/PaperPDF/TP_100015_2017060615282812121.pdf

E. Gati, M. Garst, R. S. Manna, U. Tutsch, B. Wolf, L. Bartosch, H. Schubert, T. Sasaki, J. A. Schlueter & M. Lang, *Breakdown of Hooke's law of elasticity at the Mott critical endpoint in an organic conductor*, Science Advances 2 (2016) 12, e1601646

C. F. Gauss, *Theoria attractionis corporum sphaeroidicorum ellipticorum homogeneorum methodo nova tractata*, Commentationes societatis regiae scientiarium Gottingensis recentiores 2 (1813) 355-378

C. F. Gauss, *Disquisitiones generales circa superficies curvas*, 1827, in: *Ges. Werke*, Vol. IV (Wahrscheinlichkeitsrechnung und Geometrie), 1888; reprint: Olms Verlag, 1981; German: *Allgemeine Flächentheorie*, Leipzig: Engelmann 1889 (Ostwalds Klassiker 5)

C. F. Gauss, *Über ein neues Grundgesetz der Mechanik*, Crelles J. IV (1829); in: *Ges. Werke*, Vol. V

C. F. Gauss, *Allgemeine Lehrsätze in Beziehung auf die im verkehrten Verhältnisse des Quadrats der Entfernung wirkenden Anziehungs- und Abstoßungs-Kräfte* (1839/1840), Leipzig: Engelmann [3]1912 (Ostwalds Klassiker 2)

J.-P. Gazeau, *Coherent States in Quantum Physics*, Berlin: Wiley-VCH 2009

H. Genz, *Nichts als das Nichts. Die Physik des Vakuums*, Weinheim: Wiley-VCH 2004

W. Gerlach & O. Stern, *Der experimentelle Nachweis der Richtungsquantelung im Magnetfeld*, Z. Physik 9 (1922) 349-352

W. Gerlach & O. Stern, *Das magnetische Moment des Silberatoms*, Z. Physik 9 (1922) 353-355

S. Gerlich, S. Eibenberger, M. Tomandl, S. Nimmrichter, K. Hornberger, P. J. Fagan, J. Tüxen, M. Mayor, M. Arndt, *Quantum interference of large organic molecules*, Nature commun. 2 (2011) 263

R. Geroch, *Geometrical Quantum Mechanics*, 1974, phy.syr.edu/ salgado/geroch.notes/geroch-gqm.pdf (05.07.2013)

J. D. McGervey, *Quantum Mechanics. Concepts and Applications*, San Diego *etc.*: Academic Press 1995

P. Ghose & M. K. Samal, *Lorentz-invariant superluminal tunneling*, Phys. Rev. E 64 (2001) 036620

J. W. Gibbs, *On the Equilibrium of Heterogeneous Substances*, Trans. Conn. Acad. III (1875/76) 108-245, (1877/78) 343-524; reprint in: *The Scientific Papers of J. Willard Gibbs*, 1961, Vol. I., pp. 55-353 (No. III)

J. W. Gibbs, *Elementary Principles in Statistical Mechanics Developed with Especial Reference to the Rational Foundation of Thermodynamics*, New York: Scribner 1902 (Yale Bicentennial Publ.); archive.org/details/elementaryprinci00gibbrich (17.06.2014)

J. W. Gibbs, *The Scientific Papers of J. Willard Gibbs*, New York: Dover 1961; archive.org/details/scientificpapers01gibbuoft

V. L. Ginzburg & L. D. Landau, *On the Theory of Superconductivity*, in: L. D. Landau, *Collected papers*, 1965, pp. 546-568; Russ.: В. Л. Гинзбург & Д. Ландау, *К теории сверхпроводимости*, ЖЭТФ 20 (1950) 1064ff.

R. Gilmore, *Lie Groups, Lie algebras, and Some of Their Applications*, New York: Wiley 1974, Mineola (NY): Dover 2005

N. S. Ginsberg, S. R. Garner & L. V. Hau, *Coherent control of optical information with matter wave dynamics*, Nature 445 (2007) 623-626

I. de Gispert Pastor, "What is the basic difference between classical mechanics and quantum mechanics?", 2012, linkedin.com/groupItem?setLike= gid=1892648 item=191617038 type=mem ber commentID=108631118 nogb=true trk=grp_email_like_post csrfToken=ajax%3A2389998634 603224132 ut=2jA5GxJUy-HBw1 (19.06.2012)

D. Giulini, „*Es lebe die Unverfrorenheit!"*. *Albert Einstein und die Begründung der Quantentheorie*, in: H. Hunziker (Ed.), *Der jugendliche Einstein und Aarau*, Basel: Birkhäuser 2005; arXiv:physics/0512034v1 [physics.hist-ph] 5 Dec 2005 (19.06.2006)

D. Giulini & N. Straumann, „*... ich dachte mir nicht viel dabei...".* *Plancks ungerader Weg zur Strahlungsformel*, Phys. Bl. 56 (2000) H.12, 37-41

R. Giuntini, *Quantum Logic and Hidden Variables*, Mannheim/Wien/Zürich: B. I. Wissenschaftsverlag 1991 (Grundlagen d. exakten Naturwiss. 8)

R. J. Glauber, *The Quantum Theory of Optical Coherence*, Phys. Rev. 130 (1963) 2529-2539

R. J. Glauber, *Coherent and Incoherent States of the Radiation Field*, Phys. Rev. 131 (1963) 2766-2788

M. Glazer & J. Wark, *Statistical mechanics. A Survival Guide*, Oxford: Oxford Univ. Press 2001

Th. Görnitz, *Quanten sind anders. Die verborgene Einheit der Welt*, Heidelberg · Berlin: Spektrum 1999

K. Gödel, *Über formal unentscheidbare Sätze der Principia Mathematica und verwandter Systeme, I.*, Monatshefte Math. Physik 38 (1931) 173-98; Engl.: *On formally undecidable propositions of Principia Mathematica and related systems I*, research.ibm.com/people/h/hirzel/papers/canon00-goedel.pdf (17.06.2014)

H. Goldstein, *Classical Mechanics*, Cambridge (MA): Addison-Wesley 1950

S. Goldstein, *Bohmian Mechanics*, The Stanford Encyclopedia of Philosophy (ed. E. N. Zalta), Winter 2002 Edition, plato.stanford.edu/archives/win2002/entries/qm-bohm/ (19.07.2004)

J. Goldstone, *Field Theories with Superconductor Solutions*, Nuovo Cim. 19 (1961) 154–164

J. Goldstone, A. Salam & S. Weinberg, *Broken Symmetries*, Phys. Rev. 127 (1962) 965–970

D. Gómez-Ullate, Y. Grandati & R. Milson, *Rational extensions of the quantum harmonic oscillator and exceptional Hermite polynomials*, J. Phys. A: Math. Theor. 47 (2014) 015203 (27pp); http://iop.msgfocus.com/c/15SZOP7EO53MpTcWXYZzJBw3y (20.10.2014)

M. Gondran, *The Proca equations derived from first principles*, Am. J. Phys. 77 (2009) 925-926

D. L. Goodstein & J. R. Goodstein, *Feynman's Lost Lecture. The Motion of Planets Around the Sun*, New York. London: Norton 1996

W. Gordon, *Der Comptoneffekt nach der Schrödingerschen Theorie*, Zs. Physik (1926) 117-133; itep.ru/theor/text/Gordon.pdf (19.06.2004)

M. A. de Gosson, *The Principles of Newtonian and Quantum Mechanics: The Need for Planck's constant, h*, London: Imperial Coll. Press 2001

M. A. de Gosson, *Introduction to Simplectic Mechanics: Lectures I-II-III*, 2006, https://www.ime.usp.br/ piccione/Downloads/LecturesIME.pdf (28.09.2017)

M. A. de Gosson, *Symplectic Geometry and Quantum Mechanics*, Basel: Birkhäuser 2006 (Operator Theory: Advances and Applications 166)

M. A. de Gosson, *The Angular Momentum Dilemma and Born-Jordan Quantization*; de.arxiv.org/abs/1502.04998v3, 22 Feb 2015; extended version: *Reconsidering the Schrödinger Picture of Quantum Mechanics*, March 23, 2015 (priv. commun.)

M. A. de Gosson, *From Weyl to Born-Jordan quantization: The Schrödinger representation revisited*, Phys. Rep. 623 (2016) 1-58

M. A. de Gosson & B. Hiley, *Imprints of the Quantum World in Classical Mechanics*, arXiv:1001.4632v2 [quant-ph] 15 Dec 2010

S. H. Gould, *Variational Methods for Eigenvalue Problems. An Introduction to the Methods of Rayleigh, Ritz, Weinstein, and Aronszajn*, Toronto: Univ. Toronto Press 1957 (Mathematical Expositions 10); reprint: New York: Dover 1995

H. Graßmann, *Das Top Quark, Picasso und Mercedes-Benz oder Was ist Physik?*, Berlin: Rowohlt 1997

J. E. Gray & A. D. Parks, *Some Implications of the Curie Symmetry Principle in Quantum Physics*, Quantum Processing Group, Code B-10, Naval Surface Warfare Center Dahlgren Division, Dahlgren, VA, 22448, https://www.researchgate.net/publication/268617379 _Some_implications_of_the_Curie_and_Rosen_symmetry_principles_in_quantum_physics

G. Green, *An Essay on the Applicability of Mathematical Analysis on the Theories of Electricity and Magnetism*, Nottingham 1828; German: *Ein Versuch die mathematische Analysis auf die Theorien der Elektricität und des Magnetismus anzuwenden*, Leipzig: Engelmann 1895 (Ostwalds Klassiker 61)

J. Gregory, *Vera circuli et hyperbolae quadratura*, Patavii (Padua) 1667

J. Gregory, *Exercitationes geometricae*, London 1668, https://books.google.ca/books?id=ZtRYqgyD5YsC (19.11.2015)

J. Gregory, *Geometriae pars universalis*, Patavii (Padua) 1668

W. Greiner, *Quantentheorie. Spezielle Kapitel* (*Theoretische Physik* 4A), Thun · Frankfurt am Main: Deutsch [3]1989

W. Greiner & J. Reinhardt, *Feldquantisierung* (*Theoretische Physik* 7A), Thun · Frankfurt am Main: Deutsch 1993

A. Griffin, D. W. Snoke & S. Stringari (Eds.), *Bose-Einstein condensation*, Cambridge: Cambridge Univ. Press 1995; catdir.loc.gov/catdir/samples/cam031/94027795.pdf (21.06.2014)

A. T. Grigorjan, *The initial stage of the development of classical mechanics*, in: Исследования по истории физики и механики. 1985, Moscow: "Nauka" 1985, pp. 176-223 (in Russ.)

A. T. Grigoryan & B. D. Kovalev, *Daniil Bernulli*, Moscow: "Nauka" 1981 (in Russ.)

D. J. Gross, *Gauge Theory – Past, Present, and Future?*, Chin. J. Phys. 30 (1992) 7, 955-972

E. P. Gross, *Structure of a quantized vortex in boson systems*, Nuovo Cimento 20 (1961) 3, 454–457

D. Guéry-Odelin & T. Lahaye, *Classical Mechanics Illustrated by Modern Physics*, London: Imperial College Press 2010

G. Guerrerio, *Kurt Gödel. Logische Paradoxien und mathematische Wahrheit*, Heidelberg: Spektrum 2002 (Spektrum der Wissenschaft. Biographie 1/2002)

N. V. Gulia, *Inertia*, Moscow: Nauka 1982 (in Russian); publ.lib.ru/ARCHIVES/G/GULIA_ Nurbey_Vladimirovich/Gulia_N.V._Inerciya.(1982).[djv-fax].zip (09.09.2011)

S. N. Gupta, *On the elimination of divergences from classical electrodynamics*, Proc. Phys. Soc. L. 64 A (1951) 50-56

S. N. Gupta, *Quantum Electrodynamics*, New York: Gordon & Breach 1977

G. Györgyi & J. Révai, *Hidden Symmetry of the Kepler Problem*, Sov. Phys. JETP 21 (1965) 5, 967-968; http://jetp.ac.ru/cgi-bin/dn/e_021_05_0967.pdf (09.09.2015)

D. ter Haar, *Quantentheorie. Einführung und Originaltexte*, Berlin: Akademie-Verlag / Oxford: Pergamon / Braunschweig: Vieweg ^{2}1970 (WTB 56)

H. Haken, *Laserstrahlung – ein neues Beispiel für einen Phasenübergang*, Festkörperprobleme X (1970) 351-365

H. Haken, *Quantenfeldtheorie des Festkörpers*, Stuttgart: Teubner 1973

F. D. M. Haldane, *Fractional statistics in arbitrary dimensions – A generalization of the Pauli principle*, Phys. Rev. Lett. 67 (1991) 937-940

B. C. Hall, *Holomorphic methods in analysis and mathematical physics*, arXiv:quant-ph/9912054

P. R. Halmos, *The Legend of John von Neumann*, Am. Math. Monthly 80 (1973) 382-394

P. R. Halmos, *I Want to Be a Mathematician*, New York: Springer 1985

G. Hamel, *Mechanik I: Grundbegriffe der Mechanik*, Leipzig · Berlin: Teubner 1921 (Aus Natur und Geisteswelt 684)

G. Hamel, *Theoretische Mechanik. Eine einheitliche Einführung in die gesamte Mechanik*, Berlin · Göttingen · Heidelberg: Springer 1949 (Grundlehren der mathematischen Wissenschaften LVII)

W. R. Hamilton, *Theory of Systems of Rays*, Trans. Roy. Irish Acad. 15 (1828) 69-174; emis.de/classics/Hamilton/PtFst.pdf (09.11.2011)

W. R. Hamilton, *On a General Method of Expressing the Paths of Light, and of the Planets, by the Coefficients of a Characteristic Function*, Dublin Univ. Rev. Quart. Mag. 1 (1833) 795-826; emis.de/classics/Hamilton/CharFun.pdf (D. R. Wilkins, ed., 1999) (09.11.2011)

W. R. Hamilton, *On the Application to Dynamics of a General Mathematical Method Previously Applied to Optics*, Report 4th Meeting Brit. Ass. Adv. Sci., Edinburgh 1834, London: Murray 1835, pp. 513-518; emis.de/classics/Hamilton/BARep34A.pdf (09.11.2011)

W. R. Hamilton, *On a General Method in Dynamics, by which the study of motions of all free systems of attracting or repelling points is reduced to the search and differentiation of one central relation or characteristic function*, Phil. Trans. Roy. Soc. L. 124 (1834) pt. II, 247-308; corr. on http://www.emis.de/classics/Hamilton/GenMeth.pdf (09.11.2011)

W. R. Hamilton, *Second Essay on a General Method in Dynamics*, Ibid. 125 (1835) pt. I, 95-144; in: A. W. Conway & A. J. McConnell (Eds.), *The Mathematical Papers of Sir William Rowan Hamilton, Vol. II: Dynamics*, Cambridge: Cambridge Univ. Press 1940; corr. on http://www.emis.de/classics/Hamilton/SecEssay.pdf (09.11.2011)

W. R. Hamilton, *The hodograph or a new method of expressing in symbolic language the Newtonian law of attraction*, Proc. Roy. Irish Acad. III (1846) 344-353; in: *Mathematical Papers of Sir William Rowan Hamilton* (A. W. Conway & A. J. McConnell, Eds.), Cambridge: Cambridge

Univ. Press 1940, Vol. 2, pp. 287–92; http://www.emis.de/classics/Hamilton/Hodo.pdf (D. R. Wilkins, Ed., 2000) (09.11.2011)

W. R. Hamilton, *Applications of Quaternions to Some Dynamical Questions*, commun. 1845, publ. 1847 in: Proc. Roy Irish Acad., vol. 3, Appendix, pp. xxxvi-l; http://www.emis.ams.org/classics/Hamilton/DynQue.pdf (D. R. Wilkins, Ed., 2000) (09.11.2011)

Th. Hapke, *100 Jahre Ostwald's Klassiker der exakten Wissenschaften 1889 - 1989*, 2003; https://www.tuhh.de/b/hapke/ostwklas.html (17.08.2013)

W. A. Harrison, *Solid State Theory*, New York: McGraw-Hill 1970; corr. reprint: New York: Dover 1980

W. A. Harrison, *Electronic structure and the properties of solids*, San Francisco: Freeman 1980

Th. E. Hartman, *Tunneling of a Wave Packet*, J. Appl. Phys. 33 (1962) 3427-3433

D. R. Hartree, *The wave mechanics of an atom with a non-Coulomb central field. Part I. Theory and methods*, Proc. Camb. Phil. Soc. 24 (1927) 89-110, http://karin.fq.uh.cu/qct/extras/Hartree-Fock/hartree28.pdf; *Part II. Some results and discussion*, 111-132, http://karin.fq.uh.cu/qct/Tema_07/07.00.PostHF.%20Correlaci%f3n%20electr%f3nica/hartree28a.pdf (both 03.04.2018); *Part III. Term values and the intensities of series in optical spectra*, 24 (1928) 426-437

H. Haug & S. W. Koch, *Quantum Theory of the Optical and Electronic Properties of Semiconductors*, Singapore *etc.*: World Scientific [3]1994

E. H. Hauge & J. A. Støvneng, *Tunneling times: a critical review*, Rev. Mod. Phys. 61 (1989) 917-936

S. Hawking, *Godel and the End of the Universe*, 2002, hawking.org.uk/godel-and-the-end-of-physics.html (29.06.2015)

S. Hawking (Ed.), *On The Shoulders Of Giants. The Great Works of Physics and Astronomy*, Philadelphia: Running Press 2003; German: *Die Klassiker der Physik*, Hamburg: Hoffmann und Campe 2004[7]

O. Heaviside, *A gravitational and electromagnetic analogy. Part I*, The Electrician 31 (1893) 281-282, *Part II, Ibid.* 359; http://serg.fedosin.ru/Heavisid.htm (27.10.2013); in: O. D. Jefimenko, *Causality Electromagnetic Induction, and Gravitation*, 2000 (conversion of formulas to modern notation by Jefimenko), https://de.scribd.com/document/194185341/A-Gravitational-and-Electromagnetic-Analogy-by-Oliver-Heaviside (9/4/2013; retriev. 03.04.2018)

H. van Hees, *Diskrete Symmetrietransformationen*, theory.gsi.de/ vanhees/faq/zeitum/zeitum.html (29.05.1998; retriev. 20.07.2013)

G. W. F. Hegel, *Wissenschaft der Logik*, Nürnberg: Schrag 1812-1816; deutschestextarchiv.de/book/show/hegel_logik0101_1812, deutschestextarchiv.de/book/show/hegel_logik0102_1813, deutschestextarchiv.de/book/show/hegel_logik02_1816 (all retriev. 03.04.2018)

E. Heifetz & E. Cohen, *Toward a thermo-hydrodynamic like description of Schr¨odinger equation via the Madelung formulation and Fisher information*, ; https://arxiv.org/abs/1501.00944

W. Heisenberg, *Über quantentheoretische Umdeutung kinematischer und mechanischer Beziehungen*, Z. Phys. 33 (1925) 879-893; reprint in: Ludwig, *Wellenmechanik*, 1970, pp. 193-

---

[7] *Giganten des Wissens* (Augsburg: Weltbild 2005) has only 256 out of 1068 p.

210; Engl.: *Quantum-Theoretical Re-Interpretation of Kinematic and Mechanical Relations*, in: van der Waerden, *Sources of Quantum Mechanics*, 1968, paper 12, pp. 261-276

W. Heisenberg, *Mehrkörperproblem und Resonanz in der Quantenmechanik*, Z. Phys. 38 (1926) 411-426

W. Heisenberg, *Über die Spektren von Atomsystemen mit zwei Elektronen*, Z. Phys. 39 (1926) 499-518; reprint in: Hindmarsh, *Atomspektren*, 1972, pp. 276-304

W. Heisenberg, *Schwankungserscheinungen und Quantenmechanik*, Z. Phys. 40 (1926) 501-509

W. Heisenberg, *Über den anschaulichen Inhalt der quantentheoretischen Kinematik und Mechanik*, Z. Phys. 43 (1927) 172-198; Engl. in: Wheeler & Zurek (Eds.), *Quantum Theory and Measurement*, 1983, pp. 62-84

W. Heisenberg, *Die physikalischen Prinzipien der Quantentheorie* (Lectures held in Chicago, 1929), Stuttgart: Hirzel 1958; reprint: Heidelberg · Berlin: Spektrum 2001

W. Heisenberg, *Die Geschichte der Quantentheorie*, in: *Physik und Philosophie*, Frankfurt *etc.*: Ullstein 1977, pp. 15-27

H. Hellmann, *Einführung in die Quantenchemie*, Leipzig and Wien: Deuticke 1937; reprint (D. Andrae, Ed.): Springer 2015; Russ.: *Квантовая Химия* (transl. by J. Golovin, N. Tunitskij & M. Kovner), Moscow and Leningrad: ONTI 1937

H. Helmholtz, *Über die Erhaltung der Kraft*, Berlin: Reimer 1847; in: *Wissenschaftliche Abhandlungen I*, Leipzig: Barth 1882, pp. 12-75; extended reprint: Leipzig: Engelmann 1889 / Leipzig: Geest & Portig 1982 (Ostwalds Klassiker 1); partial reprint in: Brush, *Kinetische Theorie I*, 1970, No. 7; https://books.google.de/books?id=NXcLAAAAMAAJ (09.11.2012)

H. Helmholtz, *Über Integrale der hydrodynamischen Gleichungen, welche den Wirbelbewegungen entsprechen*, J. Reine Angew. Math. (Crelle's J.) 55 (1858) 25-55; *Ges. Abh.* I, p. 101 (Ostwald's Klassiker 79)

H. Helmholtz, *Über die Thatsachen, die der Geometrie zum Grunde liegen*, Nachr. Ges. Wiss. Gött. (1868) 9, 193-221; in: *Wissenschaftliche Abhandlungen II*, Leipzig: Barth 1883, pp. 618-639

H. v. Helmholtz, *Über die physikalische Bedeutung des Prinzips der kleinsten Wirkung*, J. reine angew. Math. 100 (1887) 137-166, 213-222

H. v. Helmholtz, *Prinzipien der Statik monocyclischer Systeme*, J. reine angew. Math. 97 (1884) 111-140, 317-336

H. v. Helmholtz, *Reden und Vorträge*, Braunschweig 1884, Vol. 1, https://ia802606.us.archive.org/7/items/vortrgeundreden02helmgoog/

H. v. Helmholtz, *Einleitung zu den Vorlesungen über Theoretische Physik* (*Vorlesungen über Theoretische Physik*, Vol. I/1, ed. by A. König & C. Runge), Leipzig: Barth 1903; reprint of pp. 1-50 in: R. Rompe & H.-J. Treder, *Zur Grundlegung der theoretischen Physik*, 1984, pp. 11-62

H. v. Helmholtz, *Vorlesungen über die Dynamik discreter Massenpunkte* (*Vorlesungen über Theoretische Physik*, Vol. I/2, ed. by O. Krigar-Menzel), Leipzig: Barth ²1911

H. v. Helmholtz, *Wissenschaftliche Abhandlungen I*, Leipzig: Barth 1882, archive.org/details/wissenschaftlic00helmgoog (09.11.2011)

T. Henz & G. Langhake, *Pfade durch die Theoretische Mechanik 1. Die Newtonsche Mechanik und ihre mathematischen Grundlagen: anschaulich – axiomatisch – abstrakt*, Berlin · Heidelberg: Springer 2016

A. Hermann (Ed.), *Dokumente der Naturwissenschaften, Abteilung Physik*, München: Battenberg 1967

A. Hermann, *Frühgeschichte der Quantentheorie (1899-1913)*, Mosbach/Baden: Physik Verlag 1969

A. Hermann, „*Ich dachte mir nicht viel dabei*", Süddeutsche Zeitung v. 9.12.2000; roro-seiten.de/physik/extra/planck/planckartikel.html (09.11.2011)

J. Hermann, *d'investigare l'Orbite de' Pianeti, nell'ipotesi che le forze centrali o pure le gravità delgi stessi Pianeti sono in ragione reciproca de' quadrati delle distanze, che i medesimi tengono dal Centro, a cui si dirigono le forze stesse*, Giorn. Lett. D'Italia 2 (1710) 447-467

J. Hermann, *Extrait d'une lettre de M. Herman à M. Bernoulli datée de Padoüe le 12. Juillet 1710*, Hist. l'acad. roy. sci. (Paris) 1732, 519-521

Ch. Hermite, *Sur La Théorie des Formes Quadratiques*, Crelles J. 47, 1853; Berlin: Reimer 1854; in: *Oevres*, I, Paris: Gauthier-Villars 1905, 1er Mém. 200-233, 2nd Mém 234-263; archive.org/details/oeuvresdecharles01hermuoft (09.11.2011)

C. Hermite, *Sur un nouveau développement en série de fonctions*, C. R Acad. Sci. Paris 58 (1864) 93-100; *Oeuvres* II 293-303

H. Hertz, *Die Kräfte elektrischer Schwingungen, behandelt nach der Maxwell'schen Theorie*, Wiedemanns Ann. 36 (1888) 1-22; corr. reprint in: Hertz, *Gesammelte Werke. II*, 2001, No. 9

H. Hertz, *Gesammelte Werke. Bd. II. Untersuchungen über die Ausbreitung der elektrischen Kraft*, Leipzig: Barth (Meiner) 21894; reprint: Vaduz: Sändig 2001

H. Hertz, *Die Prinzipien der Mechanik in neuem Zusammenhange dargestellt* (ed. by P. Lenard, with a Preface by H. v. Helmholtz), Leipzig: Barth 21910 (*Collected Papers*, Vol. III); reprint: Vaduz: Sändig 2001; partial reprints (Introducing remarks by Ph. Lenard, Preface by H. v. Helmholtz; Preface by H. Hertz; Introduction) in: Hertz, *Die Prinzipien der Mechanik. Einleitung* (introduced and commented by J. Kuczera), Leipzig: Geest & Portig 1984 (Ostwalds Klassiker 263); Rompe & Treder (Eds.), *Zur Grundlegung der theoretischen Physik*, 1984, pp. 63-124; Engl.: *The Principles of Mechanics Presented in a New Form*, London: MacMillan 1899; https://archive.org/details/principlesofmech00hertuoft (09.03.2012)

E. Hilb, *Über Reihenentwicklungen nach den Eigenfunktionen linearer Differentialgleichungen 2. Ordnung*, Math. Ann. 71 (1911) 76-87

D. Hilbert, *Mathematische Probleme* (Internat. Congress of Mathematicians, Paris 1900), Nachr. Königl. Ges. Wiss. Göttingen, Math.-Phys. Klasse, Heft 3 (1900) 253-297; Archiv Math. Physik (3) 1 (1901) 44-63, 213-237; https://www.math.uni-bielefeld.de/ kersten/hilbert/rede.html, http://www.digizeitschriften.de/dms/img/?PID=GDZPPN002498863; French: *Sur les problèmes futurs des mathématiques* (transl. by M. L. Laugel), *Compte Rendu du Deuxième Congrès International des Mathématiciens*, Paris: Gauthier-Villars 1902, pp. 58-114; Engl.: *Mathematical Problems* (transl. by M. W. Newson), Bull. Am. Math. Soc. 8 (1902) 437-479; reprint in: F. E. Brouder (Ed.), *Mathematical Developments Arising from Hilbert Problems*, Am. Math. Soc. 1976; revised transl. by D. E. Joyce: http://www2.clarku.edu/ djoyce/hilbert (1997, access 26.10.2016); see also Hilbert, *Die Hilbertschen Probleme*, 1976

D. Hilbert, *Grundzüge einer allgemeinen Theorie der linearen Integralgleichungen*, Leipzig: Teubner 1912

D. Hilbert, *Naturerkennen und Logik*, Kongress Verein. dtsch. Naturwiss. Mediziner, Königsberg 1930; mp3 file of its partial translation by radio, its transcription and the Engl. transl. of the latter by J. T. Smith: *David Hilbert's Radio Address*, Convergence, Febr. 2014; http://math.sfsu.edu/smith/Documents/HilbertRadio/HilbertRadio.pdf

D. Hilbert, *Hilbert's Problems* (P. S. Aleksandrov, Ed.), Moscow: "Nauka" 1969 (in Russian); German: *Die Hilbertschen Probleme*, Leipzig: Geest & Portig 1976 / Thun · Frankfurt/Main: Deutsch 1998 (Ostwald Klassiker 252, with 220 p. of comments)

D. Hilbert, J. v. Neumann & L. Nordheim, *Über die Grundlagen der Quantenmechanik*, Math. Ann. 98 (1927) 1-30

R. C. Hilborn & G. M. Tino (Eds.), *Spin-Statistics Connection and Commutation Relations*, AIP 2000 (CP545)

G. W. Hill, *On the Part of the Motion of Lunar Perigee Which is a Function of the Mean Motions of the Sun and Moon*, Acta Math. 8 (1886) 1-36

W. R. Hindmarsh, *Atomspektren*, Berlin: Akademie-Verlag / Oxford: Pergamon / Braunschweig: Vieweg 1972 (WTB 76)

A. R. von Hippel, *Do We Really Understand Ferroelectricity?*, J. Phys. Soc. Japan 28 (1970) Suppl., 1-6

R. Holland, *The Quantum Theory of Motion. An Account for the de Broglie-Bohm Causal Interpretation of Quantum Mechanics*, Cambridge: Cambridge Univ. Press 1993

W. Hollmann, *Die Zeitschriften der exakten Naturwissenschaften in Deutschland*, München: Zeitungswissenschaftl. Vereinigung 1937 (Zeitung und Leben 39)

R. Hooke, *A Description of Helioscopes and some other Instruments*, London: John Martyn 1676, http://www.e-rara.ch/zut/collections/content/thumbview/731058?lang=de; in: *Lectiones Cutlerianæ*, Lect. III

R. Hooke, *Lectures de potentia restitutiva, or, of Spring: explaining the power of springing bodies: to which are added some collections*, London: John Martyn 1678; in: *Lectiones Cutlerianæ*, Lect. VI

R. Hooke, *Lectiones Cutlerianæ, or A Collection of Lectures: Physical, Mechanical, Geographical, & Astronomical*, London: John Martyn 1679, https://archive.org/details/LectionesCutler00Hook; reprint in: Robert T. Gunther, *Early Science in Oxford*, Vol. 8: *The Cutler lectures of Robert Hooke*, Oxford: Oxford Univ. Press 1931, https://archive.org/details/earlyscienceinox08gunt

D. Howard, *The Early History of Quantum Entanglement, 1905-1935*, 2007, http://tam.ung.si/2007/slides/20070828_Howard.pdf (22.06.2015)

D. Howard, *What makes a Classical Concept Classical? Toward a Reconstruction of Niels Bohr's Philosophy of Physics*, www3.nd.edu/ dhoward1/Classcon.pdf (22.06.2015)

F. Hund, *Zur Deutung der Molekelspektren. I*, Z. Phys. 40 (1927) 742-764; *III. Bemerkungen über das Schwingungs- und Rotationsspektrum bei Molekeln mit mehr als zwei Kernen*, 43 (1927) 805-826

F. Hund, *Geschichte der Quantentheorie*, Mannheim: B. I. – Wissenschaftsverlag ^{2}1975

K. Husimi, *Some formal properties of the density matrix*, Proc. Phys.-Math. Soc. Japan 22 (1940) 264-314

K. Husimi, *Miscellanea in elementary quantum mechanics II*, Progr. Theor. Phys. 9 (1953) 381-402, https://academic.oup.com/ptp/article/9/4/381/1849279 (03.04.2018)

Chr. Huygens, *De motu corporum ex percussione*, London: Royal Society, Talk dated Jan. 4, 1669, publ. 1703

Chr. Huygens, *Horologium oscillatorium sive de motu pendulorum ad horologia aptato demonstrationes geometricae*, Paris 1673; in: *The complete Works of Christiaan Huygens*, Vol. XVII, Den Haag 1934

Chr. Huygens, *Traité de la Lvmière. Où font expliquées les causes de ce qui luy arrive dans la reflexion, & dans la refraction. Et particulierement dans l'etrange refraction dv cristal d'Islande. Par C. H. D. Z. [Christian Huygens de Zulichen] Avec un Discurs de la Cause de la pesantevr*, Leiden: van der Aa 1690; https://books.google.de/books?id=X9PKaZlChggC; German: *Abhandlung über das Licht*, Thun · Frankfurt am Main: Deutsch [4]1996 (Ostwalds Klassiker 20); archive.org/details/abhandlungberda00mewegoog; Engl.: TREATISE ON LIGHT In which are explained The causes of that which occurs In REFLEXION, & in REFRACTION And particularly In the strange REFRACTION OF ICELAND CRYSTAL By CHRISTIAAN HUYGENS Rendered into English By SILVANUS P. THOMPSON, Univ. Chicago Press, thegod720.com/2384177-Treatise-on-Light-by-Huygens-Christiaan-16291695.pdf; *Treatise On Light*, Macmillan 1912, archive.org/details/treatiseonlight031310mbp; reprint: Echo Library, Jan. 16, 2007 (all retriev. 09.03.2013)

L. Infeld and T. E. Hull, *The Factorization Method*, Rev. Mod. Phys. 23 (1951) 21-68

A. Yu. Ishlinskii', *Preface by the editor*, in: Gulia, *Inertia*, 1982, pp. 3-7 (in Russian)

R. Jackiw, *Introducing scale symmetry*, Physics Today 25 (1972) Jan., 23–27

C. G. J. Jacobi, *Über die Reduction der Integration der partiellen Differentialgleichungen erster Ordnung zwischen irgend einer Zahl Variabeln auf die Integration eines einzigen Systemes gewöhnlicher Differentialgleichungen*, Crelle's J. 17 (1837) 97ff.; in: Gesammelte Werke, Vol. IV, Berlin: Reimer 1886

C. G. J. Jacobi, *Neue Methode zur Integration partieller Differentialgleichungen erster Ordnung zwischen irgend einer Anzahl von Veränderlichen*, Leipzig: Engelmann 1906 (Ostwalds Klassiker 156)

C. G. J. Jacobi, *Vorlesungen über Dynamik, gehalten an der Universität Königsberg im Wintersemester 1842-1843 und nach einem von C.W. Borchart ausgearbeiteten Hefte* (A. Clebsch, Ed.), Berlin: Reimer [2]1884;, replica: Book Renaissance; 2nd, revis. ed. in: *Gesammelte Werke. Suppl.* (E. Lottner, Ed.), Berlin: Reimer 1884, https://archive.org/details/cgjjacobisvorle00lottgoog (03.04.2018)

G. Jaeger, *Quantum Information. An Overview*, Heidelberg: Springer 2007

H. A. Jahn, *Stability of degenerate electronic states in polyatomic molecules II. Spin degeneracy*, Proc. Roy. Soc. A164 (1938) 117-131

M. Jammer, *Concepts of Force*, Cambridge: Harvard Univ. Press 1957; reprint: New York: Dover 1999

M. Jammer, *Concepts of Mass in Classical and Modern Physics*, Cambridge: Harvard Univ. Press 1961; reprint: New York: Dover 1991

M. Jammer, *The conceptual development of quantum mechanics*, New York etc.: McGraw-Hill 1967, New York: AIP [2]1989

M. Jammer, *The philosophy of quantum mechanics*, New York etc.: Wiley 1974

M. Jammer, *Concepts of Space. The History of Theories of Space in Physics*, New York: Dover, 3rd ext. ed. 1993

M. Jammer, *Concepts of Time in Physics: A Synopsis*, Phys. in Perspective 9 (2007) 266-280 (Pais Prize Lecture, APS April Meeting, Jacksonville, 2007)

E. T. Jaynes, *The Gibbs Paradox*, in: C. R. Smith, G. J. Erickson & P. O. Neudorfer (Eds.), *Maximum Entropy and Bayesian Methods*, 1992, 1-22; bayes.wustl.edu/etj/articles/gibbs.paradox.pdf (26.07.2013)

J. D. Joannopoulos, S. G. Johnson, J. N. Winn & R. D. Meade, *Photonic Crystals. Molding the Flow of Light*, Princeton & Oxford: Princeton Univ. Press 22008; http://ab-initio.mit.edu/book/photonic-crystals-book.pdf (09.03.2015)

O. D. Johns, *Analytical Mechanics for Relativity and Quantum Mechanics,*, Oxford: Oxford Univ. Press 22011; http://www.metacosmos.org/mechanics/index.html (08.03.2016)

N. Jolley, *The Cambridge Companion to Leibniz*, Cambridge. New York. Melbourne: Cambridge Univ. Press 1995, books.google.de/books?id=SnRis5Gdi8gC (09.03.2015)

P. Jordan, *Über eine neue Begründung der Quantenmechanik*, Z. Phys. 40 (1927) 809-838

P. Jordan, *Die Physik des 20. Jahrhunderts*, Braunschweig: Vieweg 1947

P. Jordan & O. Klein, *Zum Mehrkörperproblem der Quantentheorie*, Z. Phys. 45 (1927) 751-765

P. Jordan & W. Pauli, *Zur Quantenelektrodynamik ladungsfreier Felder*, Z. Phys. 47 (1928) 151-173

P. Jordan & E. Wigner, *Über das Paulische Äquivalenzverbot*, Z. Phys. 47 (1928) 631-651; reprint in: J. Schwinger (Ed.), *Selected Papers on Quantum Electrodynamics*, 1958, No. 4

P. S. Kamenov & I. D. Christoskov, *Interference of classical particles on the surface of a liquid*, Phys. Lett. A 140 (1989) 1-2, 13-18

H. Kamerlingh Onnes, *Further experiments with liquid helium. D. On the change of electric resistance of pure metals at very low temperatures, etc. V. The disappearance of the resistance of mercury*, Comm. Phys. Lab. Univ. Leiden, No. 122b, 1911

I. Kant, *Gedanken von der wahren Schätzung der lebendigen Kräfte und Beurtheilung der Beweise derer sich Herr von Leibnitz und andere Mechaniker in dieser Streitsache bedienet haben nebst einigen vorhergehenden Betrachtungen welche die Kraft der Körper überhaupt betreffen*, Königsberg: Dor n 1746; https://bildsuche.digitale-sammlungen.de/index.html?c=viewer&bandnummer=bsb00074230 (09.04.2012), https://books.google.de/books?id=XcFRAAAAcAAJ (03.04.2018)

I. G. Kaplan, *Pauli Spin-Statistics Theorem and Statistics of Quasiparticles in a Periodical Lattice*, in: Hilborn & Tino, *Spin-Statistics Connection and Commutation Relations*, 2000, pp. 72-78

I. G. Kaplan, *Is the Pauli Exclusive Principle an Independent Quantum Mechanical Postulate?*, Int. J. Quantum Chem. 89 (2002) 268-276

T. Kashiwa, Y. Ohnuki & M. Suzuki, *Path Integral Methods*, Oxford: Clarendon 1997 (Oxford Science Publ.)

R. Kather, *Gott ist der Kreis, dessen Mittelpunkt überall ist ... Von der Dezentrierung der Erde und der Unendlichkeit des Universums bei Nikolaus von Kues und Giordano Bruno*, in: Peitz (Ed.), *Der verfielfachte Christus. Außerirdisches Leben und christliche Heilsgeschichte*, 2004, pp. 17-72; akademie-rs.de/fileadmin/user_upload/pdf_archive/khr46.pdf (09.04.2014)

V. J. Katz, *The History of Stokes' Theorem*, Math. Mag. 52 (1979) 3, 146-156; https://web.archive.org/web/20150402154904/, http://www-personal.umich.edu/madeland/math255/files/Stokes-Katz.pdf (27.07.2017)

E. H. Kennard, *Zur Quantenmechanik einfacher Bewegungstypen*, Z. Phys. 44 (1927) 326-352

J. Kepler, *Astronomia Nova ΑΙΤΙΟΛΟΓΗΟΣ seu physica coelestis, tradita commentariis de motibus stellae Martis ex observationibus G.V. Tychonis Brahe*, Prague 1609; e-rara.ch/zut/content/titleinfo/162514 (09.04.2014); Engl.: *New Astronomy* (transl. by William H. Donahue), Cambridge: Cambridge Univ. Press 1992

J. Kepler, *Harmonices Mundi Libri V*, Linz: Johann Planck 1619; https://books.google.de/books?id=ZLlCAAAAcAAJ (09.04.2014); Engl.: E. J. Aiton, A. M. Duncan & J. V. Field (transl.), *The Harmony of the World*, Philadelphia: American Philosophical Society 1997; in: Hawking, *On the Shoulders of Giants*, 2003, pp. 533-628

W. Ketterle, *Experimental studies of Bose-Einstein condensation*, Phys. Today 52 (1999) No. 12, 30ff.

W. Ketterle, *When Atoms Behave as Waves: Bose-Einstein Condensation and the Atom Laser* (Nobel Lecture 2001), nobel.se/physics/laureates/2001/ketterle-lecture.html (09.04.2005)

P. Kayupe Kikodio & Z. Mouayn, *New coherent states with Laguerre polynomials coefficients for the symmetric Pöschl-Teller oscillator*, J. Physics A: Math. Theor. 48 (2015) #21

G. Kirchhoff, *Zur Theorie der Lichtstrahlen*, Sitzungsber. Königl. Preuß. Akad. Wiss. Berlin (1882) 641-670; Ann. Phys. 18 (1883) 663-695, bibliothek.bbaw.de/bibliothek-digital/digitalequellen/schriften/anzeige?band=10-sitz/1882-2&seite:int=00000063 (09.04.2004)

W. S. Kirsanow, *The development of the notion of potential in L. Euler's work*, in: A. N. Bogoljubow (Ed.), *Mechanics and Physics of the Second Half of 19th Century*, 1978, pp. 141-147 (in Russ.)

C. Kirsten & H.-G. Körber (Eds.), *Physiker über Physiker. Wahlvorschläge zur Aufnahme von Physikern in die Berliner Akademie 1870 bis 1929 von Hermann v. Helmholtz bis Erwin Schrödinger*, Berlin: Akademie-Verlag 1975 (Studien Gesch. AdW DDR 1)

C. Kirsten & H.-G. Körber (Eds.), *Physiker über Physiker II. Antrittsreden / Erwiderungen bei der Aufnahme von Physikern in die Berliner Akademie / Gedächtsnisreden 1870 bis 1929*, Berlin: Akademie-Verlag 1979 (Studien Gesch. AdW DDR 8)

C. Kirsten & H.-J. Treder (Eds.), *Albert Einstein in Berlin 1913-1933 Teil I. Darstellung und Dokumente*, Berlin: Akademie-Verlag 1979 (Studien Gesch. AdW DDR 6)

J. R. Klauder & B. S. Skagerstam, *Coherent States – applications in physics and mathematical physics*, Singapore: World Scientific 1985

S. C. Kleene, *Mathematical Logic*, New York: Wiley 1967; reprint: Dover 2002

O. Klein, *Die Reflexion von Elektronen an einem Potentialsprung nach der relativistischen Dynamik von Dirac*, Zs. Physik 53 (1929) 157-165; users.physik.fu-berlin.de/kleinert/files/klein.pdf (26.04.2014)

H. Kleinert, *Path Integrals in Quantum Mechanics, Statistics, Polymer Physics, and Financial Markets*, Singapore: World Scientific [5]2009; users.physik.fu-berlin.de/kleinert/kleinert/?p=booklist&details=11 (26.04.2014)

H. Kleinert, *Gauge Fields in Condensed Matter*, Vol. I. *Superflow and Vortex Lines*, Singapore: World Scientific 1989

A. Kneser, *Das Prinzip der kleinsten Wirkung von Leibniz bis zur Gegenwart*, Leipzig: Teubner / Wiesbaden: Springer Fachmedien 1928

D. H. Kobe, *Lagrangian Densities and Principle of Least Action in Nonrelativistic Quantum Mechanics*, arXiv:0712.1608v1 [quant-ph] 10 Dec 2007; https://arxiv.org/pdf/0712.1608.pdf

W. Kohn & D. Sherrington, *Two kinds of bosons and Bose condensation*, Rev. Mod. Phys. 42 (1970) 1-11

A. N. Kolmogoroff, *Über die analytischen Methoden der Wahrscheinlichkeitsrechnung*, Math. Ann. 104 (1931) 415ff.

A. N. Kolmogoroff, *Grundbegriffe der Wahrscheinlichkeitsrechnung*, Berlin: Springer 1933

M. Komma, *Mathematische Behandlung der Schrödingergleichung*, 2003, mikomma.de/qph3htm/qph3a.htm (20.08.2013)

M. Komma, *Moderne Physik mit Maple. Von Newton zu Feynman*, Internatl. Thomson Publ. 1996; Redline 1998; mikomma.de/fh/embuch.html (20.08.2013)

L. Koopman, W. H. Kaper & A. L. Ellermeijer, *Understanding Student Difficulties in First Year Quantum Mechanics Courses*, physik.uni-mainz.de/lehramt/epec/koopman.pdf (20.08.2011)

Y. Kosmann-Schwarzbach, *The Noether Theorems. Invariance and Conservation Laws in the Twentieth Century*, New York *etc.*: Springer 2011

F. Krafft (Ed.), *Lexikon großer Naturwissenschaftler. Vorstoß ins Unbekannte*, Weinheim: Wiley-VCH 1999

N. v. Kues, *De docta ignorantia*, Strasbourg 1488; Engl.: Bond (ed.), *Nicholas of Cusa* 1997; German: *Von der Wissenschaft des Nichtwissens*, zeno.org/Philosophie/M/Nicolaus+von+Cues/Von+der+Wissenschaft+des+Nichtwissens (24.07.2015)

M. Kuhlmann, *Sein oder Nichtsein? Felder, Teilchen, Tropen – die Quantenfeldtheorie im Dialog zwischen Philosophie und Physik*, Physik Journal 15 (2016) Juni, 29-35

W. Kuhn, *Ideengeschichte der Physik. Eine Analyse der Entwicklung der Physik im historischen Kontext*, Springer Spektrum [2]2016

J.-L. Lagrange, *Nouvelles recherches sur la nature et la propagation du son*, Misc. Taurinensia II (1761-1762) 11-172; reprint in: J. A. Serret (Ed.), *Oeuvres de Lagrange*, Paris: Gauthier-Villars 1867, vol. 1, pp. 151-316, https://books.google.de/books?id=3TA4DeQw1NoC (04.04.2018)

J.-L. Lagrange, *Application de la méthode exposée dans le mémoire précédent à la solution de différents problèmes de dynamique*, Misc. Taurinensia II (1760-1761); in: *Œuvres*, Vol. I, Paris 1867

J.-L. Lagrange, *Mécanique Analytique*, Paris: La Veuve Desaint 1788; reprint of the 2nd revised ed. 1811-1815: Cambridge: Cambridge Univ. Press, 2009; https://gdz.sub.uni-goettingen.de/dms/load/toc/?PPN=PPN308899466 (Tomes 11 & 12) (20.08.2012)

J.-L. Lagrange, *Abhandlungen zur Variationsrechnung*, Leipzig: Engelmann 1894 (Ostwalds Klassiker 57)

M. G. G. Laidlaw & C. M. DeWitt, *Feynman Functional Integrals for Systems of Indistinguishable Particles*, Phys. Rev. D 3 (1971) 1375-1378

T. Lalescu, *Introduction à la théorie des équations intégrales* (with a preface by É. Picard), Paris: Hermann 1912

T. Lancaster & S. J. Blundell, *Quantum Field Theory for the Gifted Amateur*, Oxford: Oxford Univ. Press 2014

C. Lanczos, *The Variational Principles of Mechanics*, Toronto: Toronto Univ. Press [4]1970; reprint: New York: Dover 1986

C. Lanczos, *The Poisson bracket*, in: A. Salam & E. P. Wigner (Eds.), *Aspects of Quantum Theory: Papers in honour of P. A. M. Dirac*, Cambridge: Cambridge Univ. Press 1972, #10, 169-178, reprint 2010; German: *Die Poissonsche Klammer in der Quantenmechanik* (transl. and introd. by H. Rechenberg), Phys. Bl. 31 (1972) 7, 301-308; http://onlinelibrary.wiley.com/doi/10.1002/phbl.19750310705/pdf (20.08.2015)

L. D. Landau, *Das Dämpfungsproblem in der Wellenmechanik*, Z. Phys. 45 (1927) 430-441

L. D. Landau, *Diamagnetismus der Metalle*, Zs. Physik 64 (1930) 629-637

L. D. Landau, *Über die Bewegung des Elektrons im Kristallgitter*, Phys. Zs. Sowjetunion 3 (1933) 664-665

L. D. Landau & E. M. Lifschitz, *Lehrbuch der Theoretischen Physik, Vol. I. Mechanik*, [9]1976; *Vol. III. Quantenmechanik*, [9]1979; *Vol. V. Statistische Physik*, [3]1971; *Vol. VIII. Elektrodynamik der Kontinua*, 1967, Berlin: Akademie-Verlag

L. Landau & R. Peierls, *Erweiterung des Unbestimmtheitsprinzips für die relativistische Quantentheorie*, Z. Phys. 69 (1931) 56-69

L. D. Landau, *Collected papers* (D. ter Haar, Ed.), Oxford: Pergamon Press 1965

M. Lange, *A Tale of Two Vectors*, dialectica 63 (2009) 397-431

M. Lange, *Why do forces add vectorially? A forgotten controversy in the foundation of classical mechanics*, Am. J. Phys. 79 (2011) 380-388

P. S. de Laplace, *Traité de mécanique celeste*, Paris 1799-1825 (5 Vols.)

P. S. de Laplace, *Mémoire sur les intégrales définies et leur applications aux probabilités, et spécialement à la recherche du milieux qu'il faut choisir entre les résultats des observations*, Mém. Cl. Sci. Math. Phys. Inst. France 11 (1811) 279-347; O. C. 12, 357-412[8]

P. S. de Laplace, *Essai Philosophique sur la Probabilité*, Paris 1814; Engl.: *Philosophical Essay on Probabilities* (transl. by Andrew I. Dale from [5]1825), Springer 1995 (Sources Hist. Math. Phys. Sci. 13); *A Philosophical Essay on Probabilities* (transl. by F. W. Truscott & F. L. Emory from [6]1840), New York: Wiley / London: Chapman & Hall 1902; archive.org/details/philosophicaless00lapliala; openlibrary.org/books/OL7124417M/A_philosophical_essay_on_probabilities (02.10.2013); reprint: 2007

J. Larmor, *Aether and Matter: A Development of the Dynamical Relations of the Aether to Material Systems on the Basis of the Atomic Constitution of Matter, Including a Discussion of the Earth's Motion on Optical Phenomena, Being an Adams Prize Essay in the University of Cambridge*, Cambridge: Cambridge Univ. Press 1900

M. v. Laue, *Geschichte der Physik*, Frankfurt/M.: Ullstein [4]1959

R. B. Laughlin, *Anomalous Quantum Hall Effect: An Incompressible Quantum Fluid with Fractionally Charged Excitations*, Phys. Rev. Lett. 50 (1983) 1395-1398

---

[8] Bibl. data corrected and extended according to download.springer.com/static/pdf/223/bbm%253A978-1-4614-5725-1%252F1.pdf?auth66=1380911739_4e8650c639b45ad9af36f49c047de65f ext=.pdf (02.10.2013)

R. B. Laughlin, *Fractional Quantization* (Nobel Lecture 1998); in: G. Ekspong (Ed.), *Nobel Lectures in Physics 1996-2000*, 2002, pp. 264-286; nobelprize.org/nobel_prizes/physics/laureates/1998/laughlin-lecture.html (02.10.2003)

A. V. Lavrinenko, J. Lægsgaard, N. Gregersen, F. Schmidt, Th. Sø ndergaard, *Numerical Methods in Photonics*, Boca Raton: CRC Press 2015

D. K. Lazarou, *Interpretation of Quantum Theory. An overview*, arxiv.org/ftp/arxiv/papers/0712/0712.3466.pdf (02.11.2013)

P. G. L. Leach & G. P. Flessas, *Generalisations of the Laplace–Runge–Lenz Vector*, J. Nonlin. Math. Phys. 10 (2003) 3, 340–423; http://arxiv.org/abs/math-ph/0403028 (02.10.2015)

T. Lee, F. E. Low & D. Pines, *The Motion of Slow Electrons in Polar Crystals*, Phys. Rev. 90 (1953) 297ff.

A. M. Legendre, *Exercises de calcul integral*, Paris 1825-1832 (3 Vols.)

G. W. Leibniz, *Nova Methodus Pro Maximis & minimis, itemque tangentibus qua nec fractas nec irrationales quantitates moratur, & singulare pro illis calculigenus*, Acta Erud. 1684; German: *Neue Methode der Maxima, Minima sowie der Tangenten, die sich weder an gebrochenen, noch an irrationalen Größen stößt, und eine eigentümliche darauf bezügliche Rechnungsart*, Leipzig: Engelmann; Deutsch ^{3}2007

G. W. Leibniz, *Brevis demonstratio erroris memorabilis Cartesii et aliorum circa legem naturae, secundum quam volunt a Deo eandem semper quantitatem motus conservari, qua et in re mechanica abutuntutr*, Acta erudit. Lipsiens. 1686

G. W. Leibniz, *Discours de métaphysique*, 1686; Engl.: *Discourse on Metaphysics*, anselm.edu/homepage/dbanach/Leibniz-Discourse.htm (02.11.2013)

G. W. Leibniz, *SPECIMEN DYNAMICUM pro admirandis Naturae Legibus circa corporum vires et mutuas Actiones detegendis et ad suas causas revocandis*, Acta erudit. Lipsiens. April 1695; Engl.: *Essay in Dynamics showing the wonderful laws of nature concerning bodily forces and their interactions, and tracing them to their causes* (transl., ed. and commented by J. Bennett, June 2006), earlymoderntexts.com/pdf/leibessa.pdf (02.11.2013); Latin & German: *Specimen Dynamicum* (ed. and transl. by H. G. Dosch, G. W. Most & E. Rudolph, with elucid. by J. Aichelin, H. G. Dosch, P. Keller, H. Lichtenberger, H. J. Maul, G. W. Most & E. Rudolph), Hamburg: Meiner 1982 (Philos. Bibl. 339)

G. W. Leibniz, *Nouveaux essais sur l'entendement humain*, 1704, in: *Œ uvres philosoph. latines et françoises de feu Mr. de Leibnitz*, Amsterdam. Leipzig 1765; German: Halle 1778-1780

G. W. Leibniz, *Essais de Theodicée sur la bonté de Dieu, la liberté de l'homme et l'origine du mal*, Amsterdam 1710; German: Hannover 1720; Engl.: *Making the Case for God in terms of his Justice, which is Reconciled with the rest of his Perfections and with all his Actions*, earlymoderntexts.com/pdf/leibcd.pdf (J. Bennett, July 2007)

G. W. Leibniz, *5th paper to Clarke*, 8.viii.1716, No. 47; earlymoderntexts.com/pdf/leibclar.pdf, pp. 35f. (J. Bennett, Ed.; April 2007, retriev. 02.11.2013)

G. W. Leibniz, *Philosophical Papers and Letters* (transl. and ed. by L. Loemker), Dordrecht: Reidel 21969

L. M. Leinaas & J. Myrheim, *On the theory of identical particles*, Nuovo Cim. 37b (1977) 1-23

W. Lenz, *Über den Bewegungsverlauf und die Quantenzustände der gestörten Keplerbewegung*, Zs. Physik 24 (1924) 197-207

J. A. J. (Hans) van Leunen, *Quaternionic versus Maxwell based differential calculus*, 12.05.2016, http://www.e-physics.eu/QuaternionicVersusMaxwellBasedDifferentialCalculus.pdf

Lexikon der Naturwissenschaftler (Digitale Bibliothek 85)

G. N. Lewis, *The conservation of Photons*, Nature 118, Pt. 2 (1926) 874-875; http://www.nobeliefs.com/photon.htm (02.11.2010)

S. Lie, *Theorie der Transformationsgruppen*, Leipzig: Teubner 1888…1893 (3 Vols.)

S. Lie, *Vorlesungen über Differentialgleichungen mit bekannten infinitesimalen Transformationen* (G. Scheffers, Ed.), Leipzig: Teubner 1891; http://gdz.sub.uni-goettingen.de/dms/load/toc/?PPN=PPN578426625 (02.11.2015)

S. Lie, *Die Grundlagen für die Theorie der unendlichen kontinuierlichen Transformationsgruppen. I*, Leipz. Ber. III (1891) 316ff.; *Ges. Abh.* vol. 6, Art. XI, pp. 300-330; Engl.: *The foundations of the theory of infinite continuous transformation groups – I.* (transl. by D. H. Delphenich), http://neo-classical-physics.info/uploads/3/0/6/5/3065888/lie-_infinite_continuous_groups_-_i.pdf (02.11.2015)

Th. M. Liebling & L. Pournin, *Voronoi Diagrams and Delaunay Triangulations: Ubiquitous Siamese Twins*, Documenta Math. Extra Vol. (2012) 419-431, http://www.math.uiuc.edu/documenta/vol-ismp/60_liebling-thomas.pdf

D.-E. Liebscher, *Relativitätstheorie mit Zirkel und Lineal*, Berlin: Akademie-Verlag 1977 (WTB 197)

M. Ligare & R. Olivieri, *The calculated photon: Visualization of a quantum field*, Am. J. Phys. 70 (2002) 58-66

J. V. Lill, M. I. Haftel & G. H. Herling, *Mixed state quantum mechanics in hydrodynamical form*, J. Chem. Phys. 90 (9) (1989) May, 4940-4950, https://www.researchgate.net/publication/234927644_Mixed_state_quantum_mechanics_in_hydrodynamical_form

J. V. Lill, M. I. Haftel & G. H. Herling, *A time-dependent variational principle and the time-dependent Hartree approximation in hydrodynamical form*, J. Chem. Phys. 90 (9) (1989) May, 4933-4939

H. J. Lipkin, *Quantum Mechanics. New Approaches To Selected Topics*, Amsterdam · London: North Holland / New York: Elsevier 1973; Russ.: Г. Липкин, *Квантовая Механика. Новый Подход К Некоторым Проблемам*, Москва: «Мир» 1977

F. London, *On the Bose-Einstein condensation*, Phys. Rev. 54 (1938) 947-954

M. Longair, *Theoretical Concepts in Physics. An alternative view of theoretical reasoning in physics*, Cambridge: Cambridge Univ. Press ²2003

G. López, X. E. López & G. González, *Ambiguities on the Quantization of a One-Dimensional Dissipative System with Position Depending Dissipative Coefficient*, Int. J. Theor. Phys. 46 (2007) 149-156

H. A. Lorentz, *La théorie électromagnétique de Maxwell et son application aux corps mouvants*, Leiden: Brill 1892; Arch. néerl. sci. exact. natur. 25 (1892) 363-552; in: *Collected Papers*, 1936, Vol.2, pp. 164-343; archive.org/details/lathorielectrom00loregoog; ia801408.us.archive.org/35/items/lathorielectrom00loregoog (02.11.2010)

H. A. Lorentz, *Versuch einer Theorie der elektrischen und optischen Erscheinungen in bewegten Körpern*, Leiden: Brill 1895; reprints: Leipzig: Teubner 1906, Elibron Classics w/o. year

(elibron.com);
https://de.wikisource.org/wiki/Versuch_einer_Theorie_der_electrischen_und_optischen_Erscheinungen_in_bewegten_Körpern (04.04.2018)

H. A. Lorentz, *Electromagnetic phenomena in a system moving with any velocity less than that of light*, Proc. Acad. Sci. Amsterdam 6 (1904) 809-832; German: *Elektromagnetische Erscheinungen in einem System, das sich mit beliebiger, die des Lichtes nicht erreichender Geschwindigkeit bewegt*, in: Lorentz, Einstein, Minkowski & Weyl, *Das Relativitätsprinzip*, 1913, pp. 6-26; de.wikisource.org/wiki/Elektromagnetische_Erscheinungen (02.11.2010)

H. A. Lorentz, *The Theory of Electrons and its Applications to the Phenomena of Light and Radiant Heat*, Leipzig: Teubner 1909; reprint of the 2nd ed.: New York: Dover 1952

H. A. Lorentz, A. Einstein, H. Minkowski, H. Weyl, *Das Relativitätsprinzip*, Leipzig: Teubner [4]1922 / Stuttgart: Teubner [7]1974; Engl.: *The Principle of Relativity*, New York: Dover 1923

G. Ludwig, *Wellenmechanik. Einführung und Originaltexte*, Berlin: Akademie-Verlag / Oxford: Pergamon / Braunschweig: Vieweg [2]1970 (WTB 55)

G. Ludwig & G. Thurler, *A New Foundation of Physical Theories*, Berlin · Heidelberg: Springer 2006

H. Lübbig, *Das Wirkungsprinzip von Maupertuis und Feynmans Wegintegral der Quantenphase*, in: H. Hecht (Ed.), *Pierre Louis Moreau de Maupertuis. Eine Bilanz nach 300 Jahren*, Berlin: Berlin Verlag 1999, pp. 505ff.

H. Lübbig & D. Suisky, *From the origin of forces to the origin of mass. Euler's algorithm for the definition of inert mass reconsidered*, DPG-Jahrestagung 2007, Heidelberg, AKPhil 9.1; dpg-tagungen.de/program/heidelberg/akphil9.pdf (02.11.2008)

J. Lützen, *Heinrich Hertz: Classical Physicist, Modern Philosopher*, in: Baird, Hughes & Nordmann, *Heinrich Hertz: Classical Physicist, Modern Philosopher*, 1998, pp. 103-122

S. Lundqvist (Ed.), *Nobel Lectures in Physics 1971-1980*, Singapore: World Scientific 1991

A. M. Lyapunov, *The General Problem of the Stability of Motion*, London: Taylor & Francis 1992 (transl. by A. T. Fuller from E. Davaux's French 1907 transl. of the orig. Russian dissertation 1892)

E. Mach, *Die Mechanik in ihrer Entwicklung* (R. Wahsner & H.-H. v. Borzeszkowski, Eds.), Berlin: Akademie-Verlag 1988; Engl.: *The Science of Mechanics. A Critical and Historical Account of its Development*, Chicago. London: Open Court [2]1919; archive.org/stream/scienceofmechani005860mbp#page/n0/mode/2up (02.11.2010)

G. W. Mackey, *The Relationship Between Classical and Quantum Mechanics*, in: Contemporary Mathematics 214, Providence (RI): Amer. Math. Soc. 1998, pp. 91-110

E. Madelung, *Eine anschauliche Deutung der Gleichung von Schrödinger*, Naturwiss. 14 (1926) 45, 1004–1004

E. Madelung, *Quantentheorie in hydrodynamischer Form*, Z. Phys. 40 (1927), 3-4, 322-326; Engl.: *Quantum Theory in Hydrodynamical Form* (transl. by D. H. Delphenich), http://neo-classical-physics.info/uploads/3/0/6/5/3065888/madelung_-_hydrodynamical_interp..pdf (30.01.2017)

N. Malebranche, *De la recherche de la vérité. Où l'on traite de la Nature de l'Esprit de l'homme, et de l'usage qu'il en doit faire pour éviter l'erreur dans les Sciences*, Paris: Michel David [6]1712; fr.wikisource.org/wiki/De_la_recherche_de_la_v%C3%A9rit%C3%A9 (05.09.2013); Engl.: *Father Malebranche his treatise concerning the search after truth...* (transl. by Th. Taylor),

London 1700; archive.org/details/fathermalebranch00male (05.09.2013); *The Search after Truth* (transl. by T. M. Lennon & P. J. Olscamp), Columbus: Ohio State Univ. Press 1980

N. Malebranche, *Entretiens sur la métaphysique*, Rotterdam: Leers 1688; Engl.: *Dialogues on Metaphysics and Religion*, earlymoderntexts.com/f_maleb.html (J. Bennett, Ed.), 2007 (03.10.2013)

M. A. Man'ko & V. I. Man'ko, *Properties of Nonnegative Hermitian Matrices and New Entropic Inequalities for Noncomposite Quantum Systems*, Entropy 17 (2015) 2876-2894; mdpi.com/1099-4300/17/5/2876 (02.11.2015)

A. A. Markov, *Распространение закона больших чисель на величины, зависящие друг от друга* [Extension of the law of large numbers to quantities depending one on another], Изв. Физ.-мат. общ. Казан. унив. [2] 15 (1906) 135-156; Engl.: *Extension of the limit theorems of probability theory to a sum of variables connected in a chain*, in: R. Howard, *Dynamic Probabilistic Systems, vol. 1: Markov Models*, New York: Wiley 1971 (Decision and Control 1), Appendix B; reprint: New York: Dover 2007

A. A. Markov, *An Example of Statistical Investigation of the Text Eugene Onegin Concerning the Connection of Samples in Chains* (Lecture at the Royal Academy of Sciences, St. Petersburg, Jan. 23, 1913), Science in Context 19 (2006) 591-600; journals.cambridge.org/production/action/cjoGetFulltext?fulltextid=637500 (02.11.2010)

É. L. Mathieu, *Mémoire sur Le Mouvement Vibratoire d'une Membrane de forme Elliptique*. J. Math. Pure Appl. 13 (1868) 137-203

P. L. M. de Maupertuis, *Accord de différentes Loix de la Nature qui avoient jusqu'ici paru incompatibles*, Lu à l'Académie des Sciences de Paris le 15 avril 1744; in: *Œuvres de Maupertuis*, Vol. IV, Lyon: Bruyset 1756, 1768, pp. 3-28; https://fr.wikisource.org/wiki/Accord_de_différentes_loix_de_la_nature_qui_avoient_jusqu'ici_paru_incompatibles (04.04.2018)

P. L. M. de Maupertuis, *Loi de repos*, in: *Œuvres de Maupertuis*, Vol. IV, Lyon: Bruyset 1756, 1768

P. L. M. de Maupertuis, *Essai de cosmologie, Système de la nature, Réponse aux objections de M. Diderot*, prés. par F. Azouvi, Paris 1984

J. C. Maxwell, *On Faraday's lines of force*, Trans. Cambr. Phil. Soc. 10 (1855) 27ff., 1 (1856) 155ff.[9] (27.11.2014)

J. C. Maxwell, *On physical lines of force*, Phil. Mag. [4] 21 (1861) 161ff., 281ff., 338ff.; 23 (1862) 12ff., 85ff.; in: *Scient. Papers*, I, 451ff.; https://en.wikisource.org/wiki/On_Physical_Lines_of_Force (27.11.2014)

J. C. Maxwell, *A Dynamical Theory of the Electromagnetic Field*, Trans. Roy. Soc. CLV (1965), Pt. III; reprint (with a Preface and an Introduction by Th. F. Torrance, 1982, and an appreciation by A. Einstein, 1931): Eugene (OR): Wipf and Stock 1996; en.wikisource.org/wiki/A_Dynamical_Theory_of_the_Electromagnetic_Field (27.11.2014)

J. C. Maxwell, *A Treatise on Electricity & Magnetism*, Oxford: Clarendon ³1891; reprint: New York: Dover 1954; archive.org/details/electricandmagne01maxwrich (27.11.2014)

---

[9] The digitalization of many papers and books by Maxwell is available on archive.org/search.php?query=creator%3A%22Maxwell%2C%20James%20Clerk%2C%201831-1879%22

J. C. Maxwell, *Matter and Motion*, 1877; reprints: New York: Dover 1991, Cosimo Classics 2007

J. C. Maxwell, *Scientific Papers*, Cambridge: Cambridge Univ. Press 1890 (2 vols.); reprint: New York: Dover 1965 (1 vol.), http://strangebeautiful.com/other-texts/maxwell-scientificpapers-vol-i-dover.pdf (27.11.2014)

J. R. Mayer, *Bemerkungen über die Kräfte der unbelebten Natur,* Ann. Chemie Pharmacie 43 (1842) 233-240; Engl.: *Remarks on the Forces of Inorganic Nature* (transl. by G. C. Foster), Phil. Mag. [4] 24 (1862) 371; reprint in: W. F. Magie (Ed.), *A Source Book in Physics*, New York: McGraw-Hill 1935; web.lemoyne.edu/ giunta/mayer.html (27.11.2014)

M. McCloskey, *Intuitive Physics*, Scientific American 248 (1983) 4, 122-130

K. T. McDonald, *Orbital and Spin Angular Momentum of Electromagnetic Fields*, 7.2.2015; physics.princeton.edu/ mcdonald/examples/spin.pdf (27.11.2015)

J. Mehra & H. Rechenberg, *The Historical Development of Quantum Theory*, New York: Springer 1999ff.; Vol. 1: *The Quantum Theory of Planck, Einstein, Bohr and Sommerfeld: Its Foundation and the Rise of Its Difficulties 1900-1925*; Vol. 2: *The Discovery of Quantum Mechanics 1925*; Vol. 3: *The Formulation of Matrix Mechanics and Its Modifications 1925-1926*; Vol. 4/1: *The Fundamental Equations of Quantum Mechanics 1925-1926*; Vol. 4/2: *The Reception of the New Quantum Mechanics 1925-1926*; Vol. 5: *Erwin Schrödinger and the Rise of Wave Mechanics*; Vol. 6: *The Completion of Quantum Mechanics 1926-1941; Epilogue: Aspects of the Further Development of Quantum Theory 1942-1999*

L. Mensing, *Die Rotations-Schwingungsbanden nach der Quantenmechanik*, Z. Physik 36 (1926) 814-823

N. Mermin, *Copernican crystallography*, Phys. Rev. Lett. 68 (1992) 1172-1175

A. Messiah, *Quantum Mechanics*, New York: Wiley 1958 (2 Vols.); reprint: New York: Dover 1999 (1 Vol.)

K. von Meyenn (Ed.), *Eine Entdeckung von ganz außerordentlicher Tragweite: Schrödingers Briefwechsel zur Wellenmechanik und zum Katzenparadoxon*, Berlin · Heidelberg: Springer 2011

G. Mie, *Lehrbuch der Elektrizität und des Magnetismus*, Stuttgart: Enke ^{2}1941

A. Mielke & H. Tasaki, *Ferromagnetism in the Hubbard model*, Commun. Math. Phys. 158 (1993) 341-371; http://www.researchgate.net/profile/Andreas_Mielke/publication/1862686_Ferromagnetism_in_th e_Hubbard_Model–_Examples_from_Models_with_Degenerate_Single-Electron_Ground_States/links/02e7e5397010fa07e8000000.pdf (23.10.2014)

D., I., J. & M. Millar, *The Cambridge Dictionary of Scientists*, Cambridge: Cambridge Univ. Press 1996

H. Minkowski, *Die Grundgleichungen für die elektromagnetischen Vorgänge in bewegten Körpern*, Nachr. Ges. Wiss. Gött., Math.-Phys. Klasse 1908, pp. 53-111; de.wikisource.org/wiki/Die_Grundgleichungen_f%C3%BCr_die_elektromagnetischen_Vorg%C3 %A4nge_in_bewegten_K%C3%B6rpern (23.10.2010)

H. Minkowski, *Raum und Zeit*, 80. Vers. d. dtsch. Naturforscher u. Ärzte, Köln, 21.9.1908; in: Jahresber. Dtsch. Math.-Verein. 1909; Phys. Zs. 10 (1909) 104ff.; Leipzig u. Berlin: Teubner 1909; de.wikisource.org/wiki/Raum_und_Zeit_(Minkowski) (23.10.2010); Engl. in: Einstein *et al.*, *The Principle of Relativity*, 1923, pp. 75-91, with notes by Sommerfeld on pp. 92-96

V. A. Miransky, *Dynamical Symmetry Breaking in Quantum Field Theories*, Singapore: World Scientific 1994

A. K. Mishra & G. Rajasekaran, *Generalized Fock spaces, new forms of quantum statistics and their algebras*, arXiv:hep-th/9605204v1 29 May 1996

A. K. Mishra & G. Rajasekaran, *Quantum Field Theory for Orthofermions and Orthobosons*, arXiv:hep-th/0105004v1 1 May 2001

P. Mittelstaedt, *Klassische Mechanik*, Mannheim: Bibl. Institut 1995 (BI-Taschenbücher)

P. Mittelstaedt, *Rational Reconstructions of Modern Physics*, Dordrecht *etc.*: Springer 2011 (Fund. Theor. Phys. 169)

P. Mittelstaedt, *The Problem of Interpretation of Modern Physics*, Found. Physics 41 (2011) 11, 1667-1676

A. Moatti, *Gaspard-Gustave de Coriolis (1792-1843): un mathématicien, théoricien de la mécanique appliquée*, Thèse de doctorat de l'Université de Paris I, 2011; ddata.over-blog.com/xxxyyy/0/31/93/70/These-Moatti-Coriolis-Octobre-2011.pdf (23.10.2014)

L. Molnár, *An Algebraic Approach to Wigner's Unitary-Antiunitary Theorem*, arxiv.org/pdf/math/9808033v1.pdf (23.10.2011)

J. Moody, A. Shapere & F. Wilczek, *Adiabatic Effective Lagrangians*, in: Shapere & Wilczek, *Geometric Phases in Physics*, 1989, pp. 160-183

O. Morsch, *Licht und Materie. Eine physikalische Beziehungsgeschichte*, Weinheim: Wiley-VCH 2003

P. M. Morse, *Diatomic Molecules According to the Wave Mechanics. II. Vibrational Levels*, Phys. Rev. 34 (1929) 57-64, http://pearl.elte.hu/infra/morse_29PRL.pdf (19.07.2017)

N. F. Mott, *The exclusion principle and aperiodic systems*, Proc. Roy. Soc. L. A 125 (1929) 220-230

N. F. Mott, *The collision between two electrons*, Proc. Roy. Soc. L. A 126 (1930) 259-267

N. F. Mott, *Theory of excitation by collision with heavy particles*, Proc. Camb. Phil. Soc. 27 (1931) 553-560

D. Mugnai, A. Ranfagni, *Microwave Experiments on Tunneling Time*, in: J. G. MugaR. Sala MayatoÍ & L. Egusquiza (Eds.), *Time in Quantum Mechanics*, Heidelberg: Springer 2008, Ch. 12, pp. 355-397 (Lect. Notes Phys. 734)

D. Mugnai, A. Ranfagni & L. Ronchi, *The question of tunneling time duration: A new experimental test at microwave scale*, Phys. Lett. A 247 (1998) 281-286[10]

S. Mukherjee, A. Spracklen, D. Choudhury, N. Goldman, P. Öhberg, E. Andersson & R. R. Thomson, *Observation of a Localized Flat-Band State in a Photonic Lieb Lattice*, Phys. Rev. Lett. 114 (2015) 245504-245504-5; https://physics.aps.org/featured-article-pdf/10.1103/PhysRevLett.114.245504 (30.09.2015)

N. Mukunda, L. O'Raifeartaigh & E. Sudarshan, *Characteristic noninvariance groups of dynamical systems*, Phys. Rev. Lett. 15 (1965) 1041-1044

J. Naas & H. L. Schmid, *Mathematisches Wörterbuch*, Berlin: Akademie-Verlag / Leipzig: Teubner 1974 (2 Vols.)

Y. Nambu, *Quasiparticles and Gauge Invariance in the Theory of Superconductivity*, Phys. Rev. 117 (1960) 648-663; http://www.mat.unimi.it/users/gaeta/SD2/risorse/nambu.pdf

---

[10] I thank Daniela Mugnai for providing me with a copy of this and of the foregoing reference *via* ResearchGate.

Y. Nambu, *Broken Symmetry. Selected papers* (T. Eguchi & K. Nishijima, Eds.), Singapore: World Scientific 1995 (World Scientific Ser. 20th Cent. Physics 13)

W. Natanson, *Über die statistische Theorie der Strahlung,* Phys. Z. 12 (1911) 659-666; *On the statistical theory of radiation*, Bull. Acad. Sci. Cracovie (A) (1911), 134-148

L. Navarro & E. Pérez, *Paul Ehrenfest on the Necessity of Quanta (1911): Discontinuity, Quantization, Corpuscularity, and Adiabatic Invariance*, Arch. Hist. Exact Sci. 58 (2004) 97-141

T. Needham, *Newton and the Transmutation of Force*, Amer. Math. Monthly 100 (1993) 119-137

T. Needham, *Visual Complex Analysis*, Oxford: Oxford Univ. Press 1997; German: *Anschauliche Funktionentheorie*, München: Oldenbourg 2001

K. Neitmann, F. Beck, H. Kaak, F. Göse, J. Peters & W. Neugebauer, *Liselott Enders in memoriam. Das archiv- und geschichtswissenschaftliche Werk im Rückblick und im Ausblick*, Jahrbuch Geschichte Mittel- und Ostdeutschlands 57 (2011) 221-306 (with a revised bibliography by Florian Seher)

E. Nelson, *Feynman Integrals and the Schrödinger Equation*, J. Math. Phys. 5 (1964) 332-343

E. Nelson, *Derivation of the Schrödinger Equation from Newtonian Mechanics*, Phys. Rev. 150 (1966) 1079-1085; dieumsnh.qfb.umich.mx/archivoshistoricosmq/ModernaHist/Nelson%20a.pdf

J. v. Neumann, *Wahrscheinlichkeitstheoretischer Aufbau der Quantenmechanik*, Gött. Nachr. (1927) 245-272

J. v. Neumann, *Thermodynamik quantenmechanischer Gesamtheiten*, Gött. Nachr. (1927) 273-291

J. v. Neumann, *Allgemeine Eigenwerttheorie Hermitescher Funktionaloperatoren*, Math. Ann. 102 (1929) 49-131

J. v. Neumann, *Mathematische Grundlagen der Quantenmechanik*, Berlin: Springer 1932; https://gdz.sub.uni-goettingen.de/id/PPN379400774 (30.09.2013); [2]1996; Engl.: *Mathematical Foundations of Quantum Mechanics*, Princeton: Princeton Univ. Press 1955, [2]1996

J. v. Neumann, *Über einen Satz von Herrn M. H. Stone*, Ann. Math. [2] 33 (1932) 567-573

I. Newton, *De motu corporum in gyrum*, 1684; newtonproject.sussex.ac.uk/view/texts/normalized/NATP00089 (6.1.2014)

I. Newton, *De Gravitatione et aequipondio fluidorum*, unpubl. (Cambridge Univ. Lib., Ms. Add 4003); philoscience.unibe.ch/documents/kursarchiv/SS02/newton.PDF (6.1.2014); Engl.: *Descartes, Space and Body*, earlymoderntexts.com/pdf/newtdes.pdf (J. Bennett, Jan. 2013); *De gravitatione et æ quipondio fluidorum* (revised translation and interpolated comments by H. Stein), strangebeautiful.com/other-texts/newton-de-grav-stein-trans.pdf (10.08.2013); Latin & German: *Über die Gravitation ...: Texte zu den philosophischen Grundlagen der klassischen Mechanik* (transl. and introd. by Gernot Böhme), Frankfurt: Klostermann 1988 (Klostermann-Texte: Philosophie)

I. Newton, *Philosophiae Naturalis Principia Mathematica*, London 1687, http://www.gutenberg.org/ebooks/28233, http://cudl.lib.cam.ac.uk/view/PR-ADV-B-00039-00001/1; [3]1726; Engl.: *Mathematical Principles of Natural Philosophy* (transl. by A. Motte & F. Cajori), 1729; reprint in: Hawking, *On the Shoulders of Giants*, 2003 (Books I and III)

I. Newton, *The Principia. Mathematical Principles of Natural Philosophy* (A New Translation by I. Bernhard Cohen and Anne Whitman assisted by Julia Buden, Preceded by *A Guide to Newton's Principia* by I. Bernhard Cohen[11]), Berkeley *etc.*: Univ. of California Press 1999

I. Newton, *Principia* (transl. and annotated by I. Bruce), 2012, 17centurymaths.com/contents/newtoncontents.html (27.11.2013)

I. Newton, *Opticks*, London ²1704; reprint of the 4th ed.: New York: Dover 1952

E. Noether, *Invariante Variationsprobleme*, Gött. Nachr., Math-phys. Klasse (1918) 235-257; physics.ucla.edu/ cwp/articles/noether.trans/german/emmy235.html; Engl.: *Invariant Variation Problems*, physics.ucla.edu/~cwp/articles/noether.trans/english/mort186.html (all retriev. 6.2.2014)

A. Noguchi, Y. Shikano, K. Toyoda & S. Urabe, *Aharonov-Bohm effect in the tunnelling of a quantum rotor in a linear Paul trap*, Nature Commun. 5 (2014) May, Art. 3868

L. W. Nordheim, *Zur Theorie der thermischen Emission und der Reflexion von Elektronen an Metallen*, Z. Phys. 46 (1928) 833-855

S. Odake, *Recurrence relations of the multi-indexed orthogonal polynomials*, J. Math. Phys. 54 (2013) 083506; http://arxiv.org/pdf/1303.5820v2.pdf; Pt. II: http://arxiv.org/pdf/1410.8236v2.pdf (v2, 2015) (6.1.2015)

S. Odake & R. Sasaki, *Discrete Quantum Mechanics*, J.Phys. A 44 (2011) 353001ff., https://arxiv.org/abs/1104.0473v2 (16.3.2014)

Y. Ohnuki & S. Kamefuchi, *Quantum Field Theory and Parastatistics*, Berlin: Springer 1982

M. Ostrogradsky, *Démonstration d'un théorème du calcul intégral*, Paris 1826, publ. in abbrev. form in St. Petersburg 1831; Russ. in: А. П. Юшкевич & В. И. Антропова, *Неопубликованные работы М. В. Остроградского,* Историко-математические исследования 16 (1965) 49-96, Sect. Остроградский М. В.

M. Ostrogradsky, *Première note sur la théorie de la chaleur*, Mém. Acad. imp. sci. St. Pétersbourg [6] 1 (1828/1831) 129-133

W. Ostwald, *Johann Wilhelm Ritter*, 1894; reprint in: *Abhandlungen*, pp. 361-362; quoted after R. Zott, *Über Wilhelm Ostwalds wissenschaftshistorische Beiträge zum Problem des wissenschaftlichen Schöpfertums*, in: W. Ostwald, *Zur Geschichte der Wissenschaft*, 1999

W. Ostwald, *Zur Geschichte der Wissenschaft: vier Manuskripte aus dem Nachlaß,* Thun · Frankfurt am Main: Deutsch 21999 (Ostwalds Klassiker 267)

A. van Oudenaarden, M. H. Devoret, Yu. V. Nazarov, J. E. Mooij, *Magneto-electric Aharonov-Bohm effect in metal rings*, Nature 391 (1998) 768-770

P. Painlevé, *Analyse des travaux scientifiques*, 1900; reprint in *Librairie Scientifique et Technique*, Paris: Blanchard 1967, pp. 1-2; reprod. in *Oeuvres de Paul Painlevé*, Paris: CNRS 1972-1975, vol. 1, pp. 72-73 (after homepage.math.uiowa.edu/ jorgen/hadamardquotesource.html; 01.04.2014)

S. Pancharatnam, *Generalized Theory of Interference, and Its Applications. Part I. Coherent Pencils*, Proc. Indian Acad. Sci. A 44 (1956) 247-262

D. A. Park, *Resource Letter SP-1 on Symmetry in Physics*, Am. J. Phys. 36 (1968) 577-584

---

[11] Among others, this 370 (!) p. Guide contains not only historical notes, but also explanations of and completions to Newton's text.

O. Passon, *Bohmsche Mechanik – eine elementare Einführung in die deterministische Interpretation der Quantenmechanik*, Frankfurt a. Main: Deutsch [2]2010

H. Paul, *Auf dem Weg zur Quantentheorie: Die Erfindung des Hohlraums und ihre Folgen*, Schriften der Sudetendeutschen Akademie der Wissenschaften und Künste 34 (2015) 135-140

W. Paul & H. Steinwedel, *Ein neues Massenspektrometer ohne Magnetfeld*, Zs. Naturforschung A 8 (1953) 448-445

W. Pauli, *Über den Zusammenhang des Abschlusses der Elektronengruppen im Atom mit der Komplexstruktur der Spektren*, Z. Phys. 31 (1925) 765-783; reprints in: D. ter Haar, *Quantentheorie*, 1970, pp. 229-253; Ch. P. Enz & K. v. Meyenn (Eds.), *Wolfgang Pauli. Das Gewissen der Physik*, 1988, pp. 211-239

W. Pauli, *Über das Wasserstoffspektrum vom Standpunkt der neuen Quantenmechanik*, Zs. Physik 36 (1926) 336-363, http://www.chemie.unibas.ch/~steinhauser/documents/Pauli_1926_36_336-363.pdf; reprint in: Ch. P. Enz & K. v. Meyenn (Eds.), *Wolfgang Pauli. Das Gewissen der Physik*, 1988, pp. 181-200; Engl. in: van der Waerden, *Sources of Quantum Mechanics*, 1967, No. 16

W. Pauli, *Quantentheorie*, in: H. Geiger & K. Scheel (Eds.), *Handbuch der Physik*, Vol. 23, Berlin: Springer 1926, pp. 1-278; reprint in: S. Flügge (Ed.), *Handbuch der Physik*, Vol. V/1, Berlin: Springer 1958, pp. 136ff.

W. Pauli, *Zur Quantenmechanik des magnetischen Elektrons*, Zs. Phys. 43 (1927) 601-623; reprint in: Ch. P. Enz & K. v. Meyenn (Eds.), *Wolfgang Pauli. Das Gewissen der Physik*, 1988, pp. 282-305

W. Pauli, *Die allgemeinen Prinzipien der Wellenmechanik*, in: H. Geiger & K. Scheel (Eds.), *Handbuch der Physik*, Vol.24/1, Berlin: Springer [2]1933

W. Pauli, *The Connection between Spin and Statistics*, Phys. Rev. 58 (1940) 716-722; web.ihep.su/dbserv/compas/src/pauli40b/eng.pdf (6.1.2014)

W. Pauli, *Exclusion principle and quantum mechanics* (Nobel Lecture 1946); in: G. Ekspong (Ed.), *Nobel Lectures in Physics, Vol 3, 1942-1962*, 1998, pp. 27-43; nobelprize.org/nobel_prizes/physics/laureates/1945/pauli-lecture.html (6.1.2010)

W. Pauli, *Wave Mechanics* (Pauli Lectures on Physics 5), Cambridge (MA): MIT Press 1973; reprint: New York: Dover 2000

W. Pauli, *Selected Topics in Field Quantization* (Pauli Lectures on Physics 6), Cambridge, Mass.: MIT Press 1973; ext. reprint: New York: Dover 2000

H.-H. Peitz (Ed.), *Der verfielfachte Christus. Außerirdisches Leben und christliche Heilsgeschichte*, Diözese Rottenberg-Stuttgart 2004 (Kleine Hohenheimer Reihe 46), pp. 17-72; akademie-rs.de/fileadmin/user_upload/pdf_archive/khr46.pdf (6.1.2009)

L. S. Penrose & R. Penrose, *Impossible objects: A special type of visual illusion*, Brit. J. Psychology 49 (1958) 31–33

F. Penzlin, *Die Methode der Feynmanschen Graphen*, Fortschr. Phys. 25 (1962) 357-420

A. Perelomov, *Generalized Coherent States and Their Applications*, Heidelberg · Berlin: Springer 1986

S. Perlis, *Theory of Matrices*, Reading (MA): Addison-Wesley 1952; reprint: New York: Dover 1991

P. Pesic, *Beyond Geometry. Classic Papers from Riemann to Einstein*, New York: Dover 2007

I. Pikovski, M. Zych, F. Costa & Č. Brukner, *Universal decoherence due to gravitational time dilation*, Nature Physics, June 15, 2015

L. P. Pitaevskii, *Vortex lines in an imperfect Bose gas*, Sov. Phys. JETP 13 (1961) 2, 451–454, http://jetp.ac.ru/cgi-bin/dn/e_013_02_0451.pdf (6.6.2016); Russ.: Л. П. Питаевский, *Вихровые линии в неидельном бозонном газе*, ЖЭТФ 40 (1961) 646-651

M. Plancherel & M. G. Mittag Leffler, *Contribution a l'etude de la representation d'une fonction arbitraire par les integrales définies*, Rendiconti del Circolo Matematico di Palermo 30 (2010) 289-335

M. Planck, *Zur Theorie des Gesetzes der Energieverteilung im Normalspektrum*, Verh. dtsch. Phys. Ges. Berlin 2 (1900) 237-245 (talk dated 14.12.1900), https://archive.org/stream/verhandlungende01goog#page/n246/mode/2up; reprints in: *Die Ableitung der Strahlungsgesetze*, 1997, No. V; Ludwig, *Wellenmechanik*, 1970, pp. 107-117; Engl.: *On the Theory of the Energy Distribution Law of the Normal Spectrum*, in: ter Haar, *The Old Quantum Theory*, Oxford: Pergamon Press 1967, pp. 88ff., http://www.ffn.ub.es/luisnavarro/nuevo_maletin/Planck%20%281900%29,%20Distribution%20Law.pdf

M. Planck, *Über das Gesetz der Energieverteilung im Normalspectrum*, Ann. Phys. 4 (1901) 553-563; reprint in: Planck, *Die Ableitung der Strahlungsgesetze*, 1997, No. VI; Engl.: *On the Law of Distribution of Energy in the Normal Spectrum*, strangepaths.com/files/planck1901.pdf (5.1.2015)

M. Planck, *Das Prinzip der Relativität und die Grundgleichungen der Mechanik*, Verh. dtsch. phys. Ges. 8 (1906) 136-141; wikilivres.ca/wiki/Das_Prinzip_der_Relativit%C3%A4t_und_die_Grundgleichungen_der_Mechanik; Engl.: *The Principle of Relativity and the Fundamental Equations of Mechanics*, en.wikisource.org/wiki/Translation:The_Principle_of_Relativity_and_the_Fundamental_Equations_of_Mechanics (all retriev. 5.1.2015)

M. Planck, *Das Prinzip von der Erhaltung der Energie*, 1887; Leipzig. Berlin: Teubner [2]1908 (Wissenschaft und Hypothese VI)

M. Planck, *Vorlesungen über die Theorie der Wärmestrahlung*, Leipzig: Barth 1906; archive.org/details/vorlesungenberd04plangoog; Engl.: *The theory of heat radiation* (authorized transl. of the 2nd ed. 1913 by M. Masius), Philadelphia: Blakiston 1914 / New York: Dover 1959, 1991; books.google.de/books?isbn=0486173283 (5.1.2015)

M. Planck, *Erwiderung des Sekretars Hrn.* Planck *auf die Antrittsrede des Hrn.* Einstein *vom 2. Juli 1914 (Leibniztag)*, in: Planck, *Max Planck in seinen Akademie-Ansprachen*, 1948, pp. 22-25; Kirsten & Körber, *Physiker über Physiker II*, 1979, pp. 247f.

M. Planck, *Eine veränderte Formulierung der Quantenhypothese*, Berl. Ber. (1914) 918-923

M. Planck, *Der Ursprung und die Entwicklung der Quantentheorie* (Nobel Lecture 1919); in: G. Ekspong (Ed.), *Nobel Lectures in Physics Vol 1 1901-1921*, 1998; Engl.: *The Genesis and Present State of Development of the Quantum Theory*; nobelprize.org/nobel_prizes/physics/laureates/1918/planck-lecture.html (12.04.2014)

M. Planck, *Ansprache vom 29. Juni 1922 (Leibniztag)*, in: Planck, *Max Planck in seinen Akademie-Ansprachen*, 1948, pp. 41-48

M. Planck, *Max Planck in seinen Akademie-Ansprachen. Erinnerungsschrift der Deutschen Akademie der Wissenschaften zu Berlin*, Berlin: Akademie-Verlag 1948; planck.bbaw.de/akademieansprachen.php? (12.04.2014)

M. Planck, *Die Ableitung der Strahlungsgesetze. Sieben Abhandlungen aus dem Gebiete der elektromagnetischen Strahlungstheorie*, Leipzig: Geest & Portig / Frankfurt a. Main · Thun: Deutsch [3]1997, 2001 (Ostwalds Klassiker 206)

Planck, Nernst, Rubens & E. Warburg, *Wahlvorschlag für A. Einstein zur Aufnahme als ordentliches Mitglied in die Akademie d. Wiss. von M. Planck (Berlin, 12. Juni 1913)*, AAW Berlin, II-IIIa – Vol. 19, Bl. 36-37; Inventar A No. 1; reprints in: C. Kirsten & H.-G. Körber (Eds.), *Physiker über Physiker*, 1975, pp. 201-203; C. Kirsten & H.-J. Treder (Eds), *Albert Einstein in Berlin 1913-1933 Teil I*, 1979, pp. 95-97

G. Pöschl & E. Teller, *Bemerkungen zur Quantenmechanik des anharmonischen Oszillators*, Zs. Physik 83 (1933) 143-151

H. Poincaré, *L'équilibre d'une masse fluide animée d'un mouvement de rotation*, Acta Math. 7 (1885) Sept., pp. 259-380; http://projecteuclid.org/euclid.acta/1485888300

H. Poincaré, *Sur la dynamique de l'électron*, Comptes rendus 140 (1905) 1504-1508, fr.wikisource.org/wiki/Sur_la_dynamique_de_l%E2%80%99%C3%A9lectron_(juin); German: *Über die Dynamik des Elektrons*, 2008; archive.org/details/PoincareDynamikA (all retriev. 12.04.2004)

H. Poincaré, *Sur la dynamique de l'électron*, Rend. Circ. mat. Palermo 21 (1906) 129-176; fr.wikisource.org/wiki/Sur_la_dynamique_de_l%E2%80%99%C3%A9lectron_(juillet); Engl.: *On the Dynamics of the Electron*, en.wikisource.org/?curid=664793; German: *Über die Dynamik des Elektrons*, 2008, archive.org/details/PoincareDynamikB (all retriev. 12.04.2004)

S. D. Poisson, *Traité de Mécanique*, Paris: Courcier 1811, Bachelier [2]1833; Engl.: *A Treatise of Mechanics*, London: Longmans 1842

S. D. Poisson, *Sur la variation des constantes arbitraires dans les questions de mécanique*, Mem. Acad. sci. Inst. France (1816) T. Ier, 1-70, https://books.google.de/books?id=ASugAAAAMAAJ (12.04.2004)

D. Popov, *Coherent States of Systems with Non-Equidistant Energy Levels*, Theor. Phys. 2 (2017) 2, 97-107, https://www.researchgate.net/publication/317370275_Coherent_States_of_Systems_with_Non-Equidistant_Energy_Levels

A. Proca, *Sur la théorie ondulatoire des électrons positifs et négatifs*, J. Phys. Radium 7 (1936) 347-353

A. P. Prudnikov, Yu. A. Brychkov & O. I. Marichev, *Integrals and Series. Elementary Functions*, Moscow: "Nauka" 1981 (in Russian)

A. P. Prudnikov, Yu. A. Brychkov & O. I. Marichev, *Integrals and Series. Special Functions*, Moscow: "Nauka" 1983 (in Russian)

A. P. Prudnikov, Yu. A. Brychkov & O. I. Marichev, *Integrals and Series. Additional Chapters*, Moscow: "Nauka" 1986 (in Russian)

L. O'Raifeartaigh & N. Straumann, *Early History of Gauge Theories and Kaluza-Klein Theories, with a Glance at Recent Developments*, arXiv:hep-ph/9810524v2 5 Apr 1999 (12.04.2014)

J. P. Ralston, *Quantum Theory without Planck's Constant*, 2012, https://arxiv.org/abs/1203.5557

E. Recami, *Superluminal Motions? A Bird's-Eye View of the Experimental Situation*, Found. Phys., 31 (2001) 1119-1135, http://dinamico2.unibg.it/recami/erasmo%20docs/SomeRecentSCIENTIFICpapers/ExtendedRelativity/BIRDSEYE.PDF (08.04.2016)

E. Recami & G. Salesi, *Kinematics and hydrodynamics of spinning particles*, Phys. Rev. A 57 (1998) 98-105, http://dinamico2.unibg.it/recami/erasmo%20docs/SomeRecentSCIENTIFICpapers/StructureOfSpinningParticles/PRA98.pdf (08.04.2016)

P. Reinecker & M. Schulz, *Theoretische Physik. Elektrodynamik*, lecture script winter term 2005/06, status 11.04.2006, University Ulm 2006; P. Reineker, M. Schulz & B. M. Schulz, *Theoretische Physik II. Elektrodynamik*, Berlin: Wiley-VCH 2006

P. Renaud, *Sur une généralisation du principe de symétrie de Curie*, Comptes Rendus 200 (1935) 531-534; Engl.: *On a generalization of Curie's symmetry principle*, in: J. Rosen, *Symmetry in Physics*, 1982, pp. 26ff.

J. Renn (Ed.), *Einstein's Annalen Papers. The Complete Collection 1901-1922*, Weinheim: Wiley 2005; physik.uni-augsburg.de/annalen/history/Einstein-in-AdP.htm (08.11.2005)

G. Ricci-Curbastro & T. Levi-Civita, *Méthodes du calcul différentiel absolu et leurs applications*, Math. Annalen LIV (1900) 125-201

B. Riemann, *Über die Hypothesen, welche der Geometrie zugrunde liegen* (Habilitationsvorlesung, 1854); in: Abh. Königl. Ges. Wiss. Göttingen 13 (1868); *Gesammelte mathematische Werke*, Leipzig: Teubner ²1892, pp. 272ff.; emis.de/classics/Riemann/Geom.pdf (08.03.2015); Engl.: *On the Hypotheses That Lie at the Foundations of Geometry (1854)*, in: Pesic, *Beyond Geometry*, 2007, pp. 23-33

B. Riemann, *Ueber die Darstellbarkeit einer Function durch eine trigonometrische Reihe*, Abh. Königl. Ges. Wiss. Göttingen 13 (1867); Göttingen: Dieterichsche Buchhandlung 1867 (habilitation treatise, 1854), wwwuser.gwdg.de/subtypo3/gdz/pdf/PPN309770033/PPN309770033___LOG_0001.pdf, emis.de/classics/Riemann/Trig.pdf (08.03.2015)

S. P. Rigaud, *Historical Essay On the First Publication Of Sir Isaac Newton's Principia*, Oxford: Oxford Univ. Press MDCCCXXXVIII; reprint: Delhi: Pranava Books (254439)

W. Ritz, *Über ein neues Gesetz der Serienspektren (Vorläufige Mitteilung)*, Phys. Zs. 9 (1908) 521-529

P. S. Riseborough, P. Hänggi & U. Weiss, *Exact results for a damped quantum-mechanical harmonic oscillator*, Phys. Rev. A 31 (1985) 471-478

R. de Lima Rodrigues, *The Quantum Mechanics SUSY Algebra: An Introductory Review*, Monograph CBPF-MO-03-01 (December/2001), https://arxiv.org/pdf/hep-th/0205017.pdf (29.03.2018)

O. Rodrigues, *De l'attraction des sphérodes*, Corresp. l'École Impériale Polytechnique (Thesis, Faculty of Science, University of Paris) 3 (1816) 3, 361-385

J. Rogel-Salazar, *The Gross-Pitaevskii Equation and Bose-Einstein condensates*, Eur. J. Phys. 34 (2013) 247-258, https://arxiv.org/abs/1301.2073 (07.03.2015)

R. Rojas, *Impetustheorie des Fußballs*, heise.de/tp/artikel/41/41270/lit.html#l_2r, 19.03.2014 (retriev. 06.03.2015)

R. Rompe & H.-J. Treder, *Über Physik*, Berlin: Akademie-Verlag 1979 (WTB 107)

R. Rompe & H.-J. Treder, *Nikolaus von Kues als Naturforscher*, in: Rompe & Treder, *Über Physik*, 1979, Appendix

R. Rompe & H.-J. Treder, *Über die Einheit der exakten Wissenschaften*, Berlin: Akademie-Verlag 1982 (WTB 279)

R. Rompe & H.-J. Treder, *Zur Grundlegung der theoretischen Physik. Beiträge von H. v. Helmholtz und H. Hertz*, Berlin: Akademie-Verlag 1984 (WTB 284)

J. Rosen, *Resource letter SP-2: Symmetry and group theory in physics*, Am. J. Phys. 49 (1981) 304-319; reprint in: J. Rosen (Ed.), *Symmetry in Physics*, 1982, pp. 1-16

J. Rosen (Ed.), *Symmetry in Physics: Selected Reprints*, New York: AAPT 1982

J. Rosen, *Symmetry in Science: An Introduction to the General Theory*, New York: Springer 1995

J. Rosen, *The Symmetry Principle*, Entropy 7 (2005) 308-313; mdpi.com/1099-4300/7/4/308 (4.1.2014)

J. Rosen, *Symmetry Rules: How Science and Nature Are Founded on Symmetry*, Berlin: Springer 2008

L. Rosenfeld, *Sur la définition du spin d'un champ de rayonnement*, Bull. Acad. Roy. Belg. 28, 562 (1942); physics.princeton.edu/ mcdonald/examples/EM/rosenfeld_barb_28_568_42.pdf (06.03.2013)

B. Rothenstein & M. Costache, *From Thought Experiments to the Lorentz Transformations*, Gen. Sci. J. Jan.12, 2009, wbabin.net/ (06.03.2013)

A. E. Ruark, *The Zeeman effect and Stark effect of hydrogen in wave mechanics: the force equation and the virial theorem in wave mechanics*, Phys. Rev. 31 (1928) 533-538

C. D. T. Runge, *Praxis der Gleichungen*, Berlin. Leipzig 21921; Engl.: *Vector Analysis*, London: Methuen 1923

J. R. Rydberg, *Untersuchungen über die Beschaffenheit der Emissionsspektren der chemischen Elemente*, Leipzig: Akad. Verlagsges. 1922

G. Salesi, *Non-Newtonian Mechanics*, Int. J. Mod. Phys. 17 (2002) 347-374, https://arxiv.org/abs/quant-ph/0112052

F. C. Santos, V. Soares & A. C. Tort, *An English translation of Bertrand's theorem*, 2007, http://arxiv.org/abs/0704.2396 (05.01.2016)

S. Saunders, *On the explanation for quantum statistics*, Stud. Hist. Phil. Sci. B: Mod. Phys. 37 (2006) 192-211

G. Schaefer, *Das Elementare im Komplexen. Neue Wege zu einer fächerübergreifenden Allgemeinbildung um die Jahrtausendwende*, Frankfurt am Main *etc.*: Lang 1997

G. Scharf, *From Electrostatics to Optics*, Berlin *etc.*: Springer 1994

W. P. Schleich, *Quantum Optics in Phase Space*, Berlin: Wiley-VCH 2001

J. Schmalian, *Elastisch dank einflussreicher Elektronen*, Physik Jornal 16 (2017) 3, 24-25

P. Schmelcher, *Symmetrien diktieren nicht alles*, Physik Journal 11 (2012) 2, 16-17

M. Schmiechen, *What did Eötvös and his followers actually do? Classical theory of general relativity and gravity*, m-schmiechen.homepage.t-online.de/HomepageClassic01/grav_Eotv.pdf (19.05.2010)

E. Schmutzer, *Symmetrien und Erhaltungssätze der Physik*, Berlin: Akademie-Verlag / Oxford: Pergamon / Braunschweig: Vieweg 1972 (WTB 75)

J. Schnakenberg, *Thermodynamik und Statistische Physik*, Neuhaus: Wiley-VCH ²2002

M. M. Schneider, *Identische Teilchen und Bohmsche Mechanik*, Berlin: Logos 1997

U. E. Schröder, *Noether's Theorem and the Conservation Laws in Classical Field Theories*, Fortschr. Physik 16 (1968) 357-372; reprint in: J. Rosen, *Symmetry in Physics*, 1982, pp. 92-107

E. Schrödinger, *Quantisierung als Eigenwertproblem. Erste Mitteilung*, Ann. Phys. [4] 79 (1926) 361-376; reprints in: *Abhandlungen zur Wellenmechanik*, 1927, pp. 1-16; G. Ludwig, *Wellenmechanik*, 1970, pp. 108-122; strangepaths.com/wp-content/uploads/2008/01/schrodinger1926a.pdf; *Zweite Mitteilung*, 489-527; reprints in: *Abhandlungen zur Wellenmechanik*, pp. 17-55; G. Ludwig, *Wellenmechanik*, pp. 122-145; *Dritte Mitteilung: Störungstheorie, mit Anwendung auf den Starkeffekt der Balmerlinien*, 80 (1926) 437-490; reprint in: *Abhandlungen zur Wellenmechanik*, pp. 85-138; *Vierte Mitteilung*, 81 (1926) 109-139; reprints in: *Abhandlungen zur Wellenmechanik*, pp. 139-169, G. Ludwig, *Wellenmechanik*, pp. 174-193; copies of all on: home.tiscali.nl/physis/HistoricPaper/HistoricPapers.html; copies of Schrödinger's notebook on the construction of his time-dependent equation are available here: http://quantum-history.mpiwg-berlin.mpg.de/intranet/sourcesIntranet/fileserver/Schrodinger_1926_Notebook-18-12.pdf/V1_Schrodinger_1926_Notebook-18-12.pdf/index_html (low resolution), http://quantum-history.mpiwg-berlin.mpg.de/quantumViewer?url=http%3A//content.mpiwg-berlin.mpg.de/mpiwg/online/permanent/echo/quantum_project/SchroedingerWien/pack_18-12/index.meta pn=1 mode=texttool (high resolution) (all retriev. 08.07.2001)

E. Schrödinger, *Über das Verhältnis der Born-Heisenberg-Jordanschen Quantenmechanik zu der meinen*, Ann. Phys. [4] 79 (1926) 734-756; reprints in: *Abhandlungen zur Wellenmechanik*, 1927, pp. 62-84, G. Ludwig, *Wellenmechanik*, 1970, pp. 146-173; home.tiscali.nl/physis/HistoricPaper/HistoricPapers.html; strangepaths.com/wp-content/uploads/2008/01/schrodinger1926c.pdf; Engl.: *On the Relation between the Quantum Mechanics of Heisenberg, Born, and Jordan, and that of Schrödinger*, http://hermes.ffn.ub.es/luisnavarro/nuevo_maletin/; reprint in: *Collected Papers on Wave Mechanics by E. Schrödinger* (transl. by J. F. Shearer & W. M. Deans), London and Glasgow: Blackie & Son 1928, pp. 45-61; http://hildalarrondo.net/wp-content/uploads/2010/05/SchroedingerPapers.pdf (all retriev. 08.07.2001)

E. Schrödinger, *Der stetige Übergang von der Mikro- zur Makromechanik*, Naturwiss. 14 (1926) 664-666; reprint in: *Abhandlungen zur Wellenmechanik*, 1927, pp. 56-61; home.tiscali.nl/physis/HistoricPaper/HistoricPapers.html (08.07.2001)

E. Schrödinger, *An Undulatory Theory of the Mechanics of Atoms and Molecules*, Phys. Rev. [2] 28 (1926) 1049-1070; home.tiscali.nl/physis/HistoricPaper/HistoricPapers.html (08.07.2001)

E. Schrödinger, *Über den Comptoneffekt*, Ann. Phys. 82 (1927) 257-264; home.tiscali.nl/physis/HistoricPaper/HistoricPapers.html (08.07.2001)

E. Schrödinger, *Der Energieimpulssatz der Materiewellen*, Ann. Phys. 82 (1927) 265-272; home.tiscali.nl/physis/HistoricPaper/HistoricPapers.html (08.07.2001)

E. Schrödinger, *Energieaustausch nach der Wellenmechanik*, Ann. Phys. 83 (1927) 956-968; home.tiscali.nl/physis/HistoricPaper/HistoricPapers.html

E. Schrödinger, *Abhandlungen zur Wellenmechanik*, Leipzig: Barth [2]1927

E. Schrödinger, *Über Heisenbergs Unbestimmtheitsbeziehung*, Preuß. Ak. Wiss., Phys.-Math. Abt. XIX, 1930, 296-303; Engl.: *About Heisenberg Uncertainty Relation*, Bulg. J. Phys. 26 (1999) 193-203; arxiv:quant-ph/9903100 v2 15 Jun 2000 (08.07.2001)

E. Schrödinger, *The fundamental idea of wave mechanics* (Nobel Lecture 1933); in: G. Ekspong (Ed.), *Nobel Lectures in Physics, Vol 2, 1922-1941*, 1998, pp. 305-316; nobelprize.org/nobel_prizes/physics/laureates/1933/schrodinger-lecture.html (08.07.2001)

E. Schrödinger, *Die gegenwärtige Situation in der Quantenmechanik*, Naturwissenschaften 23 (1935) 807-812; 823-828; 844-849; http://wwwthep.physik.uni-mainz.de/ matschul/rot/schroedinger.pdf (17.09.2016); Engl.: *The Present Situation in Quantum Mechanics*, Proc. Am. Philos. Soc. 124, 323-38; reprint in: J.A. Wheeler & W.H. Zurek (Eds.), *Quantum Theory and Measurement*, Princeton (N.J.): Princeton Univ. Press 1983, Pt. I, Sect. I.11; https://www.tuhh.de/rzt/rzt/it/QM/cat.html (17.09.2016)

E. Schrödinger, *Discussion of Probability Relations between Separated Systems*, Proc. Cambridge Philos. Soc. 31 (1935) 555-563; http://www.informationphilosopher.com/solutions/scientists/schrodinger/Schrodinger-1935.pdf (10.09.2016)

E. Schrödinger, *Probability Relations Between Separated Systems*, Proc. Cambridge Philos. Soc. 32 (1936) 446-452

E. Schrödinger, *A Method of Determining Quantum-Mechanical Eigenvalues and Eigenfunctions*, Proc. Royal Irish Acad. A46 (1940) II., 9–16; http://www.jstor.org/stable/20490744 (29.03.2018)

E. Schrödinger; *Further Studies on Solving Eigenvalue Problems by Factorization*, Proc. Royal Irish Acad. A46 (1940) XIV., 183-206; http://www.jstor.org/stable/20490756 (29.03.2018)

E. Schrödinger, *The Factorization of the Hypergeometric Equation*, Proc. Roy. Irish Acad. A47 (1941) 53-54; arXiv:physics/9910003 (17.07.2004)

E. Schrödinger, *Statistical Thermodynamics*, Cambridge: Univ. Press 1948; German: *Statistische Thermodynamik*, Leipzig: Barth 1952

J. R. Schütz, *Das Prinzip der absoluten Erhaltung der Energie*, Gött. Nachr., Math.-phys. Kl. (1897) 110-123, http://dfg-viewer.de/show/?id=8071&tx_dlf%5Bid%5D=http%3A%2F%2Fapi.deutsche-digitale-bibliothek.de%2Fitems%2FABV3OUAEJBDBCVBAOXX7MDP6DVDKLVJN%2Fsource&tx_dlf%5Bpage%5D=120

B. Schulz, *A new look at Bell's inequalities and Nelson's theorem*, Ann. Phys. (Berlin) 1 (2009) 1-41

F. Schwabl, *Quantenmechanik* (Springer-Lehrbuch), Berlin *etc.*: Springer [4]1993 (*QM I*); *Quantenmechanik für Fortgeschrittene* (Springer-Lehrbuch), Berlin *etc.*: Springer 1997 (*QM II*)

B. Schwarzschild, *Physics Nobel Prize Goes to Tsui, Stormer and Laughlin for the Fractional Quantum Hall Effect*, Phys. Today 51 (1998) 17-19

K. Schwarzschild, *Zur Elektrodynamik. 1.-3.*, Nachr. Königl. Ges. Wissensch. Göttingen. Math.-phys. Klasse (1903) 126-131, 132-141, 245-278

J. Schwichtenberg, *Physics from Symmetry*, Springer 2015 (Undergraduate Lecture Notes in Physics); improved German ed.: *Durch Symmetrie die moderne Physik verstehen. Ein neuer Zugang zu den fundamentalen Theorien*, 2017 (Springer Spektrum)

J. Schwinger (Ed.), *Selected Papers on Quantum Electrodynamics*, New York: Dover 1958

C. Seligman, *Fictitious Forces*, cseligman.com/text/physics/fictitious.htm (14.05.2014)

A. Shapere & F. Wilczek (Eds.), *Geometric Phases in Physics*, Singapore *etc.*: World Scientific 1989 (Adv. Ser. Math. Phys. 5)

A. Shapere & F. Wilczek, *Classical Time Crystals*, MIT-CTP / 4347, arXiv:1202.2537v2 [cond-mat.other] 12 Jul 2012; https://arxiv.org/pdf/1202.2537.pdf

R. Shaw, *Symmetry, Uniqueness, and the Coulomb Law of Force*, Am. J. Phys. 33 (1965) 300-305, reprint in: J. Rosen, *Symmetry in Physics*, 1982, pp. 27-32

J. H. Shirley, *Interaction of a Quantum System with a Strong Oscillating Field*, PhD Thesis, CalTech, Pasadena 1963, http://thesis.library.caltech.edu/1805/1/Shirley_jh_1963.pdf (17.02.2016)

J. H. Shirley, *Solution of the Schrödinger Equation with a Hamiltonian Periodic in Time*, Phys. Rev. 138 (1965) B979-B987

B. G. Sidharth, F. Honsell & A. De Angelis (Eds.), *Frontiers of Fundamental and Computational Physics. Proc. 6th Int. Symp. 'Frontiers of Fundamental and Computational Physics', Udine, Italy, 26-29 Sept. 2004*, Dordrecht: Springer 2006

H. Sievers, *Louis de Broglie und die Quantenmechanik*, arxiv.org/pdf/physics/9807012 (17.07.2002)

M. P. Silverman, *More Than One Mystery. Explorations in Quantum Interference*, New York *etc.*: Springer 1995

K. Simonyi, *A fizika kultúrtörténete*, Budapest: Gondolat 1978, [5]2011; German: *Kulturgeschichte der Physik*, Leipzig *etc.*: Urania 1990, Frankfurt am Main: Deutsch, 3rd revis. and ext. ed. 2001; Engl.: *A Cultural History of Physics*, A K Peters/CRC Press 2012

C. Singh, M. Belloni & W. Christian, *Improving students' understanding of quantum mechanics*, https://www.compadre.org/osp/items/detail.cfm?ID=7494 (05.04.2018)

J. C. Slater, *The Theory of Complex Spectra*, Phys. Rev. 34 (1929) 1293-1322

C. R. Smith, G. J. Erickson & P. O. Neudorfer (Eds.), *Maximum Entropy and Bayesian Methods*, Dordrecht: Kluwer 1992

A. Sommerfeld, *Die Feinstruktur der Wasserstoff- und der Wasserstoff-ähnlichen Linien*, Sitzungsber. math.-phys. Klasse, K. B. Akad. d. Wiss. München, 1915, pp. 459-500 (presented Jan. 8, 1916); in: M. Eckert & A. Sommerfeld, *Die Bohr-Sommerfeldsche Atomtheorie: Sommerfelds Erweiterung des Bohrschen Atommodells 1915/16*, Berlin · Heidelberg: Springer 2013 (Klassische Texte der Wissenschaft), http://dx.doi.org/10.1140/epjh/e2013-40054-0; Engl.: *The fine structure of Hydrogen and Hydrogen-like lines* (transl. by Patrick D. F. Ion), Eur. Phys. J. H 39 (2014) 179-204

A. Sommerfeld, *Zur Quantentheorie der Spektrallinien*, Ann. Phys. [4] 51 (1916) 17, 1-94; in: *Gesammelte Schriften III*, pp. 172-265; http://gallica.bnf.fr/ark:/12148/bpt6k15353s/f7.image

A. Sommerfeld, *Atombau und Spektrallinien*, Braunschweig: Vieweg 1919-29 (2 Vols.)

A. Sommerfeld, *Wissenschafticher Briefwechsel mit Bohr, Einstein, Heisenberg u.a.*, Band I: l919-1929 (A. Hermann, K. v. Meyenn & V. F. Weisskopf, Eds.), New York/Heidelberg/Berlin: Springer 1979

A. Sommerfeld, *Vorlesungen über theoretische Physik. Vol. I Mechanik*, Leipzig 1955 / Thun · Frankfurt a. Main: Deutsch [8]1994; Engl.: *Mechanics – Lectures on Theoretical Physics, Vol. I* (transl. from the fourth German ed. by Martin O. Stern), Academic Press 1964; *Vol. III Elektrodynamik*, Frankfurt a. Main: Deutsch [4]2001; Engl.: *Electrodynamics – Lectures on Theoretical Physics, Vol. III* (transl. by Edward G. Ramberg), Academic Press 1964

Th. Sonar, *Die Entwicklung der Ballistik von Aristoteles bis Euler: Ein Beitrag zum Euler-Jahr 2007*, Abh. Braunschweig. Wissensch. Ges. LIX (2008) 203-230; digibib.tu-bs.de/?docid=00048854 (8.3.2013)

A. Speiser, *L. Euler. The principle of relativity and the foundations of classical mechanics*, in: A. N. Bogoljubow (Ed.), *Mechanics and Physics in the second half of 19ᵗʰ century*, 1978, pp. 134-140 (in Russian)

B. Średniawa, *Władys ław Natanson (1864–1937)*, The old and new concepts of physics IV (2007) 705-723; merlin.fic.uni.lodz.pl/concepts/www/IV_4_705.pdf (7.4.2012)

J. Stachel (Ed.), *Einstein's Miraculous Year. Five Papers That Changed the Face of Physics*, Princeton: Princeton Univ. Press 1998

A. A. Stahlhofen & G. Nimtz, *Evanescent modes are virtual photons*, Europhysics Lett. 76 (2006) 189–195, http://www.war-games.com/epl_76_2_189.pdf (17.03.2016)

E. V. Stefanovich, *Classical Electrodynamics without Fields and the Aharonov-Bohm effect*, https://arxiv.org/abs/0803.1326v2

E. V. Stefanovich, *Relativistic Quantum Dynamics. A Non-Traditional Perspective on Space, Time, Particles, Fields, and Action-at-a-Distance*, 2004/2014, arxiv.org/pdf/physics/0504062v17 (17.07.2016)

E. V. Stefanovich, *Relativistic Quantum Theory of Particles. I: Quantum Electrodynamics, II. A Non-Traditional Perspective on Space, Time, Particles, Fields, and Action-at-a-Distance*, LAP LAMBERT Academic Publishing 2015

E. V. Stefanovich, *Causality of the Coulomb field of relativistic electron bunches*, Eur. Phys. J. C (2016), https://www.researchgate.net/publication/299288696_Causality_of_the_Coulomb_field_of_relativistic_electron_bunches (17.07.2016)

A. Stern, *Anyons and the quantum Hall effect–A pedagogical review*, Ann. Physics 323 (2008) 204-249, pitp.physics.ubc.ca/confs/7pines2009/readings/arovas_Stern_2007.pdf (17.05.2010)

S. Stevin, *Beghinselen der Weegconst*, Leiden 1585, http://echo.mpiwg-berlin.mpg.de/ECHOdocuView?url=/mpiwg/online/permanent/archimedes/stevi_weegc_085_nl_1 586; Engl. in: *The Principal Works of Simon Stevin*, 1955; D. J. Struik (Ed.), *A source book in mathematics: 1200 - 1800*, Cambridge (MA): Harvard Univ. Press 1969

S. Stevin, *Hypomnemata mathematica* (transl. and ed. by W. Snellius), Lugduni Batavorum: Patius 1605…1608, reader.digitale-sammlungen.de/de/fs1/object/display/bsb11057828_00005.html (17.11.2013)

S. Stevin, *The Principal Works of Simon Stevin*, vol I, *Mechanics* (E. J. Dijksterhuis, Ed.), Amsterdam: Swets & Zeitlinger 1955

S. M. Stigler, *Stigler's law of eponymy*, Trans. NY Acad. Sci. [2] 39 (1980) 147-158 (Robert K. Merton Festschrift *Science and social structure*, ed. by T. F. Gieryn); republ. in: *Statistics on the Table*, Cambridge (MA): Harvard Univ. Press 2002

G. G. Stokes, *On the Effect of the Internal Friction of Fluids on the Motion of Pendulums*, Cambridge Philos. Trans. 9 (1851) 8-106, http://mural.uv.es/daroig/documentos/stokes1850.pdf, Cambridge: Pitt 1851, https://ia600406.us.archive.org/16/items/b22464074/b22464074.pdf; in: *Mathematical and Physical Papers*, Vol. III, Cambridge: Cambridge Univ. Press 1901, 1922, pp. 1-141, https://archive.org/details/cu31924004181339 (all retriev. 05.04.2018), New York: Johnson Reprint Corp. ²1966

M. H. Stone, *Linear Transformations in Hilbert Space. III. Operational Methods and Group Theory*, Proc. Natl. Acad. Sci. 16 (1930) 172-175, ncbi.nlm.nih.gov/pmc/articles/PMC1075964/ (16.11.2013)

M. H. Stone, *On one-parameter unitary groups in Hilbert space*, Ann. Math. [2] 33 (1932) 643-648; jstor.org/stable/1968538; researchgate.net/publication/200524488_On_one-parameter_unitary_groups_in_Hilbert_Space

M. Stone, *The Physics of Quantum Fields*, New York: Springer 2000

N. Straumann, *Schrödingers Entdeckung der Wellenmechanik*, 2001, arxiv.org/pdf/quant-ph/0110097.pdf; mydocs.strands.de/MyDocs/06472/06472.pdf (17.08.2013)

N. Straumann, *Gauge principle and QED*, PHOTON2005 Conf., Aug. 31 - Sept. 4, 2005, Warsaw, Poland, Acta Phys. Polon. B 37 (2006) 575-593, actaphys.uj.edu.pl/_old/vol37/pdf/v37p0575.pdf (14.11.2013)

S. Strogatz, *Einstein's First Proof*, The New Yorker, Nov. 19, 2015, http://www.newyorker.com/tech/elements/einsteins-first-proof-pythagorean-theorem (21.12.2015)

D. J. Struik, *Abriss der Geschichte der Mathematik*, Berlin: Dtsch. Verlag d. Wiss. ³1965

J. Ch. F. Sturm, *Cours d'analyse de l'Ecole polytechnique*, Paris: Gauthier-Villars 1877 (2 vols.)

J. Ch. F. Sturm & J. Liouville, *EXTRAIT D'un Mémoire sur le d éveloSement des fonctions en séries dont les différents termes sont assujettis à satisfaire à une même équation diffé rentielle linéaire, contenant un paramètre variable*, J. Math. Pure Appl. 2 (1837) 176-178, gutenberg.org/ebooks/31295 (15.09.2013)

D. F. Styer, *Notes on Relativistic Dynamics*, 28.08.2014, oberlin.edu/physics/dstyer/Modern/RelativisticDynamics.pdf (15.11.2014)

D. F. Styer, M. S. Balkin, K. M. Becker, M. R. Burns, C. E. Dudley, S. T. Forth, J. S. Gaumer, M. A. Kramer, D. C. Oertel, L. H. Park, M. T. Rinkoski, C. T. Smith & T. D. Wotherspoon, *Nine formulations of quantum mechanics*, Am. J. Physics 70 (3), March 2002, 288-297, online.kitp.ucsb.edu/online/utheory03/styer/ (12.06.2007)

D. Suisky, *Über eine Differenz in der Begründung des Wirkungsprinzips bei Maupertuis und Euler*, in: H. Hecht (Ed.), *Pierre Louis Moreau de Maupertuis. Eine Bilanz nach 300 Jahren*, Berlin: Berlin-Verlag 1999, 293-320

D. Suisky, *The Newton - Leibniz controversy on space and time and the development of mechanics by Euler and Einstein*, VIII. Int. Leibniz-Kongress "Einheit in der Vielheit", Hannover, July 24-29, 2006; nlb-hannover.de/Leibniz/Gesellschaft/Dokumente/Kongressprogramm.pdf (15.11.2013)

D. Suisky, *Zur methodologischen Bedeutung von Eulers Begründung der Mechanik*, DPG-Jahrestagung 2006, Dortmund, AKPHIL 10.1, dpg-tagungen.de/archive/2006/dortmund/akphil10.pdf (15.11.2013)

D. Suisky, *Zur Ableitung der Lorentz-Transformation mittels Ordnungsrelationen*, DPG-Jahrestagung 2006, Dortmund, AKPHIL 10.2, dpg-tagungen.de/archive/2006/dortmund/akphil10.pdf (15.11.2013)

D. Suisky, *Euler's early relativistic theory*, in: R. Baker (Ed.), *Euler Reconsidered. Tercentenary essays*, Heber City: Kendrick Press 2007

D. Suisky, *On the post-Newtonian period in the development of mechanics*, DPG-Jahrestagung 2007, Heidelberg, AKPhil 8.2, dpg-tagungen.de/program/heidelberg/akphil8.html (15.11.2013)

D. Suisky, *Euler's mechanics as a unified theory of matter and motion*, DPG-Jahrestagung 2007, Heidelberg, AKPhil 8.3, dpg-tagungen.de/program/heidelberg/akphil8.html (15.11.2013)

D. Suisky, *Euler as Physicist*, Berlin · Heidelberg: Springer 2009

D. Suisky, *Are there elements of Leibniz's theory in Newton? On the different shapes of Newton's 2ⁿᵈ Law*, Verh. Dtsch. Phys. Ges. AGPhil., Berlin 2012; Abstract: dpg-verhandlungen.de/year/2012/conference/berlin/part/agphil/session/6/contribution/4 (15.11.2013)

D. Suisky & P. Enders, *Leibniz' foundation of mechanics and the development of 18ᵗʰ century mechanics initiated by Euler*, in: H. Poser (Ed.), *Nihil sine ratione*, Proc. VII Intern. Leibniz Congress, Berlin 2001, leibniz-kongress.tu-berlin.de/webprogramm.html; information-philosophie.de/philosophie/leibniz2001.html (15.11.2005)

D. Suisky & P. Enders, *Dynamische Begründung der Lorentz-Transformation*, DPG-Jahrestagung, Berlin 2005, Poster GR18.1, orion.physik.hu-berlin.de/POSTER/DPG05_Suisky2005a2.pdf (15.11.2005)

R. H. Swendsen, *The ambiguity of distinguishability in statistical mechanics*, Am. J. Phys. 83 (2015) 545-554

I. Szabó, *Geschichte der mechanischen Prinzipien und ihrer wichtigsten Anwendungen*, Basel *etc.*: Birkhäuser ³1987 (ext. and corr. ed. by P. Zimmermann & E. A. Fellmann), corr. reprint 1996 (Wissenschaft und Kultur 32)

E. Szücs, *Dialógusok a müszaki tudományokról*, Budapest: Müszaki Könyvkiadó 1971; German: *Dialoge über technische Prozesse* (Konrad Werner, ed., István Gedeon, transl.), Budapest: Müszaki Könyvkiadó / Leipzig: Fachbuchverlag 1976

T. E. Tahko, *An Introduction to Metametaphysics*, Cambridge: Cambridge Univ. Press 2015

B. Taylor, *Methodus Incrementorum Directa et Inversa*, London: Innys 1715

P. Teller, *The Ineliminable Classical Face of Quantum Field Theory*, in: T. Cao (Ed.), *Conceptual Foundations of Quantum Field Theory*, 1999, pp. 314-323

W. Thompson & P. G. Tait, *Treatise on natural philosophy*, Cambridge: Cambridge Univ. Press 1879

J. D. Titius, *Betrachtungen über die Natur*, Leipzig 1766

O. Toeplitz, *Zur Theorie der quadratischen und bilinearen Formen von unendlich vielen Veränderlichen. I. Teil: Theorie der L-Formen*, Math. Ann. 70 (1911) 351-376

M. Tomes, K. J. Vahala & T. Carmon, *Direct imaging of tunneling from a potential wall*, OPTICS EXPRESS 17 (2009) No. 21, web.eecs.umich.edu/ tcarmon/0E347872-BDB9-137E-C85B689E7B1CC1B6_186579.pdf (24.08.2015)

E. Torricelli, *Opera geometrica*, 1644; ique.de/aufg/tm4/a4_3_ml.PDF

H. Touchette, *Legendre-Fenchel transforms in a nutshell*, 2005-2007, maths.qmul.ac.uk/~ht/archive/lfth2.pdf (17.03.2014)

R. Torretti, *The Philosophy of Physics*, Cambridge: Cambridge Univ. Press 1999 (The Evolution of Modern Philosophy)

R. J. Tykodi, *On Euler's theorem for homogeneous functions and proofs thereof*, J. Chem. Educ. 59 (1982) 557

H.-J. Treder, *Isaac Newton und die Begründung der mathematischen Prinzipien der Naturphilosophie*, Sitzungsber. AdW d. DDR 12/N, Berlin: Akademie-Verlag 1977

H.-J. Treder (Ed.), *Newton-Studien*, Berlin: Akademie-Verlag 1978

H. Tributsch, *Quantum Paradoxes, Time, and Derivation of Thermodynamic Law: Opportunities from Change of Energy Paradigm*, J. Gen. Phil. Sci. 37 (2006) 287-306

R. A. R. Tricker, *Die Beiträge von Faraday und Maxwell zur Elektrodynamik*, Berlin: Akademie-Verlag 1974 (WTB 95)

C. Truesdell, *Essays in the History of Mechanics*, New York *etc.*: Springer 1968

C. A. Truesdell, *Rückwirkungen der Geschichte der Mechanik auf die moderne Forschung*, Humanismus und Technik 13 (1969) No. 1

J. Turney & M. Riordan (Eds.), *A Quark for Mister Mark: 101 Poems about Science*, Leipzig: Faber & Faber 2000 (Faber Poetry)

F. Ueberweg, *Grundriß der Geschichte der Philosophie*, 3. Teil: *Die Philosophie der Neuzeit bis zum Ende des XVIII. Jahrhunderts* (völlig neu bearb. v. M. Frischeisen-Köhler & W. Moog), Berlin: Mittler [12]1924

A. Unsöld, *Beiträge zur Quantenmechanik des Atoms*, Thesis, LMU Munich 1927, Ann. Phys. [4] 82 (1927) 355-393

R.-M. Vetter, *Simulation von Tunnelstrukturen. Experimentelle und theoretische Untersuchungen an Systemen mit anomaler Dispersion*, Theses, University at Cologne, 2002

R. A. Vicencio, C. Cantillano, L. Morales-Inostroza, B. Real, C. Mejía-Cortés, S. Weimann, A. Szameit & Mario I. Molina, *Observation of Localized States in Lieb Photonic Lattices*, Phys. Rev. Lett. 114 (2015) 24, 245503 ff., online with suppl. material: http://arxiv.org/abs/1412.6342 (16.04.2016)

G. Vincze & A. Szasz, *Nonequilibrium Thermodynamic and Quantum Model of a Damped Oscillator*, in: M. Gorji-Bandpy (Ed.), *Recent Advances in Thermo and Fluid Dynamics*, Rijeka: Intech 2015, Ch. 3 (pp. 39-80), http://www.intechopen.com/books/recent-advances-in-thermo-and-fluid-dynamics/nonequilibrium-thermodynamic-and-quantum-model-of-a-damped-oscillator#exportas (15.02.2016)

J. Violle, *Lehrbuch der Physik*, 2 Vols. (Vol. 1: Mechanik; Vol. 2: Akustik und Optik), Berlin: Springer 1892-1893

M. Visser, *Some general bounds for 1-D scattering*, arXiv:quant-ph/9901030 (15.04.2013)

Voltaire, *Élements de philosophie de Newton, mis á la portée de tout le monde*, Amsterdam: Ledet 1738; in: *Œuvres complètes de Voltaire*, Vol. 31, De l'imprimerie de las société littéraire-typographique 1784

Voltaire, *Sammlung verschiedener Briefe des Herrn von Voltaire, die Engelländer und andere Sachen betreffend*, Jena 1747

V. Volterra, *Sulla inversione degli integrali definiti*, Rend. Accad. Lincei [2] 5 (1896) 177–185, 289–300

V. Volterra, *Sopra alcune questioni di inversione di integrali definiti*, Ann. di Math. (2) 25 (1897) 139–178

V. Volterra, *Leçons sur les équations intégrales et les équations intégro-différentielles*, Paris: Gauthier-Villars 1913; Engl.: *Theory of Functionals and of Integral and Integro-Differential Equations*, Glasgow: Blackie 1930; reprint: Mineola (N.Y.): Dover 1959 (Dover Phoenix Editions, with a preface by Griffith C. Evans, and a biography and bibliography by Sir Edmund Whittaker)

G. F. Voronoy, *Nouvelles applications des paramètres continus à la théorie de formes quadratiques*, J. reine angew. Math. 134 (1908) 198–287

B. L. van der Waerden, *Sources of Quantum Mechanics*, Amsterdam: North-Holland 1967; reprint: New York: Dover 1968, archive.org/details/SourcesOfQuantumMechanics (29.06.2015)

B. L. van der Waerden, *Eulers Herleitung des Drehimpulssatzes*, in: *Leonhard Euler 1707-1783*, 1983, pp. 271-282

H. Weber, *Über die Integration der partiellen Differentialgleichung* $\partial^2 u/\partial x^2 + \partial^2 u/\partial y^2 + k^2 u = 0$, Math. Ann. I (1869) 1-36

S. Weinberg, *The search for unity. Notes for a history of quantum field theory*, Daedalus 106 (1977) No. 4, 17-35; fafnir.phyast.pitt.edu/py3765/WeinbergQFThistory.pdf (15.04.2005)

S. Weinberg, *The Quantum Theory of Fields, Vol. I. Foundations, Vol. II. Modern Applications, Vol. III. Supersymmetry*, Cambridge: Cambridge Univ. Press 1995, 1996, 2000

C. F. v. Weizsäcker, *Über die Entstehung des Planetensystems*, Z. Astrophys. 22 (1943) 319-355, adsabs.harvard.edu/full/1943ZA.....22..319W (15.04.2013)

C. F. v. Weizsäcker, *Komplementarität und Logik*, Naturwiss. 42 (1955) No. 19/20; in: *Zum Weltbild der Physik*, 1990, pp. 281-331

C. F. v. Weizsäcker, *Zum Weltbild der Physik*, Stuttgart: Hirzel [13]1990

C. F. v. Weizsäcker, *Aufbau der Physik*, München: dtv [4]2002 (dtv 33084)

G. Wentzel, *Zur Quantenoptik*, Zs. Physik 22 (1924) 193-199

R. Westfall, *Isaac Newton*, Heidelberg *etc.*: Spectrum 1996

H. Weyl, *Über gewöhnliche Differentialgleichungen mit Singularitäten und die zugehörigen Entwicklungen willkürlicher Functionen*, Math. Ann. 68 (1910) 220–269; https://eudml.org/doc/158437

H. Weyl, *Gravitation und Elektrizität*, S.-Ber. Kgl. Preuss. Akad. Wiss. Berlin (1918) 465-480; Engl.: *Gravitation and Electricity*, in: Einstein, Lorentz, Minkowski & Weyl, *The Principle of Relativity*, 1923, pp. 201-216

H. Weyl, *Raum – Zeit – Materie. Vorlesungen über allgemeine Relativitätstheorie*, Berlin: Springer 1918; Engl.: *Space – Time – Matter*, New York: Dover 1952 (transl. of [4]1922)

H. Weyl, *Gruppentheorie und Quantenmechanik*, Leipzig: Hirzel 1928; Engl.: *The Theory of Groups and Quantum Mechanics*, New York: Dover 1950 (transl. of the 2nd rev. ed. 1931)

H. Weyl, *Elektron und Gravitation. I.* Z. Phys. 56 (1929) 330-352

J. A. Wheeler & W. H. Zurek (Eds.), *Quantum Theory and Measurement*, Princeton: Princeton Univ. Press 1983

A. Whitaker, *Einstein, Bohr and the Quantum Dilemma. From Quantum Theory to Quantum Information*, Cambridge: Cambridge Univ. Press [2]2006

E. T. Whittaker, *On the Functions associated with the Parabolic Cylinder in Harmonic Analysis*, Proc. London Math. Soc. s1-35 (1902) 417-427

E. T. Whittaker, *A Treatise on the Analytical Dynamics of Particles and Rigid Bodies*, Cambridge: Cambridge Univ. Press [4]1947

E. T. Whittaker & G. N. Watson, *A Course of Modern Analysis*, Cambridge: Cambridge Univ. Press [4]1927; new ed. 1996 (Cambr. Math. Libr. Ser.)

C. E. Wieman, *The creation and study of Bose-Einstein condensation in a dilute atomic vapour*, Phil. Trans. R. Soc. L. A 355 (1997) 2247-2257; in: *Collected Papers of Carl Wiemann*, Singapore: World Scientific 2008

A. S. Wightman (Ed.), *The Collected Works of Eugene Paul Wigner*, Vol. 4, Berlin · Heidelberg: Springer 1997

E. P. Wigner, *Über nichtkombinierende Terme in der neueren Quantentheorie. Zweiter Teil*, Z. Physik 40 (1927) 883-892

E. Wigner, *Über die elastischen Eigenschwingungen symmetrischer Systeme*, Nachr. Ges. Wiss. Göttingen, Math. Phys. Kl. 133 (1930); Engl. in: Cracknell, *Applied Group Theory*, 1968, pp. 223-236 (No. 1)

E. Wigner, *Gruppentheorie und ihre Anwendung auf die Quantenmechanik der Atomspektren*, Braunschweig: Vieweg 1931; Engl.: *Group theory and its application to the quantum mechanics of atomic spectra* (expanded and improved ed. by J. J. Griffin), New York: Academic Press 1959

E. P. Wigner, *Über die Operation der Zeitumkehr*, Nachr. Akad. Wiss. Göttingen, Math.-Physik. (1932) 546-559

E. P. Wigner, *On the Quantum Correction for Thermodynamic Equilibrium*, Phys. Rev. 40 (1932) 749–759; reprint in: A. S. Wightman (Ed.), *The Collected Works of Eugene Paul Wigner*, Vol. 4, 1997, pp. 110-120

E. P. Wigner, *Invariance in Physical Theory*, Proc. Am. Phil. Soc. 93 (1949) #7; reprint in: J. Rosen (Ed.), *Symmetry in Physics*, 1982, pp. 33-38

E. P. Wigner, *Events, laws of nature, and invariance principles* (Nobel Lecture 1963); in: G. Ekspong (Ed.), *Nobel Lectures in Physics Vol 4 1963-1980*, 1998, pp. 6-17, https://www.nobelprize.org/nobel_prizes/physics/laureates/1963/wigner-lecture.pdf

E. P. Wigner, *Symmetry and Conservation Laws*, Phys. Today (1964) March; reprint in: J. Rosen (Ed.), *Symmetry in Physics*, 1982, pp. 39-45

E. P. Wigner, *The role of invariance principles in Natural Philosophy*, Proc. Int. School Phys. "Enrico Fermi" 29 (1964) ix-xvi; reprint in: J. Rosen (Ed.), *Symmetry in Physics*, 1982, pp. 46-53

E. P. Wigner & F. Seitz, *On the constitution of metallic sodium*, Phys. Rev. 43 (1933) 804-810; 46 (1934) 509–524; reprints in: A. S. Wightman (Ed.), *The Collected Works of Eugene Paul Wigner*, Vol. 4, 1997, pp. 365-371, 372-387

F. Wilczek, *Magnetic flux, angular momentum and statistics*, Phys. Rev. Lett. 48 (1982) 1144-1146

F. Wilczek, *Quantum mechanics of fractional-statistics particles*, Phys. Rev. Lett. 49 (1982) 957-979

F. Wilczek, *Quantum Time Crystals*, 2012, https://arxiv.org/abs/1202.2539v2

F. Wilczek & B. Devine, *Longing for the Harmonies. Themes and Variations from Modern Physics*, New York · London: Norton 1988

M. N. Wise, *The Mutual Embrace of Electricity and Magnetism*, Science 203 (1979) 4387, 1310-1318, http://www.fflch.usp.br/df/opessoa/Norton-Wise-Embrace.pdf (15.04.2016)

H.-U. Wöhler, *Dialektik in der mittelalterlichen Philosophie*, Berlin: Akademie-Verlag 2006

T. T. Wu & C. N. Yang, *Concept of nonintegrable phase factors and global formulation of gauge fields*, Phys. Rev. D 12 (1975) 3845-3857

H. Wußing, *Die Genesis des abstrakten Gruppenbegriffs. Ein Beitrag zur Entstehungsgeschichte der abstrakten Gruppentheorie*, Berlin: Deutscher Verlag der Wissenschaften 1969

D. Xiao, M.-C. Chang & Q. Niu, *Berry Phase Effects on Electronic Properties*, 2009, arXiv:0907.2021v1 [cond-mat.mes-hall] 12 Jul 2009; Rev. Mod. Phys. 82 (2010) 1959-2007

L. G. Yaffe, *Large N limits as classical mechanics*, Rev. Mod. Phys. 54 (1982) 407-435

C. N. Yang, *The law of parity conservation and other symmetry laws of physics* (Nobel Lecture 1957); in: G. Ekspong (Ed.), *Nobel Lectures in Physics Vol 3 1942-1962*, 1998, pp. 393-403, https://www.nobelprize.org/nobel_prizes/physics/laureates/1957/yang-lecture.html

Y. G. Yi, *Lagrangian Approaches of Dirac and Feynman to Quantum Mechanics*, 2006, arxiv.org/pdf/physics/0005044.pdf (15.04.2013)

W. Yourgrau & S. Mandelstam, *Variational Principles in Dynamics and Quantum Theory*, New York: Dover 1979

H. Yukawa, *On the interaction of elementary particles I*, Proc. Phys.-Math. Soc. Japan [3] 17 (1935) 48-57

A. M. Zagoskin, D. Felbacq & E. Rousseau, *Quantum metamaterials in the microwave and optical ranges*, EPJ Quantum Technology (2016) 3:2 (17 p.), http://epjquantumtechnology.springeropen.com/articles/10.1140/epjqt/s40507-016-0040-x (10.05.2016)

A. Zangwill, *Modern Electrodynamics*, Cambridge: Cambridge Univ. Press 2013

E. Zauderer, *Partial Differential Equations of Applied mathematics*, New York *etc.*: Wiley ²1989

A. Zee, *Fearful Symmetry. The Search for Beauty in Modern Physics*, Princeton: Princeton Univ. Press ²1999

A. Zee, *Quantum Field Theory in a Nutshell*, Princeton: Princeton Univ. Press 2003; Reprint: Hyderabad: Univ. Press (India) 2005

A. Zeilinger, *Einsteins Schleier. Die neue Welt der Quantenphysik*, München: Beck ²2003

Ya. B. Zel'dovich, *The quasienergy of a quantum-mechanical system subjected to a periodic action*, Sov. Phys. JETP 24 (1967) 1006-1008, http://www.jetp.ac.ru/cgi-bin/dn/e_024_05_1006.pdf (13.04.2014)

C. Zeng & V. Elser, *Numerical studies of antiferromagnetism on a Kagomé net*, Phys. Rev. B 42 (1990) 8436-8444

R. K. P. Zia, E. F. Redish & S. R. McKay, *Making Sense of the Legendre Transform*, Am. J. Phys. 77 (2009) 614-622, http://arxiv.org/pdf/0806.1147.pdf; www3.nd.edu/ powers/ame.20231/zia.pdf (30.04.2015)

R. Zott, *Über Wilhelm Ostwalds wissenschaftshistorische Beiträge zum Problem des wissenschaftlichen Schöpfertums*, in: W. Ostwald, *Zur Geschichte der Wissenschaft*, 1999, pp. 10-39

D. Zwillinger, *Handbook of Differential Equations*, San Diego *etc.*: Academic Press ³1998, https://books.google.de/books?isbn=0127843965 (30.04.2016)

SUBJECT INDEX

www.ingramcontent.com/pod-product-compliance
Lightning Source LLC
Chambersburg PA
CBHW050802220326
41598CB00006B/97